P9-ELW-690

RANCHER, FARMER, FISHERMAN

Also by Miriam Horn

*Earth: The Sequel: The Race to Reinvent Energy
and Stop Global Warming* (coauthored with Fred Krupp)

*Rebels in White Gloves: Coming of Age with
Hillary's Class—Wellesley '69*

RANCHER, FARMER, FISHERMAN

CONSERVATION HEROES OF
THE AMERICAN HEARTLAND

MIRIAM HORN

W. W. NORTON & COMPANY
Independent Publishers Since 1923
New York • London

Copyright © 2016 by Miriam Horn

Mississippi River System map on pp. x–xi © Jeffrey L. Ward; photograph on p. xx courtesy
of Discovery Channel from the film *Rancher, Farmer, Fisherman* / Beth Alala; photographs
on pp. 80 and 208 courtesy of Discovery Channel from the film *Rancher, Farmer, Fisherman* /
DP Buddy Squires; Louisiana land loss map on p. 171 courtesy of the Coastal Protection
and Restoration Authority; photograph on p. 254 © Otis Dobson.

For information about permission to reproduce selections from this book,
write to Permissions, W. W. Norton & Company, Inc.,
500 Fifth Avenue, New York, NY 10110

For information about special discounts for bulk purchases, please contact
W. W. Norton Special Sales at specialsales@wwnorton.com or 800-233-4830

Manufacturing by RR Donnelley Harrisonburg
Book design by Ellen Cipriano Design
Production manager: Julia Druskin

ISBN 978-0-393-24734-3

W. W. Norton & Company, Inc.
500 Fifth Avenue, New York, NY 10110
www.wwnorton.com

W. W. Norton & Company Ltd.
15 Carlisle Street, London W1D 3BS

2 3 4 5 6 7 8 9 0

for Francesca

CONTENTS

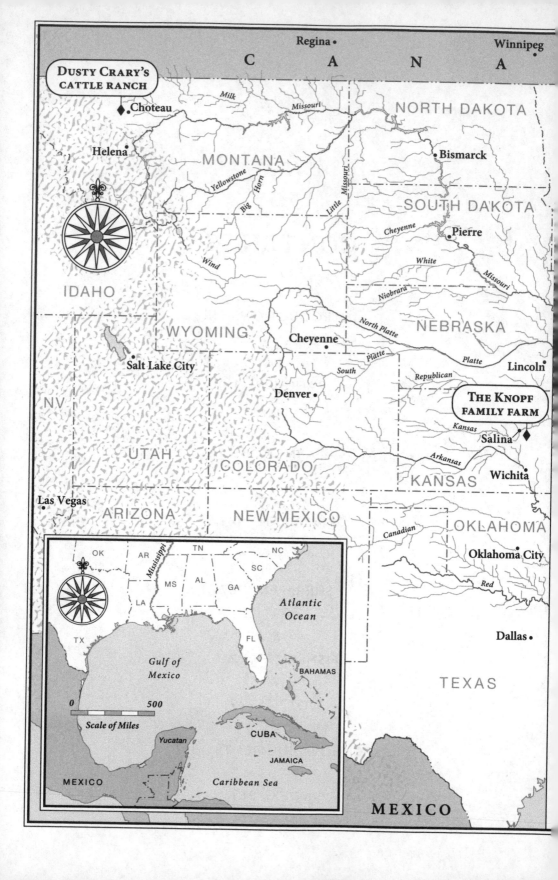

Mississippi
River System

0   500

Scale of Miles

CANADA

Lake Superior

MINNESOTA

St. Paul

Minneapolis

Mississippi

WISCONSIN

Milwaukee

Lake Michigan

MICHIGAN

Detroit

Lake Huron

Toronto

Lake Ontario

NY

Lake Erie

Allegheny

IOWA

Des Moines

Omaha

Chicago

Illinois

OHIO

Indianapolis

Cincinnati

Ohio

Monongahela

Pittsburgh

PA

MISSOURI

Missouri

Kansas City

ILLINOIS

St. Louis

Wabash

INDIANA

Louisville

Ohio

Charleston

WV

Richmond

**MERRITT LANE'S
CANAL BARGE
COMPANY**

Cairo

Cumberland

KENTUCKY

Nashville

VIRGINIA

NORTH
CAROLINA

Raleigh

Arkansas

TENNESSEE

Memphis

Tennessee

Little Rock

ARKANSAS

Mississippi

Greenville

Atlanta

Birmingham

SOUTH
CAROLINA

MS

Shreveport

LA

Jackson

Vicksburg

Natchez

Red

Baton Rouge

ALABAMA

GEORGIA

Charleston

*Atlantic
Ocean*

Houston

New Orleans

**SANDY NGUYEN'S
SHRIMPING
COMMUNITY**

Jacksonville

Leeville

Venice

FLORIDA

**WAYNE WERNER'S
FISHING GROUNDS**

Gulf of Mexico

© 2016 Jeffrey L. Ward

INTRODUCTION

*The land belongs to the future . . . that's the way it seems to me
. . . I might as well try to will the sunset over there to my brother's
children. We come and go, but the land is always here. And the
people who love it and understand it are the people who own it—
for a little while.*

—WILLA CATHER, *O PIONEERS!*

WITH HIS envoys in Paris negotiating to purchase Louisiana,
Thomas Jefferson implored his friend and fellow farmer James Mon-
roe to join them, to "secure our rights and interest in the Mississippi"
River and surrounding territories. "On the event of this mission," the
president wrote in January 1803, "depends the future destinies of this
Republic."

Jefferson's imminent concern was possible war with France, but
his words would prove prophetic across centuries. The Mississippi
River watershed—an immense funnel spun of 7,000 tributaries reaching from the Rockies to the Appalachians and draining 40 percent of
the continental United States—is central to the American story. The
third largest in the world (behind only the Amazon and the Congo),
this basin holds most of the nation's natural wealth and produces
most of its minerals and food: metals and coal from its mountains,
meat from its northern grasslands, grains and beans from its central

plains, fish and black oil from its delta. The connectivity provided by its thousands of miles of waterways—linking the heartland to the rest of the nation and the world—has been critical to America's rise and reign as a global economic power. The nation's politics, too, have been crucially shaped in these middle reaches, for two hundred years the place to sort out such fundamental questions of democracy as the proper balance between federal and local authority. Most important have been the values born here: on this iconic terrain—these mountain majesties, fruited plains, shining seas—explorers and cowboys, pioneers and riverboat captains, forged the American identity. It is not by chance that the Mississippi provided the setting for two of America's founding journeys: Lewis and Clark's up from St. Louis to the Missouri headwaters and across the whole of the Louisiana Purchase Jefferson sent them to explore, and Huck Finn's down the river, to freedom and an understanding of the common human purpose.

America depends on these grand working landscapes, and they in turn depend on a small number of people: the families who live by harvesting their bounty. Farmers and ranchers make up just 1 percent of the U.S. population but manage two-thirds of the nation's land; agriculture has greater impacts on water, land and terrestrial biodiversity than any other human enterprise. That's true everywhere, making this region a model for the world. Half of Earth's ice-free land is in pasture or farms. Crops now cover an area the size of South America and livestock graze an expanse as big as Africa; together they use 70 percent of all fresh water. Fishermen have an equally enormous impact, harvesting 90 million metric tons of fish annually—equivalent, as author Paul Greenberg calculates, to pulling the human weight of China out of the sea every year.

As these productive landscapes grow increasingly precarious—overgrazed, overtilled, overfished; threatened by invasive species, development, ill-conceived feats of engineering, and extreme weather—it is the families who run the tractors and barges and fishing boats who are stepping up to save them. Theirs are the most consequential efforts to restore America's grasslands, wildlife, soils, rivers, wetlands and fisheries—the vast, rich bounty that shaped our national character and sustains our way of life.

In the still half-wild frontier of the northern Rockies, near the headwaters of the Missouri 4,000 miles upstream from the Mississippi's mouth, Montana cowboy Dusty Crary has gathered an improbable band of longtime enemies—cattlemen, fishermen, federal land managers, outfitters, hikers, hunters, environmentalists—to protect the epic ranches and untamed wilderness and elk and grizzlies and trout they all love. On the Kansas prairie, Justin Knopf is using "industrial-scale" farming to restore depleted soils cultivated by his family since homestead days. On the Mississippi itself and its sultry delta, Canal Barge CEO Merritt Lane—scion of an old aristocratic Southern family—has joined an unprecedentedly ambitious effort to reestablish the river's natural land-building functions, to protect his mariners and New Orleans. On the Louisiana bayou, Sandy Nguyen is building alliances to rescue the estuaries that harbor the shrimp and oysters and crabs her community relies on. And in the deep blue waters of the Gulf of Mexico, beyond the river's mouth, commercial fisherman Wayne Werner is tangling with fisheries regulators to bring back red snapper and keep his and his buddies' small businesses afloat. The challenges they face are nearly as daunting as those met by their forebears when they settled the frontier, founded companies in the depths of the Depression, or fled war and Communism in tiny fishing boats adrift on vast seas. But like those ancestors, they draw on deep reservoirs of courage, ingenuity, optimism and resolve.

All are conservationists because their livelihoods and communities will live or die with these ecosystems, but also because they love these land- and river- and seascapes where nature's elemental forces remain vivid in their beauty and danger; where lives of self-creation, self-reliance and liberty remain possible; where the ideas of home and homeland remain strong. All bear a sense of moral responsibility to both the future and the past, a determination to pass on to their children and grandchildren a heritage often generations deep: the family memories imprinted on this land, the seasonal rhythms and traditions built around the bounty they reap. Many acknowledge something sacred here—larger than human understanding or will, a gift to be tended and revered.

Those imperatives aren't new but continue a long (if presently obscured) American tradition. Teddy Roosevelt called conservation "a great moral issue, for it involves the patriotic duty of insuring the safety and continuance of the nation." Frank Capra animated that belief in his "boy ranger" Jefferson Smith, a "big-eyed patriot" gone to Washington to fight for a place he loves, "the prairies, wind leaning on the tall grass, cattle moving down the slope against the sun," a place where boys can "learn something about nature and American ideals." Richard Nixon created the EPA and signed more major environmental laws than any president before or since, including the Endangered Species Act, which the Senate passed 92–0 and the House by a vote of 355 to 4. Ronald Reagan signed the Montreal Protocol, the first global agreement to protect the atmosphere. "What is a conservative after all but one who conserves?" he asked in a 1984 speech. "And we want to protect and conserve the land on which we live—our countryside, our rivers and mountains, our plains and meadows and forests. This is our patrimony."

Linked by those traditional values, these conservation heroes are caring for their families and resources and communities not by digging into ideological trenches or warring to protect their own narrow interests but by coming together like neighbors used to do when raising barns or bringing in the wheat—and often with people very different from themselves. Their fortunes, they know, are entirely intertwined, not least by the river itself. Wayne Werner and Justin Knopf both recognize that the red snapper harvest depends on water quality in the Gulf and therefore on choices Justin makes about fertilizing his wheat a thousand miles upstream. Still physically connected to the earth and weather and the complex web of life, they have come to see diversity as paramount for survival: a diversity of grasses and grazers and predators, of seeds and crops and soil microbes and pollinators, of water and mud, fresh and salt, and—most crucially—of people.

What is new are the alliances they are forging with nature, recognizing there too an irreducible interdependency, and discovering that in emulating nature or hitching a ride on her immense forces they can create more safety and wealth for people. So Justin farms like the prairie: keeping his soils sheltered under living plants and messy mats of fallen stalks and leaves to sustain their hidden microbial universe, knowing those bacteria and fungi care for his crops in ways no one can yet understand, let alone improve upon. So Merritt and partners recognize the muddy Mississippi's power to build storm protection and fish nurseries beyond engineers' wildest imaginings. As Canal Barge captain John Belcher has learned in nearly forty years on the Mississippi, "There's only so long you can levee the world out."

Their collaborations with other species and geologic forces overturn the tragic zero-sum story: that meeting human needs invariably requires sacrificing nature, and vice versa. These ranchers, farmers and fishermen haven't just figured out how to minimize the damage, leave a lighter mark on the land. Their productive work has itself become the path to restoration: Dusty's livestock the curative for stripped grasslands and invasive weeds; Justin's commodity crops the tonic for soil health. Which is not to say they have found the perfect way to fish or farm; they would be the first to acknowledge that there is no such ideal. Rather, their heroism lies in the depth of their commitment to consider the largest implications of what they do, across geographic and generational lines; to forever listen more intently, weigh each choice for the impact it will have on their neighbors and all of life, challenge themselves to do better as they understand more and the world changes around them.

In that work, and their view of the world, they turn out to have little in common with the cartoon versions of heartland citizens regularly trotted out to serve this or that political end. Dusty, Justin, Merritt, Sandy, Wayne and their many partners tell a far more interesting story about what "real Americans" care about and believe.

RANCHER, FARMER, FISHERMAN

1

RANCHER

DUSTY CRARY was just a skinny little boy when he decided to be a rodeo cowboy. He'd been on a horse since age five, riding into the mountains with his dad to the log cabin built by his great-grandfather. By age ten he was helping gather up the family's cattle from their summer pastures on the Blackfeet reservation, where they'd "be scattered clear over creation," or following the Indian kids, "bareback and doubled up on their broom-tailed horses," down to the river for a swim. But Dusty had bigger plans, so while his father put up hay, he'd be rigging up ways to climb atop half-ton steers that didn't want him there, or "waspy" horses that bucked the second he touched their backs. "Those big strong horses, when you're a kid, you'd get pitched off: zzzzt, zzzzt," he says. "You take some terrible beatings learning how to stay on, and *after* you learn how. But the men that were looked up to on an outfit rode the rough string. I wanted to be able to stand up a little straighter and say I'm not just somebody who bought a hat." If he went in the house bawling, his mom would shoo him back outside. "I'd tell him, 'It's not near your heart,'" recalls Bonnie, who is tiny and now white-haired but sustains the same bemused outlook on life. "'You'll be all right.'"

When he was in ninth grade, Dusty got up on his first—and to

his mother's relief, last—bull. He hung on well enough to the bucking horses to be named the 1979 Rodeo Rookie of the Year and go
on to the national high school rodeo finals. By nineteen he was a
professional, his then longish, curly hair bushing out from under his
Texas-made Resistol hat, and by twenty-five a two-time Montana state
saddle bronc champion. For the next decade he kept "going pretty
hard," traveling with his best friend and future best man Lane Yeager
to the big rodeos in Denver, Cheyenne, Houston, Fort Worth, San
Antonio, El Paso, and Tucson, the "swamp rodeos" in Lafayette and
Baton Rouge and the Calgary Stampede.

Dusty saw bull riders killed and more broken bones than he can
remember; Bonnie remembers watching frozen more than once as
crazed horses reared up and over backward on him in the chutes. He
still walks a bit stiffly, and when he turns his square jaw and thick
neck, his blocky shoulders turn too. (He also talks out of the lower
right corner of his mouth, as if chewing tobacco, though the scar
under his chin tugging at the skin came not from a bucking horse but
at the wheel of a truck in high school, when he and a neighbor kid in
a second truck coming his way both got caught in the same rut in the
snow and headlonged into each other.)

Dusty now writes off those rodeo years as "mostly a waste of time
. . . not much more than running around wild." Yet those itinerant
years set the course for his life.

He had never thought it unusual, growing up, to live in a community where ranches and ranching families were still holding on. Like
most kids, he took his world for granted. He figured there were lots of
places a kid could go check cattle down on river pastures and spook
up elk or maybe a grizzly bear; lots of places a kid could round up
horses turned out on the range while a herd of antelope grazed nearby,
watch eagles swoop in to snag the placenta from a newborn calf, or
listen to tales of the frontier from people who'd lived it, like his great-
grandmother Emily, who at ninety-nine could bring to life their town

of Choteau's early days, when men were still publicly hanged and she ran the silent picture show.

But in those two decades spent traveling to Denver and other rodeo towns, Dusty saw that his Montana backyard might in fact be, as the state calls itself, "the last best place." Though those towns shared the dramatic setting he'd grown up with—on the Front Range, where the Rocky Mountains rise abruptly from the prairie—they were now encircled by endless suburbs. The ranches, and the families and wildlife they'd once sustained, were gone. The wide-open home on the range their forebears had thought limitless and inexhaustible had in fact been used up, and Dusty saw that the same danger shadowed the land his family had worked for four generations. "When you got more shingles than grass," he says, shaking his head, covered as always with a rain- and mud-stained cowboy hat, "it's too late, pardner. You're not going to get that back."

So Dusty Crary decided to spend his life where it began, in the "Crown of the Continent": the last intact multimillion-acre landscape in the lower forty-eight, stretching from the Scapegoat Wilderness in the south to Glacier National Park in the north, and from there across the border into Alberta and British Columbia. Straddling the Continental Divide, the Crown includes national forests and tribal lands, the headwaters of the Missouri, Columbia and Saskatchewan rivers draining into the Gulf, the Pacific and Hudson Bay, and the Bob Marshall Wilderness, at 1,577 square miles the fifth largest wilderness outside Alaska.

From Dusty's ranch on the banks of the Teton River, which runs east to the Marias River and from there into the Missouri, he can look up to the snowy summits of "the Bob," as it's known to its friends. The cattle he runs with his mom and wife and three sometimes-home-schooled children live side by side with the biggest population of grizzlies south of Canada as well as wolves, mountain lions, Canada lynx, wolverines, swift foxes, black-footed ferrets and countless other

species that have been here since long before Lewis and Clark found
Blackfeet hunting bison on this northern prairie. The last healthy
populations of native westslope cutthroat trout and Arctic grayling
still hide in the rivers. The air overhead fills at times with hundreds of
thousands of snow geese and tundra swans, peregrine and gyrfalcons,
ibis and avocets, hawks, owls and the songbirds Meriwether Lewis
loved, including the ruby-crowned kinglet, western meadowlark and
warbling vireo.

The wildness has survived, in part, because of this land's unrelenting toughness. The topography, shaped by tectonics and glaciers,
is as severe as the Sahara, with barren dunes of ice and stone and
white dolomite cliffs (here called "reefs") rising vertically out of the
prairie. Winter temperatures regularly drop to 30 or 40 below zero.
Slicing winds roll off the mountains at 50, 80, even 100 miles an
hour. (David Letterman, who owns a ranch a few miles from Dusty's
and married his girlfriend of twenty-three years in the Teton County
courthouse, jokes that every so often they have to go retrieve their
lawn furniture from Canada.)

The Rocky Mountain Front's abrupt transition from mountain to
prairie is a boon for wildlife, putting within easy commuting distance
both the high-altitude meadows and forests many animals favor in
summer and the rolling limber pine savanna and sheltered river bottoms they need to survive the brutal winter. Both the Teton River,
which slices through the Crary ranch, and the nearby Pine Butte "fen"
(a flowing bog, the largest in the western U.S.) are "primo habitat,"
Dusty says, for grizzlies. "They use the river as a safe migratory corridor. It's their main highway. The steep coulees and thick brush in the
river bottom provide them with good cover. And they love the choke-
cherries and serviceberries, tubers and roots and moths." Dusty also
takes pleasure in spotting long-billed curlews, North American shore-
birds that summer here before their long journey south to Mexico.

The Front is just as wonderful for people—at least for those tough enough to flourish here. Up in the ragged peaks and cool, moist alpine valleys, Dusty spent his childhood camping and hunting with his father and uncles amid fir, spruce, white-bark pine and wildflowers: low-growing bitterroots with their pink tutu of petals, silvery wolf willow filling the air with sweetness, white bear grass as fluffy-headed as Dr. Seuss's Truffula trees. Just a few miles east, where Dusty's wife Danelle grew up farming, the dry northwest Great Plains open out to infinity. The rich rough fescue bunch grasses of the savanna that feed Dusty's cattle and also elk and pronghorn (the second fastest land mammal on earth, often called "antelope," the name of its old-world cousin) give way to the needlegrass and wheatgrass of the northern mixed grass prairie, extending hundreds of miles.

This dramatic landscape provided the setting for an equally dramatic human history, the relics of which Dusty loves to find. Bouncing in his beat-up truck across a pasture to a ridgetop from which he can admire the Bob stretched out before him, he points out a prehistoric pictograph he calls Stickman Butte. This is the Old North Trail, where the first humans to cross from Siberia into North America found passage south between the mountain glaciers and the continental ice sheet. (Superior predators, those Pleistocene hunters contributed to the extinction of the mammoths and mastodons then abundant here; the supersized bison survived by evolving into a smaller beast.) He and friends also frequently find relics of more recent centuries, including teepee rings and eagle-catching sites, "where the Indians used to lie down in a pile of rocks and put brush over the top and maybe a buffalo liver so the eagles would come down and they'd reach up and grab their legs and pull feathers for their headdresses." Dusty points out Frazier and Werner peaks, named for two members of Lewis and Clark's Discovery Corps, which spent more time in Montana than anywhere else on its three-year journey. "They were so glad to get back here after wintering on the Columbia River eating fish and dead whales," he says. "Everywhere you looked here there were bison and elk and sheep. That's why there were so many grizzlies on the upper Missouri when they came through; the bison and elk would fall through the ice trying to cross, and die. It was just a smorgasbord in spring when the bears come out of hibernation and here's all these wonderful nice rank carcasses. Bear love carrion; they'll go past fresh meat to get the maggoty one, the riper the better." (That taste for carrion can earn grizzlies a bum rap: a rancher finding a bear scavenging a dead calf will sometimes wrongly accuse it of murder.)

Meriwether Lewis described river bottoms abounding with "anamals of the fur kind" and the "lofty and open forrests . . . the habitation of miriads of the feathered tribes who salute the ear . . . with their wild and simple, yet sweet and cheerfull melody." Lewis also committed here, in July 1806, the first act of bloodshed against western Indians by Americans. Camping for a night with eight Blackfeet warriors, Lewis unwisely bluffed, telling them that the U.S. had brokered a peace between the Shoshone and Nez Perce and were rewarding both tribes with guns. The Blackfeet, frightened at the prospect of their enemies being armed, tried to steal the Discovery Corps' horses and guns but Lewis woke and killed one, leaving a medal around his neck "that they might be informed who we were."

"Montana was some of the last to be settled, in the twilight of the West," says Dusty. "With the Blackfeet so hostile and moving through here into Idaho to steal horses, it was a challenge for settlers to establish a foothold. Custer's deal was 1876 and there was still nothing out there. It wasn't until steamboats started navigating the upper Missouri clear to Fort Benton for fur trading, that this part took off. Then there was more and more white encroachment, and the Indian, in terms of being stand-alone and defending their territory, was pretty much done." The steamboats gave hunters a way to capitalize on Europe's craze for beaver hats, buffalo robes and delectable buffalo tongue.* In just a few decades' time they had emptied this landscape, a history Dusty and his friends frequently cite as warning of how a free-for-all can devastate even the most abundant life. Just twenty-six years after Lewis marveled at "the innumerable herds of buffalo attended by their shepperds the wolves," painter George Catlin, standing at the mouth of Dusty's Teton River, railed against "the deadly axe and desolating hands of cultivating man" and at the moral justifications offered for that desolation, "that God's gifts have no meaning or merit until they are appropriated by civilized man." John James Audubon's 1843 Missouri River journals described prairies blanketed with buffalo skulls and the pervasive smell of rotting meat. Turning their attention here after the Civil War, Generals William T. Sherman and Philip Sheridan championed still more wholesale slaughter, to starve the Indians into submission. By century's end an estimated forty million of what Teddy Roosevelt called "the lordly buffalo" were reduced to fewer than a hundred animals, many of them huddled pitifully in Yellowstone. The generals' plan worked: In the winter of 1883, a quarter of the Blackfeet died of starvation.

Coming home between rodeos in the 1980s with his renewed understanding of how easily both ranching communities and wild things could be erased, Dusty saw the same pressures bearing down on the place and the life he loves.

Though its remoteness has spared Choteau the worst of the real estate frenzy, Dusty watched as a few big ranches—important both for their history and as habitat—were split up and sold off. He found opportunistic weeds invading rangelands degraded by a century of overgrazing, crowding out native grasses. Most ominously, he found oil and gas companies staking claims across the hundreds of thou-

* Like the pronghorn, bison are often called by the name of their Old World cousins.

sands of acres of national forest that sits between the Bob Marshall Wilderness and still wide-open private lands; these included claims staked directly to the west of Dusty's ranch. Those leases had helped rekindle decades-old antagonisms among neighbors, with those eager for oil and gas revenues fighting others alarmed at what drilling would do to the recreation economy on which most locals depend. Other conflicts also brewed: between cattlemen fearful of grizzlies and wolves and biologists who see big predators as necessary to keep populations in balance, as well as the federal authorities tasked with protecting those threatened species; families who love four-wheeling in the mountains pitted against outfitters and hunters who know that wild game flee whenever ATVs come in; wilderness advocates who believe nature deserves some places where it is not entirely at man's mercy versus those who think wilderness locks away resources that rightfully belong to the people.

Dusty's head was swimming with all those conflicts when in 1994, at age thirty-four, he quit the rodeo and came home for good. Two years later, newly married and with a three-month-old baby girl, Dusty witnessed something on the ranch that locked in his fierce protective instinct toward this place and all who live here.

Usually, when Dusty tells a story, he keeps at his chores—shoeing horses, with Arby the Border collie darting in every so often to grab sliced-off bits of hoof for a snack; braiding a crown splice in a rope; hosing down his middle son's 4H "gilt" (a girl pig, who grunts affectionately and rubs her nose on the pen every time Dusty walks by); pulling a randy donkey off a shrieking mule or covering another mule's eyes with his yellow rain slicker so he can quickly burn into her rump the brand his great-grandfather bought in the 1920s from the U.S. Cavalry. But for this story, he stops work and stands nearly still.

He and his father, Doug, were at work together as usual—loading some Charolais bulls onto an old gooseneck trailer. "The bulls were fighting and raising hell so we let 'em out to reshuffle 'em," he says.

"I was out in the alley locking some up and my dad was on the scale loading another and I saw that bull walking by him just turn, kind of cock his head and, bang, nail him"—he punches his fist into his palm—"up against the side of the rack. My dad jumped up on the fence and I thought, 'Aw, he roughed him up but it's not too bad.'

"But then my dad was standing there with his feet on the rail saying, you know, 'Go on,' and that bull come again and swept dad's feet out from under him so he fell down on that concrete. And then I was getting over there, and the bull got his head down and just worked him over"—Dusty butts and swings his head—"crunching him against the side. He was just being a bull; they use their heads, you know. Gosh, I don't know how many times I've been smacked, rolled in a ball.

"That bull was fuzzed up and dad shoulda not been in there with him. But the bull wasn't trotting and bouncing off the walls and blowing snot everywhere. It wasn't like he ran across the field and smacked dad like a rodeo bull. They're just so huge and strong and heavy, a ton or more, and on that concrete against that rack, dad was just crushed.

"I hollered at that bull and got him off and shut the gate on him. Dad stood up and man, he's not feeling good. He kept saying, 'Man, it's getting hard to breathe. Dusty, go get me some oxygen from the shop.' I called 911. It tore his aorta loose is what it did. He was filling up with blood. The ambulance took forever, and he died just going to town there."

Doug Crary was sixty-three. Dusty, his only son, gives an almost imperceptible shake of his head at the memory. "My kids never did get to know him. But my dad absolutely was petrified of becoming an old man. He'd see some little old feeble half-with-it guy teetering along and he'd go, 'Oh my God, I don't want to get like that.' Life is not without risk, and there's something to be said for going down with your boots on, working alongside your son. I think he'd just as soon gone out with a bang than die in a rest home eating mush and not knowing where he was. Hopefully my mom can fall over on her rake deader'n a wedge.

"My dad had always said, 'Them goddamn undertakers preying on these grieving widows, you don't let them. You haul me down in the pickup.' The coroner's there at the hospital and says, 'Whenever you're ready,' and I said, 'I'm going to take him to Great Falls myself.' And he said, 'Well, that's highly irregular.' So we got a foam pad and put it in the back of my dad's pickup. He was in a body bag so we carried him out, and you're pretty numb right then. And the cremation guy said, 'We have a cardboard box,' and I said, 'That's perfect.'"

"Dusty said it wasn't so hard taking him down," recalls Bonnie, also matter-of-fact in the telling. "What was hard was to leave him there." The first thing Dusty had to do when he got home was turn loose the bulls, who had stood in the trailer all day. They put his father's ashes up on the hill, where "he could keep an eye on things." It was, as Dusty said a few years later, "a little transition . . . It made me stop and think. None of us are going to live forever, and how do I want this to be when I'm gone? It was . . . a realization that we have to pass things on."

Just two years had passed since Dusty had been a traveling man living the half-reckless life of a rodeo cowboy. Now he felt the full weight of his responsibility: to protect his family's livelihood, the Crary heritage, his childrens' future, and the wildness of the mountains his father had first taught him to love. Cast abruptly and half-reluctantly in the role of protector, Dusty would spend uncountable hours over the next two decades leaving the animals and chores to Danelle and the kids as he forged alliances with seeming antagonists and fought battles with lifelong friends to keep this last fragment of the wild, glorious American West unspoiled.

First up was the Crary ranch. "I began visiting with a fellow by the name of Dave Carr at The Nature Conservancy [TNC]. Typically the big threats are conversion to cropland—you know, taking native ground and cultivating it, or splitting a big ranch into ranchettes." Dave explained how a conservation easement—sale of the development rights to a land trust, which would hold them in perpetuity—could ensure that the Crary land would always remain a working ranch. "There was one or two places on the Front that had easements on 'em, but no one really knew about it," Dusty recalls. "And there was tremendous animosity. You know, 'It's all a conspiracy. At some point the invisible ink's going to dry and some greenie outfit's going to own your outfit and be calling all the shots.'

"Gosh, I got criticized. My neighbors were like, 'What the hell are you doing, man? You gave away your kids' birthright. What if your kids want to develop that? You shouldn't be deciding for the next generation. Perpetuity's a long time.'" Bonnie heard similar warnings from fellow members of Montana CattleWomen. "They were against TNC," she says. "They said they couldn't just come in here and tell us how to use this land."

"But to me it was about my property right," Dusty continues. "I'm making the decision to sell those development rights in this package versus selling off 160 acres over here or cutting the whole thing up. Some do it for estate planning; that value is subtracted from the basis of the property to reduce the tax. And everything we do affects the next generation. What if I'd sold it? That's pretty permanent too. I think a lot of the skepticism comes from the view that there's cowboys and there's hippies and the two shall not get along. Folks think all environmental groups hate cattle and ranchers and there's certainly some truth to that. Dave Carr will be the first to say they made the same bad insinuations, assumed there's just no way we could have the same goals. But the real conservation-minded who do not have an agenda to exclude people realize we're as organic as everything else here, we're going to be part of this equation, and that the truly successful initiatives have always been partnerships with those working and living off the land."

Dusty was also coming to see that "protecting private lands is as crucial, if not more crucial, to this entire ecosystem than even the Wilderness." For wildlife, the islands of high rock and ice that are typically protected are simply not enough: grazers and carnivores need room to find food, amenable weather and unrelated mates, and access to lush, protected lowlands for winter grazing and spring nurseries for newborn calves, fawns and cubs. Because those bottomlands are also the best suited for agriculture, nearly all are in private hands. It is the integrity of so many farms and ranches on the Front that has allowed so many big mammals to survive here; this is the last place in the lower forty-eight where grizzlies still utilize the Plains habitat they roamed for centuries. But private lands, especially ranches and farms, presents opportunities across the nation: many of those working lands are adjacent to vast areas of protected public land, providing vital connectivity across large landscapes, and many have *higher* rates of plant and animal biodiversity than public lands.

By protecting the ranch, Dusty honored not only his father's memory but a heritage reaching back four generations in Montana, a heritage his children hope to carry on.

Dusty and Bonnie love to spread old family photos across the kitchen counter and share stories of the strong women, risk-takers and romantics that came before. Even as recently as Bonnie's childhood, taming the wilderness meant overcoming countless hardships. This land carried so many mortal threats along with its bounty that Dusty's forebears could scarcely have imagined the day he's come to see, when the roles have reversed and it is man that poses the far greater threat to the land.

Bonnie finds a picture of her paternal grandmother, Sara Hale, buttoned to the throat in a dark woolen dress, her black hair pulled severely off her face. Sara married Herman Van De Reit, a Dutch immigrant, in what the young bride called the "vast wilderness" of

Kansas. Bonnie remembers Sara as a "strong Yankee," bitter to the end at the mistreatment her father and uncles had suffered in a Confederate prison in Tyler, Texas. (Fed once a day "like hogs" with corn shoveled through the window, one uncle fell ill and died; Sara's father's hands shook till the end of his life, showing up blurred in every family picture.) In 1906, Sara packed the family into a "prairie schooner" and moved to Montana, settling in Great Falls. "My grandfather had only finished the fourth grade," says Bonnie, her own undimmed can-do nature evident in the brightness of her blue eyes and the little affirmative refrain ("mm-huh") she tacks onto the end of most sentences. "But he taught himself enough to be bookkeeper for the famous Strain Brothers' store." Cowboy painter Charlie Russell was a neighbor. As a boy, Bonnie's father Harry, then eleven, "would hold Russell's horse, and he'd repay him with a little sketch," she says. Those sketches, alas, have all been lost.

The family's early years in Montana were terribly hard. "Most of those early homesteads were short-lived," Dusty says. "This is brittle farming: lots of glacial moraine; not much soil. You had to be just a little grittier to hang on." In 1909, diphtheria broke out, killing Harry's four-year-old cousin. The next winter was so harsh the Van De Reits paid to have hay shipped in, but when they got to Choteau to claim it they found the boxcars ransacked by desperate stockmen. In 1917, Harry joined the war; gassed in France, he suffered from lung illness the rest of his life. Bonnie's sister remembers their mother, Bessie—whose own mother was among the first women in Montana to file on a homestead—washing dozens of his bloody handkerchiefs daily, without complaint. Bess and Harry had nine children, only six of whom survived infancy. "All that loss," says Bonnie, "certainly wore her down."

Living for a time on a ranch inside the Blackfeet reservation, near the Canadian border, Harry and Bess bootlegged and trapped coyotes for the government bounty, feeding their meat to the dogs. "They'd keep their moonshine down in this old stone house," says Dusty. "And hang green coyote pelts up in front. Nobody wanted to pick through and get fleas jumpin' all over them, so if the revenuers showed up that was a good way to deter too much investigation." Bonnie's uncle Guy did get caught making a delivery and went to prison at Deer Lodge. "Yep, he had to go sit in shade for a year," says Dusty, chuckling and shaking his head.

Bonnie's brother Ray, whose chaps Dusty borrowed the first time he ever rode a bull, wrote a fond reminiscence of those years. "An Indian family named Bearwalker worked on the ranch. Their youngest boy Joe was a little older than I, and when we would try to follow the older boys . . . they would tie Joe to the horse manger in the barn by his long braid. We'd have to go have our mothers untie him and the older boys would be gone by then." In 1929, outlaw Charlie Gannon (a.k.a. Hillary Loftis, a.k.a. Tom Ross, a.k.a. Frank Hale), who'd rustled cattle and (according to family lore) murdered at least four men, hid out in their bunkhouse. "My mother said he was a perfect gentleman," says Bonnie. "She danced with him at one of the parties in the big house." Soon after, as the law closed in, Gannon retreated into the bunkhouse and shot himself.

In 1932, the year Bonnie was born, the family moved to the Sun River Valley, where a little milk and mail train, the "Galloping Goose," ran by their hay meadow daily. Bonnie's older brothers would race it on horseback or play on the trestle and hitch rides in the smoky old engines. "I remember the engineers giving us coffee that tasted like lye," wrote Ray, "but we were afraid not to drink it because they might not let us ride next time."

On his father's side, Dusty has an equally intrepid pioneer past. The first to come west was his great-great-grandfather Oliver Crary, born in Connecticut in 1819. At age twenty-nine, Oliver set off by buggy for the territories, finally settling in Iowa "with not a house in sight" where nights were "made hideous by the howling of wolves." At age forty-five Oliver married an eighteen-year-old Norwegian immigrant; their ten children included Dusty's great-grandfather Elisha John (E. J.).

A stalwart Whig turned Republican, "Oliver became a county judge and a man of some means," says Dusty, "and insisted his kids go to college. Back then children listened to their fathers and respected their wishes." (He says this with exaggerated seriousness, signaling that he's sort-of kidding, though when his seventeen-year-old daughter Charlotte, called Chottie, asks if she can go to the lake, he gives her stern instructions to stay away from wild kids and jet skis and clearly means what he says.) E. J. complied, earning his dentistry degree at Northwestern University and settling in Leeds, North Dakota. There he met Emily Sturgeon—who would become a towering presence in Dusty's life—when at age fourteen (in 1900) she came to him for a bad toothache. "My Doc Crary was the 1st dentist in Leeds," she wrote in a vivid memoir begun in her nineties. "I forgot about the pain and let the Dr. set me in a chair and pull it out."

Like the Crarys, Emily's family had come to America before the Revolution. Her father, Richard Sturgeon, was a depot agent with sleek dark hair and a handlebar moustache. Emily's earliest memories include eating mince pies and head cheese, traveling 150 miles alone on a train under care of the engineer and fireman and, at age eight, "how scared we all were when Coxey's Army, the early Labor Movement Organization, seized the train heading for St. Paul, after putting all the regular trainmen off and shooting through the depot over father's head while he sat telegraphing headquarters."

Emily briefly attended college, playing basketball in bloomers. In 1906, twenty and stylish, boarding at Miss Telfte's and working for Hillman Mercantile Co., she made out the lease for Doc Crary to rent a dental office upstairs. She invited him on a hayride and by Christmas they were engaged. They had just one child ("her duty and done," says Bonnie), a beautiful boy they named Lyall, whom Emily dressed like a doll in white lace, short pants and Little Lord Fauntleroy hair.

Elisha John soon grew restless: a boyhood spent with dime-store westerns had filled him with dreams of a cowboy's life in Montana. Emily agreed to move, but her "ultimatum—and she was that strong," says Bonnie, giving a little salute, was that there would be a motion picture theater. In 1913, they arrived in Choteau. "He hung out his shingle and started pulling teeth," says Dusty, "so that he could buy her the town's first picture show." Doc, as everyone now called him, paid $1,000 for the theater and an old Edison Kinetoscope; Emily named it the Royal and began showing three silent reels each evening, playing the piano accompaniment herself.

Choteau was then little more than a few dirt streets and wooden houses scattered around an 1894 jail and a county courthouse built in 1905 and still standing today. Nevertheless, when the Great Western Railroad arrived, Emily's family joined her there. (Her brother Robert died in town from an unexplained gunshot wound a few years later.) Emily proved a canny businesswoman, "undaunted," says Bonnie, by anything, including social expectations of women. "If there was something to try: full speed ahead." In 1926, Emily went to the convention of the National Association of Theatre Owners in Los Angeles, "in a beaded dress she'd sent for from Paris," Bonnie says admiringly, "escorted by opera star John McCormack." A picture from the same year shows her crossing the Sun River on horseback, water halfway up her horse's chest.

In 1932 Doc finally got to be a cowboy, buying the land Dusty now ranches and his first cattle. Several years earlier he'd built his mountain cabin, named (during Prohibition, says Dusty, when they had homemade whiskey but were always running out) the Empty Jug. Pictures from that time reveal a family deeply in love with their new Western identity. One finds Lyall—who'd gone back east to become both a dentist and a doctor, married a Chicago socialite and returned to set up medical practice in Great Falls—sitting sprawl-legged in the cabin, a rifle across his lap and a pistol in his hand, hat on the table, his chiseled face smudged with dirt and as handsome as Gary Cooper. Children in ringlets, lace bonnets, gingham smocks and white ruffled socks play in bare dirt or carry stick fishing poles. Emily wrote of their dozens of guests, of Doc cooking breakfast hotcakes for any who hadn't made it back to town after a rowdy night, of her 1937 Christmas surprise for him: a bull she'd bought for $500, decorated with red ribbon and bows.

Life changed abruptly when E. J. died in 1944 from a cerebral hemorrhage. The next "eight long years I had to take over management of the ranch and it wasn't easy," wrote Emily. She sold the picture show, leased the ranch for thirty cents an acre to oil prospectors (though they never drilled) and went back to town. She'd already helped found the library (and treasured a note she'd received from Montana's most illustrious novelist, Choteau resident A. B. Guthrie, Jr., who thanked her for "the mighty gracious service you rendered"). She now became president of the County Republican Women's Club, receiving in 1948 another cherished note, this time from Congressman Wesley D'Ewart: "I am sure we need have no fears for the future of the Republican Party so long as we have energetic and effective leaders such as yourself." By the 1960s, she had deeded the ranch to Dusty's father, her grandson Doug, and begun traveling: by car down the Pan-American Highway to Mexico, by train with the National Federation of Republican Women to "heartily endorse" Eisenhower–Nixon, by ship to Alaska and Hawaii and finally by air to Europe, several times. At eighty-nine she was named the only "lady" director of the Choteau bank, at ninety she enjoyed a "swell dinner" at the Tick Tock in Hollywood and at ninety-eight she rode in Choteau's Fourth of July parade as Pioneer Queen; her king was J. C. Salmond, whose family would become one of Dusty's prime antagonists and critics.

Dusty's uncles Lyall Jr. (called "Ditto") and Duke had followed their father's path to medical school. But his father Doug was called to active duty in Korea before he finished high school; Doug's best friend Gene went too and was killed the day they hit the beach. A photo taken at war shows Doug shirtless, his chest crisscrossed by bandoliers, rifle in hand, helmet tipped and a cigarette dangling from the side of his mouth. Back from the war, he married his high school sweetheart Bonnie and settled in the original ranch house, built in 1889. Bonnie had grown up with woodstoves, gas lanterns and outhouses (electrification was delayed by the war), so was unfazed by ranch life, whether the den of skunks under her kitchen or the untold times she had to help move cattle or "pull somebody out of a mud-hole, even if there was bread rising." She and Doug had two daughters and one son, all delivered by their grandfather Lyall. The littlest, born in 1960, she nicknamed Dusty after Roy Rogers's son.

Doug built a feedlot on the ranch and expanded the operation to a thousand head. At that size they needed hired hands, so Bonnie would go down to the old Johnson Hotel in Great Falls to scoop up a few drifters, "who always called me ma'am." Little Dusty loved to follow them around the ranch, watching them shoe or brand; he remembers especially "an old campaigner, a World War I vet, who would come out to milk. Them old winos could do a lot drunk, not like these new drunks. They all knew how to calve. They'd be down and out and broke. You'd get thirty days out of 'em and then they'd be ready to tie one on again."

Dusty can't remember a meal without a few hired men at the table, Bonnie cooking for all. To this day, if you show up at mealtime, you'll be fed: thick buttered slices of fresh-baked banana bread, homemade stew, or burgers made from Crary-raised beef. Bonnie recalls dropping fresh-killed chickens into scalding water so she could pluck them, raising beans and beets and peas and radishes and rutabagas, canning cherries and watermelon pickles. ("I foundered on those when I was a kid," says Dusty, though his son Carson found the notion of canned watermelon baffling: "Wow, how big's that jar?") They shipped the cows off for slaughter but butchered the pigs on the farm. "We'd skin 'em out and sell the hide to Pacific Hide and Fur," says Bonnie, "but always saved the liver and heart. Ooh yes, delicious."

At the same time as Dusty was sorting out the easement on his family ranch, a century-long, intermittent hunt for oil and gas was heating up again.

Montana has been a center of mining since the 1870s' discovery of copper in Butte, soon famous as "the richest hill in the world." The Powder River Basin, straddling the Montana–Wyoming border, accounts for 40 percent of U.S. coal production. The Bakken shale formation in eastern Montana and western North Dakota was one of the first to be fracked and remains a top oil and gas producer. And a fight has been underway for years over plans for vast new strip mines in southeast Montana, and rail lines to serve them.

The Front itself had seen a brief gold rush in the nineteenth century, but exploration for fossil fuels and minerals began in earnest with the 1973 oil crisis and even more feverishly with the approach of the 1984 deadline for staking new claims in federally designated wilderness (the 1964 Wilderness Act had provided for twenty years of exploration). Dusty was on a rodeo scholarship at the University of Montana (a short-lived experiment: "I didn't get two nickels' worth of my folks' money") when he learned in a biology class about oil companies' plans to seismograph the entire Bob Marshall Wilderness. "They called it 'bombing the Bob.'"

Back home, the threat roused into action two guys Dusty had known all his life, who began calling themselves the Friends of the Front.

Friend number one, Gene Sentz, a soft-spoken, gray-bearded former Peace Corps volunteer, had moved to Choteau from West Virginia coal country when Dusty was ten to work as a wilderness ranger and teach fourth grade; Dusty's daughter Chottie was one of his students. Though Dusty calls the seventy-seven-year-old a "liberal old beatnik," his admiration for Gene's gentle manner and iron will is evident. "Gene has hiked every inch of that wilderness. And he lives an exemplary life of frugality and conservatism. He is humble and thrifty, the Gandhi of the Front."

Friend number two, Roy Jacobs, had worked sometimes for Dusty's dad. "We'd ride, move cattle together. He's like my big brother." Taught to hunt at age seven by his own father, who owned a Choteau saloon, Roy had spent years guiding in Alaska and the Northwest Territories for caribou, moose and bear before opening a taxidermy shop in Pendroy. Having since lost the taste for mounting hunters' trophy kills, he now works on David Letterman's ranch. Dusty calls his old friend an "easy keeper," a quality that endears him to Dusty's children. They tease him about his "happy-go-lucky" ways, like the time Roy lost several mules off a pack train and realized it only miles later, when the kids caught up with him, pack animals in tow.

Both men had seen mining's impacts at close range. As a boy in West Virginia, Gene had hunted fox squirrels a few miles upriver from where outsiders had "offered the hillbilly farmers a deal they couldn't refuse," as he wrote in *High Country News* in June 1995. "Landowners could keep farming just as always. All those city slickers wanted was a hypothetical thing called a 'mineral right' . . . Then one day, the company men showed up . . . And the hill people began to learn the meaning of the word 'undermine.' Slag heaps developed and some of them still burn. The rivers ran black . . . Old-timers were saddened to watch the land undergo a kind of 'progress' they couldn't control . . ."

Roy spent two years on a seismograph crew, witnessing "what they were doing to the country I love. We were 'dozing roads anywhere we wanted to go; if the hill was too steep we'd hook the trucks onto a Cat and pull them up. We destroyed it; you can still see the tracks today.

Then I worked on oil rigs in the Blackleaf. You put all those toxic chemicals down, 'the mud,' and it comes up into a pit in the ground with no liner. We were supposed to pump it into trucks to be hauled away; instead, at night, we pumped it over the bank right into a fen. I was hot. I turned 'em in. We've got sustainable income in wilderness, outfitters, recreational equipment. Just the hunters generate ten million a year on the Front. That's not a boom-bust; it can be here forever. But once you destroy the habitat, and the oil's gone, what do you got: the moon?"

In 1981, when Dusty was at the peak of his professional rodeo career, the Friends won protection for the Bob, but at a terribly high price. "Our congressman, Pat Williams, finagled a deal in Congress that said, 'Nope, we're not going to let you drill the Bob,'" recalls Gene. "But in return he had to agree to open up everything outside the wilderness. They began drilling holes up there and pumped out some gas. They never hit anything big, but Roy and I figured we'd lost the Front."

Over the next decade, as Gene and Roy grew increasingly hopeless, their voices seemingly unheard, the Forest Service and the Bureau of Land Management (BLM) accelerated leasing, ultimately putting more than 150,000 acres on the Front into the hands of oil and gas companies. When Dusty came home for good in 1994 and found that one of those leases, held by Calgary-based Startech, was to drill in the BLM's Blind Horse Outstanding Natural Area, on a high rock bench below Choteau Mountain just west of his family's ranch, he joined them in the fight. They used the most basic levers of democracy: writing plain-spoken but persistent letters to their representatives and newspaper editors, speaking up quietly but fervently at every public meeting held by the state or the BLM, leaving behind their cattle and jobs to travel repeatedly to Washington, D.C. Out of that came lifelong, abiding bonds but also, for Dusty, irreparable rifts with a few of his oldest friends.

Two more neighbors, Stoney Burk and Karl Rappold, filled out their posse. Stoney is a lawyer who grew up in a logging family in the big timber of northwest Montana and flew F-4 Phantom fighter-bombers on two tours in Vietnam, an experience that left him cynical about government (he was bombing Cambodia while Kissinger and Nixon were denying any U.S. presence there) and with a black view of human nature. "We kill or destroy everything beautiful for self-indulgence or money. All of us, we're not victims, we're purposeful administrators of devastation." But he also gained at war a "greater appreciation for the wild country and wild things, and the realization each of us has to come to, that we have to fight for the things we love." Mixed in with Stoney's darkness is a wide sentimental streak; his hobby is painting bucolic landscapes populated by sweet-faced moose and bighorn sheep. "I look at a mule deer and just revel in its beauty."

"You never know what planet Stoney is on," says Dusty. In a single conversation, he'll quote Fox News and Al-Jazeera, complain about the Everglades being overrun by immigrants and how "fifty years of giving poor people more money just made us fifty years worse," then rail against Rick Santorum for saying waterboarding is not torture: "I'd like to waterboard you, you bastard." He also amazes Dusty with his feats of slobbery. "You get in his truck and there's an old spilled milkshake basically epoxied to the dash and licorice nibs all over the seat. He'll invite you to help yourself."

Karl Rappold owns the only private ranch directly bordering the Bob, 20,000 magnificent acres just south of Glacier National Park. Karl met Dusty when the Rappolds switched from Herefords, whose pink udders were getting burned by the high-altitude sun reflecting off spring snow—"they'd kick their calves off and you had to go rope 'em and grease their bags"—in favor of Black Angus, some of which he sold to Dusty's dad. (Karl also galloped as an extra through numerous Hollywood productions, including *Heaven's Gate* and *Return to Lonesome Dove*.)

Some of the proposed leases would have torn right through Karl's ranch. Sitting on the front porch of his weatherbeaten little house, he points toward Walling Reef, one of the dizzying swoops of rock that give the northern Front its distinctive beauty. "One of their first proposed wells was right in that gallop up there against the cliff. And they were going to build a road all the way up through me to get there. Once you tear up that native sod it can never be restored for future generations."

"Nobody's anti-energy; I run diesel pickups and tractors, and there's nothing makes jet engines fly like kerosene," says Dusty. "And I'm guilty of being a NIMBY like anybody else. If oil is your heritage, well good on you, but I don't want an oil patch here. There aren't any more million-acre blocks of pristine land. To get one resource, we're going destroy the rest? Destroy the last wild places for a few days of gas? We don't have dinosaurs anymore; we're not making more of this stuff. At some point we're going to run out. So, we're either going to leave that land untouched and run out, or tear it up and run out. And it's the beauty and wildlife that produce most of our economic value."

Gathered around Bonnie's dining table snacking on Crary beef jerky, Dusty and his buddies are a motley crew. Though they've spent hundreds of hours together, "at every flippin' meeting, sometimes two and three a week," their politics are all over the place, a fact they joke about when they touch it at all. Roy and Gene are mostly liberal; Karl is a lifelong Republican, Stoney a wild card, Dusty the most conservative of all. Though he scorns those who "blindly follow a party line," he concedes, "I wind up voting for way more Republicans than Democrats and on most things land very conservative. I think we do need a balanced budget amendment. I think a lot of our entitlement programs have not helped people but done some terrible things to our society. I think you need private ownership and pride of trying to achieve something on your own." He can't abide Obama, which gives Gene opportunity to tease him about showing up on TV at the

White House standing right behind the president, for the launch of the America's Great Outdoors initiative. Listening to Dusty recount the posse's "bickering and arguing and shouting," Danelle would think: "How amazing that such diverse people could come together, but how could you ever agree on anything?" That they are hilarious, with the timing of a comedy team, no doubt helps; when Dusty asked his daughter what she wanted to do for her eighteenth birthday, she told him, "I just want *your* friends to come over."

For the 1990s' fight against the oil and gas leases, this ragtag posse joined forces with an even more diverse group that included Trout Unlimited and the National Rifle Association (NRA). With Gene quietly leading the way, they pushed for full Environmental Impact Studies and commissioned a study of their own, which calculated the damage the projected forty-three wells and 100 miles of new road would do: mucking up streams with eroded sediments, poisoning fish with toxic chemicals and fragmenting the habitat of game species like elk and deer, which would reduce their reproductive success. The potential gains, they showed, would not begin to offset those losses. Though "the Lewis Overthrust running from Alberta into Montana . . . the Disturbed Belt, some call it . . . is such a classic visible example of the earth's shiftings and buryings that it has inspired generations of hydrocarbon seekers and visionaries," *Field and Stream* editor Hal Herring wrote in *High Country News*, "with a few exceptions . . . the result has been lots of dry and marginal holes." Analysis by the Wilderness Society of U.S. Geological Survey data found that roadless areas across the entire Front held oil reserves sufficient to meet U.S. demand for just three weeks, and the Blackleaf near Dusty less than a day's worth of natural gas and fifteen minutes' worth of oil. "If this little pocket of energy here on the Front is going to save the nation, we're in a world of hurt," Karl told one of the many journalists he and the others took the time to talk to, always at the cost of interrupted work.

Montana's history again provided them a cautionary tale. Writers at the turn of the twentieth century described how the copper industry, which boomed in the new age of electricity, left towns "crushed and sullen," their extracted riches vanished into Wall Street accounts. Dashiell Hammett set his 1927 detective story *Red Harvest* in Personville (called Poisonville by his fictional locals) based on his own experiences at the Pinkerton Agency in Butte: "an ugly city . . . between two ugly mountains that had been all dirtied up by mining. Spread over this was a grimy sky that looked as if it had come out of smelters' stacks." Jared Diamond devotes the first chapter of his 2005 book *Collapse* to Montana, detailing the kinds of natural resource degradation he believes cause the downfall of human societies. Diamond notes that ranchers sued Anaconda Copper a century ago for poisoning their cows, and that many of Montana's 20,000 abandoned mines still leak acid and toxic metals. "The mining companies sell these dreams," says Stoney. "Then they leave you with the junk, taxpayers clean it up and you lost the wildlife."

"And it's not just the impact on the ground," says Dusty. "It doesn't seem like the culture of that industry has evolved. It's still this boom-and-bust rush that puts these communities on their ear. A guy I buy canvas from in Sidney, Montana, in the Bakken, had a small leather manufacturing business but shut his doors when he couldn't get help for less than $35. And I asked a friend who works in oil fields around the West to name one town that's had oil and gas development and is now better off than before the energy companies came in. He couldn't name a single one. It's a one-shot deal. Houston and Calgary is where all the money ends up. The best thing we can do is keep our landscape intact. It gives us more options."

The fight had been underway for more than a decade, "with numerous times where we all thought we'd give up," recalls Gene, when a new player appeared on the scene. Gloria Flora, one of the first women ever to oversee a National Forest, stepped in as the new Lewis and Clark Forest supervisor in 1995. Over the next two years,

Gene wrote Flora so many letters that her office had to create a dedicated Sentz file. "Gene is the most polite gentleman you'd ever want to meet," she recalls, "politely counseling me on the Front and what I ought to do. It was good counsel." By 1997, Flora had decided: she would use her administrative authority to put 350,000 acres along the Front off-limits to new oil and gas leases for a decade. "It was scary," she says. "I realized I could lose my job. But I remember sitting next to the Smith River and thinking, 'So what? What does my career mean in relation to this landscape?'" At the announcement, to which she brought U.S. marshals "in case people got overwrought," she spoke of the "incomparable human relationship to this landscape over time." Afterward, she gave the signing pen to Gene. That same year, Gene took his young daughter to D.C. and talked the chief of the Forest Service into a "hard-rock withdrawal" on the Front, blocking 104 claims filed by a Wyoming prospector. (Per the 1872 Mining Act, public land is open to mining claims unless officially "withdrawn.")

Flora received a flood of thank-you letters. "People sent me pictures of their grandchildren to say, 'I want you to see who you're making this decision for.'" She was also sued by the oil and gas industry, with appeals all the way to the Ninth Circuit, which upheld her decision. "We think progress is measured by how many board feet, barrels of oil, cubic feet of gas," Flora reflects. "That's a piece, but joy, experiences, memories . . . the real expansion of the human spirit is that relationship to landscape and place."

Though it bought them a ten-year reprieve, Flora's administrative action was not the end of the oil and gas story. The leases that predated her decision remained in place, and in 2000 Startech came back with a request to road and drill their Blind Horse lease near Dusty's ranch. By now the public was fully mobilized. "There just became a very loud outcry," recalls Dusty, with 50,000 letters opposed to drilling pouring into the Department of the Interior and protestors vowing to physically block any rigs that tried to roll in. With that wave behind them,

Dusty and posse went to their Republican senator, Conrad Burns, a conservative favorite with a habit of blurting out controversial things but also top ratings from the U.S. Chamber of Commerce, the American Land Rights Association and the NRA. They persuaded Burns to draft a bill withdrawing all federal lands on the Front from oil and gas leasing. Passed in 2006 with the help of Democratic senator Max Baucus, the law turned Flora's provisional administrative decision into an Act of Congress. It also provided for outstanding leases to be bought back, as long as that was not done with taxpayers' money. "So we got private donors in," says Dusty. "It was kind of a shell game; we didn't know what was there. But the oil companies sold, and I don't blame them. I'm a capitalist, free-market guy. We didn't jerk the leases out from anybody. I don't believe in overpowering people with a legitimate right."

Still, the community was, and remains, split, including over a lease held by a Louisiana-based company on land north of Karl's, the source of the two streams that feed the Blackfeet reservation (Badger Creek and the Two Medicine), with sacred significance to the tribe. "The county commissioners were for the drilling, as were many in town," says Dusty. "I don't think they're idiots and boneheads." In a region that has seen its population, per capita income and budgets for services like schools in steady decline, "it probably could provide economic stimulus. I just think there're many more reasons not to."

Most painful for Dusty was the rift that opened between him and his old rodeo buddy and best man, Lane Yeager. Lane's dad, Harold Yeager—a "good acquaintance" of Dusty's dad—had led the charge for drilling. "Thinking we were going to be a Bakken 2," says Roy, Yeager and a partner pulled together landowners willing to lease more than 100,000 acres of private land to Primary Petroleum, which invested millions in exploration before ultimately pulling out. Yeager told journalist George Black that he shared Dusty's worries about "the havoc the oil and gas boom had inflicted on the Bakken—the man camps, sexual assaults, crystal meth," towns like Sidney becoming a "cesspool." But he and his allies were confident that Choteau "would put the brakes on before that ever happened." Dusty responded that "it won't be up to them. There's nothing like the raw power of petroleum, and these things get out of control real fast."

When such fights become intimate—with friends and neighbors turning on one another—they often become a battle over whose history is deepest and therefore who has the most authentic claim on the land and its future. That's especially true here, where so much of the past persists, with stockmen and outfitters still on horseback, moving across remarkably unchanged land. Harold Yeager's great-grandfather homesteaded in 1876, the year of Custer's Last Stand; Harold himself remembers, he told Black, trailing 1,600 head of cattle over four days up to the reservation. "Your eyes would be like mud pies in the morning." Next to that, Dusty acknowledged, his family are newcomers, having arrived only in 1913 and even then settled in town. "My sister and I are the first in the family actually born and raised on the ranch," he joked, "so I guess we're kind of bastard fourth-generation." Those who arrived before 1900, he grants, do merit special respect. "That's a little different, and rightfully so. Those guys came here when it was really tough, and started from scratch. There's a responsibility to honor all the hardships they went through to build a ranch and community."

But in these heavyweight heritage championships, Dusty's team has its own contender.

Karl Rappold's grandfather, also named Karl, came to America from Germany in 1872; unable to feed their children, his desperate parents had put the boy and his fourteen-year-old sister on a boat to New York. Separated soon after landing, the siblings never saw each other again.

All alone in the world and just twelve years old, "all my grandpa could think about was coming west, to the Montana territory," says the younger Karl, his handsome weathered face and thick moustache shadowed by his dusty black cowboy hat. The boy spent four years making his way, finally coming up the Missouri River to Fort Benton, where he spent six more years "swamping out saloons" or doing any work he could find, getting his grubstake together and waiting to be of legal age to homestead. In 1882, he found a piece of land newly cleared of Indians, who had been moved to the reservation three miles to the north ("from Muddy Crick to Birch Crick," as Karl says). "My grandpa wasn't real fond of people and there wasn't anybody else here, so this is where he stayed." It was October and too late to build a cabin, so Karl Sr. put canvas around the wheels of his wagon and lived underneath it all winter, through deep snows, forty-below-zero nights and the 100-mile-an-hour winds that regularly roll off the summits. "He got along great with the Indians, who'd come through here on horseback and camp, even still when I was a kid. Other than when he almost killed one that first winter. He had an old 45-70 he bought off a soldier out of the Civil War. When an Indian looked underneath the wagon he shot at him." Fortunately, with several thousand Blackfeet just a few miles away, he missed, and he and the Indians became lasting friends.

By 1894 Karl was ready to start a family and sent away for a Norwegian mail-order bride. A year later she died in childbirth, leaving him to raise their baby daughter alone. In 1896, he married a second mail-order bride, Gundhilde, who had been abandoned at the Choteau station by a fellow frontiersman who (apparently) lost his nerve. "They had my dad, whose name was John, and another daughter, and then twin boys who died at a year old. The oldest daughter, from the first marriage, died in childbirth when she was twenty-one, and that baby died too. It was a rough life. Out there with no help, they just bled to death."

Like Karl Sr., John Rappold preferred the hard, solitary life of a mountain man.

He was almost sixty by the time Karl Jr. was born, but even with a family he turned his back on modern conveniences. "He didn't believe in backhoes. He dug the irrigation by hand with a horse and slip; he and my brother dug the footings of the house with a shovel, through 'bout solid rock, and I hand-dug two 26-foot wells. Electricity came up the road here in the 1940s, but dad didn't bring it in until 1959. We didn't have running water; we heated buckets on the stove once a week for a bath. We kept food in an icehouse: we'd cut big blocks of ice in winter and bury it with straw, and it'd last near all summer. When it melted you'd cut green willows and smoke the quarters of beef so flies would stay off."

John expanded the ranch to several thousand acres, buying out "probably forty homesteaders who were moving out; they couldn't make it." Contrary to the promises made by the railroads eager to lure settlers west to fill their railcars, this rocky, arid landscape was anything but a new Eden. To this day, abandoned homesteads dot the prairie and mountains in picturesque solitude, each one holding sorrowful stories. A memoir by homesteader Pearl Price Robertson, who arrived in 1911, recounted how she "sobbed aloud to shut out the dreary soughing of the wind" and of grasshoppers covering the crops with a "gray, slimy, creeping cloud." *Great Falls Leader* editor Joseph Howard recalled families freezing with no way to get to town or even fetch coal from the pile without the crusted snow fatally gashing their horses' legs, of those "splendid farmers—somewhere else" who here "failed and starved."

John Rappold kept his family afloat by trapping beaver, mink and muskrat for the European market and paying cash for land others abandoned, never more than twenty-five cents an acre. He also trapped a lot of grizzlies. "You got to realize times were a lot different back then," says Karl. "Those guys were trying to scrabble out a living on a five-dollar steer; they had to run every blade of grass they could get just to keep their families going, and if a bear killed the family milk cow that was it. Then the federal government dumped all them Yellowstone garbage bears up here along the Front. Those bears didn't have any idea how to make a living so they went to killing livestock.

"Everybody come to my dad saying, 'Would you trap those bears?' He bought big leg-hold traps, made for elephants; I've still got his Newhouse #6, the largest trap ever made. They had grizzly hunting seasons then, in fall, but the bears were killing cattle in spring when they were coming out of hibernation. It was illegal to trap 'em in the spring so my dad'd dump 'em over into a coulee. He killed his last bear in 1959, a big old rogue grizzly the federal trappers had been after for three years. That bear killed one of our calves, and my dad ran him down in a snowstorm and killed him."

By then, there were almost no grizzlies left. Of the 50,000 that had roamed the Great Plains in the 1800s, fewer than 1,000 remained, hemmed into just 2 percent of their original range; in 1975, they were listed as a threatened species. "My dad died in 1986 at age ninety, ranching up to the very end. But before he died, one of the last things he said to me was that he felt he had a big part of why the grizzly bear had declined. And that was one of the biggest regrets of his life. He wanted me to promise him that the big bears would always have a home on this ranch."

The fight to keep drilling rigs from crossing his family's ranch was one way Karl set about keeping that promise. The second was entering into a conservation easement, though that was a tougher decision. "When you own all your rights and then start selling your development rights, you feel like you're giving up control of your ranch. And good friends and cousins and some of my closest neighbors refused to talk to me; they said I'd gone over to the greens." Luke Salmond, descendant of the first white family to settle the Front (whose father was Fourth of July king to Emily Crary's queen) sees easements as an effort to clear working people off the land for the benefit of the elite. "We wouldn't move because they would never offer what the land was worth, they wanted to steal our place," he told the *Fairfield Sun Times*. "They would show up and want us to sell them an easement . . . Whenever the economy turns hard on ranchers, the environmentalists are on your doorstep . . . ready to lend a hand in exchange for control of your land. . . . They act like this area is suddenly the Serengeti. No one is supposed to be on 'the Front' except the rich and famous."

Karl sees in easements the opposite—a tool that allows someone like him, who scrapes along, to secure and grow his ranch for his children. "I'm now working on my sixth deal, and have gone from 5,000 to 20,000 acres. And when my neighbors seen that the government didn't come in and take my ranch away and that I grew it four times, paying cash: now lots of them have put their land in easements too."

Karl now sees grizzlies all the time. In early 2003, one air survey over his land spotted fifteen adults, and in 2014 he had "half a dozen that would tip the scale at a thousand pounds." Remarkably, the bears don't touch his cattle; that 1959 depredation was the last on his land. "Bears were killing calves because this country was pretty well used up. First by the Native Americans; when you've got 8,000 head of horses at one camp, it's not a paradise. Then the white man came along and ran it harder still. But a bear's not a carnivore. Ninety-plus percent of their diet is grazing. If you leave grass for them, run your cattle light enough that they can get a free meal, they're not going to spend the energy to kill a calf. They'd much sooner munch on grass, roots and berries." And, like Dusty's grizzlies, on dead stuff. In 2001, Karl described to *National Geographic* how he "watched a bear picking up the remains of a cow carcass and slamming it down on the ground like a professional wrestler. I couldn't figure it out until I saw it was breaking bones to get at the marrow."

He pulls out pictures of a grizzly sow and her three cubs eating grain side by side with young calves. "These here were taken with a little digital cell phone. That's in the corral. This old girl here, she raised a set of three cubs every two years. I used to wean the calves at Labor Day and feed 'em in cell feeders. She got to where she knew exactly when Labor Day was, and her and her cubs would come across the meadow and move right in and eat with the calves." When pellets would fall down into the cracks, the mama bear would turn the feeders over to shake them loose. "She'd stay there till I shipped 'em in December. As soon as the calves were gone she'd head up and hibernate.

"So with these conservation easements I've done what my dad asked. Those big bears will always have a home here."

Their family heritage secured, Dusty and Karl signed on for a more far-ranging mission. In the course of working out their easements, both had joined The Nature Conservancy's advisory council. The group told them that demand from landowners for easements far outstripped TNC's resources, and they wanted to bring in U.S. Fish and Wildlife funds.

"We were like, 'Are you nuts?'" Dusty recalls, sitting with Roy, Gene and Stoney. "The only thing we'd ever had with U.S. Fish and Wildlife was the hammer of endangered species. When the grizzly recovery program was in full swing, it was the firing squad, massive fines or jail if you killed a bear. A guy in Dupuyer, near Karl; he shot a bear *inside* a pen eating his sheep and they didn't spare the rod. We'd also met with state Fish, Wildlife and Parks. It didn't go well. Later I went, 'You know, if our dads woulda been there they'd a drug the one guy out on Stoney's porch and kicked the shit out of him.'"

"They were arrogant," says Roy, "came in like, 'We know it all. You guys just sit and listen to us.'"

"Some of those bureaucrats, it's the only life they've known," says Stoney. "But when you get into an independent-spirited group, they don't like being told what to do by somebody's who's never had a callus on their hand. It took us years to overcome that hostility in the community."

"But there was a ranch for sale, critical winter mule deer habitat, that we were trying to keep from getting subdivided," says Dusty. "So we were like, 'Well, you can invite this guy but we don't have to listen to the son of a bitch or like him.' They brought in this Gary Sullivan from the U.S. Fish and Wildlife real estate program and, man, we hit it off. He didn't say, 'Well, this is what we want to do; will you guys help?' He saw how much passion we all had for this landscape and said, 'I want to help you guys do what you need to do.' Big difference. We wound up authorized for public-funded easements on 295,000 acres, approved the same day as the oil and gas withdrawal. Yeah, we were pretty full of ourselves that day."

With most of the important private ranchlands protected, Dusty and crew turned to the last remaining piece of the puzzle: the 400,000 acres of still mostly roadless federal land that sits between those intact ranches and the Bob Marshall Wilderness. That public land is as much a home to Dusty as his family ranch, and as important to his livelihood. In summer and fall, he packs campers and hunters into the back country. He often works with another local, Chuck Blixrud, from whom he recently bought the outfitting arm of the 7LazyP, a guest ranch with a history in the Teton River Valley nearly as old as ranching. More than a century ago, the Great Northern Railroad booked guests directly, using their promotional brochures and traveling Wild West shows to fan the Eastern romance with cowboy country.

The stakes, this round, were the highest yet. As Chuck told the makers of a short film about the Front, *Common Ground*: "Once the wildness is taken away, you can't get up some morning and say, 'Man, I made a mistake here,' because it's done." Put a road up through it, says Stoney, "and it'd be gone and there's no more, never, it's gone."

Though they got lambasted for starting out with just "hippies and greenie ranchers," Dusty, Karl, Stoney, Gene and Roy sat down first with wilderness advocates. "They were kind of, 'Goddamn, wilderness is the only true protection,'" says Dusty. "We had to help 'em see that, politically, we're not going to get wilderness over all of this. And it wouldn't be appropriate if we could. You put it everywhere and it cheapens the real stuff. When I pass that sign I want to know there's real risk; that you can die if you're stupid and need to rise to the challenge. And we don't want this to be a place where the only people you see are in Patagonia underwear. We said, 'Let's protect the grazing permits there now, so the people that hold 'em feel it's theirs and take care of it. But let's not leave any of this out, hangin' in the breeze. It's pretty near perfect; let's just keep it the way it is.'

"We had knock-down drag-out fights; it got pretty rowdy. But we spent seven years at it, bringing more and more people in, sitting around kitchen tables, thrashing it out acre by acre, drainage by drainage, drawing and redrawing boundaries. We always had someone who'd actually been to the different places, and could say, 'No, that won't work.' We'd get off-center sometime but everyone at core never lost sight of the bottom line, that it's not about what I want but about what's best for this place.

"The best thing I've ever done was getting away from just my own kind—from hanging out in the feed store or coffee shop which I never do, or the bar which is the final word, you know—and talking with people different than me. They're not rabid, you know, and I can have some effect on their thought process and they've shaped mine too." He sustains that commitment to listen even—or perhaps especially—to the ranchers who deem him a traitor: "They don't care about this any less than I do." As he said to Black, regarding the Yeagers, "It's not for me to judge what someone else gets in spirituality from wild country. No one side is holier than the other. Humans are what humans are, on all sides of the spectrum, and I've never yet not learned more and tuned my perspective by listening to people who are polar different than me."

By summer 2008 the group had drafted a piece of federal legislation, the Rocky Mountain Front Heritage Act, that would add 67,000 acres to the Bob Marshall Wilderness and designate another 208,000 acres a Conservation Management Area, a homegrown designation that would keep intact all existing grazing, woodcutting and motorized uses but add no new roads or leases. Dusty likens it to a big easement on public land: letting families continue the work they've always done on the land, but blocking any new uses that would alter its historic character.

Dusty and Karl had already traveled to Washington, D.C., several times to discuss oil and gas and easements, but now became regulars. "Lots of ag people don't even want to go to *town* and talk," says Dusty. "But if you don't want outsiders to make all the decisions, you got to get involved; if you don't create your own destiny, somebody else will. And to be involved in the process at the national level; if you want to you can. This government's pretty frustrating but if you work at it there is access and you can affect it." Karl is equally enthusiastic. "I love D.C. People are nice to me. I think a lot of it has to do with this." (He taps the brim of his cowboy hat.) "Dusty and I was wandering around there and we had more people lead us around, say, 'I know you're not from here and you look lost.' We'd say, 'You got 'er right.' Everybody bitches about the government not working. Well, if all you do is sit on your ass and bitch, that's the way it's going to be."

In March 2012, Dusty testified before Congress: "This legislation was not generated at the federal level and sent down for comment. If ever there was a start-from-scratch, kitchen-table proposal, this is it . . . My kids are the fifth generation of Crarys growing and working on the ranch my great-grandfather started. This is our Homeland Security Bill."

With critical support from Jennifer Ferenstein at the Wilderness Society (who looked like a little girl when she first showed up, recalls Gene, like "maybe she was selling Girl Scout cookies," but was "our glue, a real gem," says Stoney), as well as the Montana Wilderness Association and Back Country Horsemen, the group organized dozens of public meetings. Again, Dusty's neighbors, including Lane Yeager, stood up in opposition. "We're losing our young people, our economy. We're dying down here, bleeding to death," said one. "I understand we want to save the Front," said another. "But how much land do we need to save?" Far more spoke up in support, in public forums or privately. At the end of 2013 the Heritage Act was unanimously approved by the Senate Energy and Natural Resources Committee.

With Baucus's departure to be ambassador to China, however, and paralysis settling over Washington, the bill languished through most of 2014, feeding Dusty's cynicism about American politics. "Honest to God, our whole political system right now, it's like two really bad cable packages. There's maybe one thing worthwhile on each side and then there's a lot of crap you have to take. Neither party stands for anything. It's all politically motivated, all about the pound of flesh. Honesty and goodness can't survive there. You just get ate alive."

The Heritage Act was also held hostage by noisy minority opposition at both ends, Dusty says, "those who think, 'If you can't drive in there, what's it good for,' and those who want us out of there entirely." The former group included Citizens for Balanced Use, who denounced the bill as far too restrictive of motorized use. The latter included the Alliance for the Wild Rockies, which criticized national environmental groups that signed on to the Heritage Act for giving too much away. Others excoriated "collaborationists" more broadly, charging that they "capitulate to power for a cut of the action." Though she was often the object of their ire, Ferenstein believes all these groups "fill a valuable role. They set the outermost markers; their plans were a kind of threat that made our proposal much more appealing."

Most contentious was the issue of grazing, with protest both from those bent on getting all livestock off public lands and from ranchers fearful that the Heritage Act would put an end to a cherished tradition: moving cattle to the high country in summer and back down in fall.

For three decades, the most outspoken critic of grazing has been

Jon Marvel, who founded the Western Watersheds Project to end livestock's "negative impacts on 250,000,000 acres of western public lands." Dusty agrees with Marvel that public range is underpriced. Ranchers with BLM or Forest Service leases paid just $1.69 in 2015 to graze a cow and calf pair for a month, far below the $20–40 that Dusty (who has no federal leases) has to pay on private land. (Gene Sentz calls the leases "federal welfare grazing allotments.") Dusty also agrees that oversight of federal leases is terribly lax, with just a single "range con" overseeing the whole Lewis and Clark Forest, and that "public land grazing practices remain antiquated and need improvement, which maybe should be funded from increased fees."

He also sees "plenty of rotten ranchers, whose permits shouldn't be sacred." None has infuriated Dusty more than Cliven Bundy, the Bunkerville, Nevada, rancher who for twenty years refused to pay fees on his 154,000-acre federal grazing lease and in April 2014 staged a weeklong armed standoff with the BLM, casting himself (with the aid of national media) as the voice of the freedom-loving Western rancher. (His sons reprised the conflict in January 2016, staging an armed occupation of a federal bird sanctuary in Oregon in an ostensible act of solidarity with local ranchers, many of whom disavowed their aims and methods.) Bundy is in fact an outlier, both in the vastness of his allotment and in his refusal to pay. Of those roughly 16,000 ranchers who do graze cattle on 155 million acres of BLM land, less than 1 percent have ever fallen more than two months past due on their bills; Bundy's million-dollar debt was four times the amount owed by every other rancher in America combined. "You always hear about the ones in a standoff with the feds and going to go down shooting," says Dusty. "But Bundy's refusal to pay is a slap in the face not only to ranchers that don't have federal grazing permits but a slap to the many that do and want to do it right, not just pay their fees but try and work with sometimes cumbersome and not-very-light-on-its-feet bureaucracies." (Sitting at Bonnie's table discussing the Bunkerville standoff, Roy, as usual, cracks them all up with his account of the parody YouTube video promoting "Bundyfest 2014": a kind of Burning Man 2 but liberated from the need for permits and fees—or clothes and toilets—thanks to their hero "declaring the entire area surrounding his Nevada ranch a completely rules-free zone.")

For all his concerns about federal grazing, however, Dusty's prime allegiance has always been to the families with leases; he spoke up for them repeatedly in the years of hashing out the Heritage Act. "We had some core principles from the get-go, and one was that this couldn't disrupt anyone's livelihood or way of life. It was never, well, if one guy gets bulldozed that's for the greater good, and those are the casualties of war, you know?" He underscored that commitment in his testimony to Congress. "When we were working on the boundaries, we talked to all the guides and livestock operators in advance to make sure that the Heritage Act would not hurt their ability to make a living off the land. . . . Federal land grazing is important to many of these multi-generational operations. Protecting these grazing permits has been the highest priority throughout this entire process. It is paramount to the integrity of the entire system that these large ranches remain intact. Keeping them economically viable is the best way to insure that. This legislation will in no way jeopardize any grazing permit and in fact provides additional language emphasizing their importance and safeguarding their continued use."

Despite years of hammering by some neighbors, his loyalties never changed. "For a lot of those families, who have leases for just a few dozen cows for a few months, by the time they move 'em up there and check 'em and fix fence, they could do it cheaper at home. But it's a family tradition. And an important cultural aspect of this area, that if it was gone, would make the whole thing different."

Still, extremism kept getting in the way of progress.

On the left, the "narrow-line" environmentalists' indifference to his neighbors' histories, Dusty says, just got everyone dug in. The "litigious vironmental groups are so far removed from this landscape and what its traditional uses mean to families: how it's part of a culture set down over generations, the oral history that ties us to this country, gives us pride of place.

"And when you keep hammering that older generation, that 'you guys did it wrong and were destructive, that it was just stupidity and greed that drove it,' those old guys just get pissed and ready to fight. Yeah, we didn't do everything right: a lot of disingenuous extractive industries that use federal resources needed reined in and better practices. Yeah, our great-granddads killed all the bears and wolves and had the federal government's blessing to do it. Yeah, farmers like Danelle's dad, who came of age when technology jumped way ahead—big tractors, 35-cent diesel—broke thousands of acres of sod to plant to wheat. The guys doing it with a horse had to pick the best soil to farm, but if you get enough horsepower and cheap fertilizer you can farm these rocks." (As he talks, he keeps an eye on a bunch of new horses and mules he's turned out on the range to meet his pack string, who are now all galloping toward him. "Here comes my thunderin' herd.")

"Yeah we didn't do everything right. OK, so let's move forward from here." Because if the effort to drive ranchers off public lands winds up hardening them against the larger project, far more will be lost than gained. "If all you want to do is get cows off Forest Service land—if you're not trying to make this whole thing, including private land, better—you're not doing much."

Dusty and crew grew equally frustrated with the "right-wing types" insistent on exercising their freedoms regardless of the impact on other people's uses of the land. "They want cattle through the whole thing," says Stoney, "oil and gas rigs and roads in every draw and ridge and develop everywhere and have it private. They want fracking and pipelines and if it destroys the aquifer, well that's your family; I don't care as long as I'm making money."

A particular nemesis has been state representative Kerry White.

White, whose grandfather came to Montana in 1864, testified in a 2013 U.S. House field hearing on behalf of motorized use as a "traditional pastime." He has also opposed federal designations of endangered species and extension of the Clean Water Act to the shallow wetlands called "prairie potholes" (which hunters prize as the nation's "Duck Factory"), and joined other Western Republicans in demanding the transfer of federal land to state jurisdiction. "He wants everything open to ATVs," says Roy, "even if it drives out game and tears up the land and wrecks it for everyone else. He says you're a greenie if you're against that. Though of course he doesn't want to pay to maintain the roads; he wants that subsidized."

Dusty's crew was also maddened by the sometimes clueless, sometimes intentional, spread of misinformation. "Our meetings got pretty hot," says Karl. "Ranchers'd get up and say that wilderness locks you out of the land; that it's for everything except people. They'd say if wilderness butts up against your land it hurts your property rights; that the Forest Service will refuse to fight fires and all. The funny part is most of these guys pack into the Bob Marshall every fall to hunt and haven't been kicked out and it's been there since 1940. And I've been bordering the wilderness my whole life and nothing's happened yet."

Coming off a Heritage Act call with congressional staffers in July 2014, Dusty sounded disheartened. "Golly, it just needs to get another hearing, but it's such a cesspool back there. They're getting the usual grief from a few locals who think it endangers their grazing permits, that it'll bring additional scrutiny to any that wind up in wilderness and bring out the forces that sue. But there's nothing to stop those groups from suing now. Jon Marvel has sued lots of permits not on wilderness; those permits are in jeopardy every day. The bill actually adds a firewall . . . with congressional grazing guidelines they'll be more protected.

"I have lots of old-school Western thoughts and belief systems, fiscally and socially. But so much of the time people shove into one or other pigeonhole: 'Well, I want to be with them so I got to oppose conservation.' But should you be *all* preservationist or *all* developer? You going to sue every grazing or timber permit, designate everything wilderness? Or maybe you're for leaving no stone unturned for oil and gas? Are those really the only two options? The best we can do? Some folks, I think, would just rather fight than win, because if somebody's not getting smashed or broke or been sued, jeez, their group's not relevant. That's where that organized opposition's coming from: it's because the Heritage Act *doesn't* smash anybody. Resolution means they might have to do something beyond agitating, blogging, slamming somebody, bitching."

Whatever the self-interest at play, two deep and abiding arguments also surfaced in the fight over the Heritage Act.

The first is about man's rightful relationship to nature. One view of that relationship, rooted in the Bible, is that man has dominion over creation, the power and duty to shape the Earth and shepherd all other living things. That concept of dominion inspired the American pioneer, who saw a moral duty to tame the wilderness, redeem the savagery of the land. By turning trees to lumber, prairies to farms, he would bring God's order and fruitfulness to a dangerous wasteland.

That idea of dominion still inspires much of the most effective stewardship in America. But for some, in Dusty's view, the sense of grave responsibility to creation has evaporated, reducing nature to a resource and nothing more, valuable only to the degree it can be owned and exploited. In that "drill it and kill it crowd," as he calls them, he includes a number of his relatives. "There's a biblical verse you always hear around here about how man shall have dominion over the earth and animals and fish, which some interpret to mean that God just put it all here for man to use. But I'm thinking that's God saying, 'Now lookit, you got a lot of power. You *do* have dominion. So what are you going to do with it? Are you just going to be the crush-everything guy, because you can? Or are you going to do it wisely?' To me it's a warning, not a permission slip."

Set against the view of man as the measure of all things are those who believe animals and plants have inherent value independent of their utility, neither "conferred nor revocable" by man, as biologist Michael Soule wrote in a famous 1985 essay. Author David Quammen, who lives in Bozeman, laments what he calls the "most ominous strategic error" he and others ever made in defending other species on grounds that humans might someday need their DNA for medicine or agriculture or our own adaptive capabilities. "A snail darter is useless but also something far more precious than a floppy disc storing genes for a rainy day," he wrote in an essay that first appeared in *Audubon* magazine in 1982. "It isn't easy to say what without gibbering in transcendental tones." He then gives way to those gibberings, describing Montana's threatened Arctic grayling as beautiful until lifted from its stream into the air, when all its brilliant turquoise, aquamarine and red-orange iridescence drains away. "The magic vanishes like a dreamed sibyl when you pull it toward you."

That view of something beyond material value impels Dusty and friends. Wilderness connects them to the experience of their trailblazing ancestors: "taking your own shelter and provender," as Dusty says, "for a week or more to allow that [he gestures toward the mountains] to come over us"; braving the inky darkness and silence, heart pounding at the snap of a branch on a black night. And to something larger still: the huge, vast forces that over eons have shaped the earth—deforming the walls of white rock, turning thousand-year-old trees, exposed in high open parks, into stunted, twisted bonsai. Stoney loves Montana's most famous novel, A. B. Guthrie, Jr.'s 1947 *The Big Sky*, for its evocation of "a pushing inside of [a man] . . . a reaching for things he couldn't single out" when he rides alone under the stars. "It's spiritual, you see," says Stoney. "This landscape is just awe-stricken." Karl, too, rejects a purely material view of this world. "Not every ounce of land needs to produce a dollar."

A corollary to that reverential view, often, is that man is so destructive—"we dropped off the biotic team," as historian Roderick Nash wrote in *Wilderness and the American Mind*—that he must be banished or tightly restrained for those other sacred creatures to survive. "Hardcore wilderness advocates see us ranchers as enemies, are disgusted with humans," Dusty says. Stoney often shares that loathing, though it drives him not to defeatism but to a fiercer protective stance. "Who speaks for the little baby fawn? Who speaks for the streams and the beautiful cliffs? It's you and I."

Dusty, as usual, finds a way to integrate both views. He is straightforward that nature is useful and ought to be used: "We're not trying to put a glass dome over this place. This is a working landscape and that's how it ought to stay." But he also, shyly, touches on its meaning beyond the material. "This country has made such a profound effect on our lives. We don't talk about it, but this is all I ever wanted. I've got the Teton River and you look up there at those mountains, and it's all you need to keep going."

He also still counts himself a member of life's team, not more but not less, an animal among animals. He watches a coyote trot across his land with an unconcerned shrug. "Aw, he's just trying to make a living." He strives not to interfere with that coyote's livelihood but believes humans deserve the same opportunity. "Everything's connected, and I'm not synthetic; I'm part of it. Like the deer and the grizzlies. I'm part of this landscape."

It may seem paradoxical that Dusty's feelings about hunting and slaughter—the killing of wild or domestic animals—best express that integration.

He tells the story of joining the last legal grizzly hunt in Montana, on his own land. Until 1991, the state had sold a few grizzly tags every fall, "but every year they'd hit the mortality quota before we ever got to hunting season: either a biologist had killed 'em by doping 'em or they'd get hit on the tracks or accidental shot. This state biologist was wanting to take out a few older aggressive males, to make room for young males, so he got approval for a spring hunt. My friend put in and got a tag, came out early April and asked me and another guy to come along. One day we got nice wet snow, good tracking snow, so we go over acrost here to the river and sure enough there's a great big set of bear tracks. Kelly, he guided on Kodiak for brown bear guys, so we figured he's the candidate to go in the brush. He goes in and keeps bumping into this bear, who pretty soon got sick of it, turned around and woofed at him. So Kelly pulled out and we start in [Dusty is now laughing], and that brush is just thick, thick, thick. And that grizzly jumps up about twenty feet away. Instead of heavier calibers with scopes we'd brought our old carbine lever-action 30-30s, lighter, iron-sighted, nimble. The bear jumped up, we all shot and that was it. It was an old, old bear, with no teeth; he wasn't going to be living much longer."

Dusty killed a second grizzly in 1997, helping a state biologist with a "management take" of a bear that had been killing calves. "He had a tracking device, so it was pretty much an assassination. We were crawling on our hands and knees and found him laid down, curled up like a dog and covered with snow. I took the safety off my rifle, he got up and turned around, looked right at me, and I shot him." He brings out a picture of himself kneeling alongside the dead bear, a huge beauty with lustrous fur and stunningly long, curved claws, like a Chinese empress. "He was just gorgeous, gorgeous.

"I don't need to kill another bear," he says quietly. "I don't need to kill another anything, really. But I think I appreciate and understand wildlife better because I have killed. I'm a meat eater, so I should do it: killing forces you to confront yourself and your own emotions, you can examine one more position, how does that play into my whole . . .

"And if taking those two bears, who knows for sure, but if that allowed three young males who were running into a brick wall of people on one side and aggressive males on the other, who will kill them or harass them to the point where they're here in my yard looking for something to eat and end up shot or transplanted or screwed with; if killing those two big old males gave those young ones a chance to go unmolested and establish themselves, then maybe it was a good thing.

"What does offend me is the modern hunting channels. They kill a deer and the first thing they're showing is the logo of the product they used to kill it. It's so disrespectful to the life you just took. Is that all that's about is just selling more crap?" When booking guests for their outfitting business, they screen out the guys that "just want to kill something," says Danelle. "One guy said a prayer over the elk he'd shot. He thought long and hard and understood that it's a big decision to take a life."

Dusty's sense of the meaning of hunting, and its deformation by commercialization, is shared by Roy and Stoney. Roy quit taxidermy "because I hate what hunting's turned into—the trail cameras, guns that shoot 1,500 yards, with a smart scope that compensates for everything, the wind, uphill, downhill. It does everything but pull the trigger; how does a deer stand a chance? The animals stand out there 400 yards and feel safe. And they're slaughtering the big males, the ones that used to breed. I'd hate to be born with balls and four legs in Montana. What are we doing to natural selection?

"If it wasn't for people like Dusty, Karl and Letterman, there wouldn't be any mature animals at all. Hunting was supposed to be about population control, not shooting the most handsome of the species. Other predators cull the weak; they don't cull the strong. But now guys are so bent on making the record book they'll poach, or shoot in closed areas. I went to Zimbabwe and watched them shoot baboons and it was like watching them shoot a human being. And the elephants, who have a grieving process that's unbelievable. Who the hell do we think we are? That we can go out and kill this stuff to hang a head in your house? And I was part of that. Killing used to mean nothing to me. I guess reality comes to you as you get older. My whole life is wildlife. I don't like 'em just the day I go out and kill 'em. I want to spend my whole life with them."

"When I grew up deer, elk, fish, grouse were our primary meat," says Stoney. "These trophy guys don't hunt for the meat, just for the big horns, to get their picture in the magazine. And they don't tell you about the ones they've gut-shot, or shot the ham off and it went off and died alone. We grew up with the ethic that if you wound something, you track it till you can put it out of its suffering."

That sense of another life to be fully honored is there, as well, when Dusty sees his cattle off to slaughter. "I don't feel sad. That animal's a food animal, whether he's domestic or in the wild and subject to predation, like elk and deer and antelope. They're a meat animal. But that doesn't mean I don't value that animal's life, that I don't want to handle him carefully and want him killed quickly. I know there's people see it differently. But that big steer in there providing 500 pounds of meat, that's a noble life. And I can respect that animal completely and yet still end its life, quickly, and make good use of it. I certainly can end his life more stress-free than four wolves killing him."

Living daily with life and death, and so close to some of the world's last real wildness, adds another layer of complexity to these questions of man's relationship to the world.

The idea of what Dusty calls "Big W Wilderness" was to preserve places, as the 1964 Wilderness Act states, "where the earth and its community of life are untrammeled" by man. From at least the time of Teddy Roosevelt, such dangerous and demanding places had been seen as critical to the creation of the American character: forging both secular virtues—courage, self-reliance, independence—and a reverence for eternal creation, "unworn of man," as TR said of a still wilder continent, "changed only by the slow change of the ages through time everlasting."

The idea of any such thing as untrammeled nature, however, has taken a beating in recent decades. First, historian William Cronon's *Changes in the Land* (1983) challenged the notion of a land untouched before Columbus, showing that humans had been altering the earth for at least 10,000 years. Six years later, Bill McKibben argued in *The End of Nature* that modern man had so broadened and accelerated that trammeling—especially by altering the atmosphere—that there remains no place on earth free of man's fingerprints. We have entered, some say, a new geological era: the Anthropocene, or age of man. That in turn has led many conservationists to argue that man must embrace his shaping role; that "we're so implicated in nature," as author Michael Pollan put it, "that you can't stop now." That's true even though well-intended interventions may go terribly wrong, as they did in Montana's Big Hole River when Fish and Wildlife planted hatchery grayling, year after year. "Five million raining down on wild grayling," wrote Quammen, "was the most disastrous thing that ever happened to the wild fish," creating "instantaneous tenement and famine conditions."

Still, nostalgia for some imagined ideal past—of nature in perfect balance, or manly freedom—retains a powerful hold on Montana. Guthrie set *The Big Sky* in the 1820s but described a paradise already lost. When seventeen-year-old Boone Caudill arrives in the country west of Choteau bent on becoming a mountain man, his Uncle Zeb tells him he's already "ten year too late. She's gone, goddam it! The whole shitaree. Gone, by God, and naught to care, savin' some of us who see 'er new . . . This was man's country onc't. Every water full of beaver and a galore of buffler any ways a man looked, and no crampin' and crowdin' Christ sake." Two years after the novel's publication, an infamous *Partisan Review* essay heaped contempt on Montanans' preoccupation with their past, mocking down-at-heel cowboys feeding on their own mythology at the picture show, "watching Gene and Roy in tailored togs . . . secure the Right."

Even as he honors his family and neighbors' heritage, Dusty himself has no patience for nostalgia, which he believes often drives bad decisions. He gets especially fired up over what he calls "the bison thing": the American Prairie Foundation's campaign to build a three-million-acre wildlife preserve in eastern Montana. "People want to recreate this snapshot in time, but it's never stayed the same ever. And it's a utopian idea to think we're going to recreate pre-Columbian prairie. You can't recreate what bison had and how they acted before there was European influence. Even if they achieve their goal, get three million acres and take out all the fences, there's still a perimeter fence, and if an old bull throws his head in the air and says, 'Jeez, I'm getting a little genetic tug to head for the cypress hills in Canada,' and runs into a fence, then all you got is a three-million-acre bison ranch. It's not that there shouldn't be a sizeable area for bison, but pumping bullshit that we're going to recreate what Lewis and Clark saw: *ten* million acres wouldn't be enough."

Dusty understands the longing behind the project. He shares the grief for a world now lost where man was not the decider on the fate of every last living thing. "But one rancher out there said to me, 'The thing that bothers me is that for them to succeed I have to fail.' What kind of story is that? What kind of success? Rather than buy all that land, those millions and millions of dollars could be spent for so much more meaningful conservation, like easements to prevent subdivision, the draining of wetlands and sod-busting." With U.S. grasslands being lost to the plow at a rate comparable to the destruction of tropical rainforests, "protecting that privately-owned sod is far more valuable than trying to have a great big buffalo pasture." Rather than wall off precious bits or create a big zoo or tableaux of what once was, "keeping it whole with a family on it with pride of place . . . that's probably the best we can do. It might be messy and patched together, but that's the only scale that counts, and has a place for people in it."

Native grazers, Dusty adds, aren't inherently lighter on the land than cattle. He recounts an evening at Pine Butte Guest Ranch, listening to historian Stephen Ambrose read by lamplight from Lewis and Clark's journals. "He's sitting there by the fire, waxing eloquent, and the lodge is full of people oohing and aahing and closing their eyes and trying to visualize stuff. And so he goes into the part where Clark's going down the Yellowstone and camped on an island and they're going to take off one morning but 'Oh, we better wait, there's some bison coming.' And every time they'd go, 'Oh, they're gone, we can get going,' another thousand bison would cross. And they waited all day for bison to cross. And Ambrose'd pause to let it soak in and everybody's trying to visualize all these bison. And I piped up with 'OK, now just picture them as cows,' and people would be ballistic. I bet that was one muddy, tore-up son of a gun for years to come, with dead bison calves floating down the river and rotting."

Beyond debates over man's place in nature, the second argument underlying the struggle over Montana's wild lands is less universal and more particular to the American project, engaging debates ongoing since the nation's founding about where private and local sovereignty should prevail over federal authority. Again, differing versions of history are invoked to justify opposing positions. And again, the struggle has played out in an intimate way between Dusty and a neighboring family.

Jim "Luke" Salmond's ranch, just south of Dusty's, is the stuff of Montana legend. Luke's great-grandmother Libby Collins is in the Cowboy Hall of Fame. Hired at seventeen to cook for a 100-wagon train linking Denver to the upper Missouri, she was attacked by wild animals and captured by Indians. Prospecting for gold in Montana in 1863, she met and married a former Virginia City vigilante and in 1881 gave birth to the first white baby born in the Teton Valley; she also adopted two orphaned Blackfeet children. As the first woman ever to take her stock to the Chicago market and superintend their sale, she was dubbed the Cattle Queen of Montana; a 1954 movie starring Barbara Stanwyck and Ronald Reagan (playing a noble U.S. Cavalryman) was loosely based on her story. Libby's daughter Carrie married Frank Salmond in 1904; Luke is their grandson.

Invoking the full weight of that history, the Salmonds were among the most outspoken critics of the Heritage Act, staking their own claim on that evocative word. In a May 2013 story in the local *Fairfield Sun Times*, headlined "Ranch Family Fights for Montana Heritage," they described the bill Dusty was championing as the latest in a long line of assaults by an overreaching government on frontier freedoms.

"If there is a single ranch . . . that serves as the front line between the Cattlemen and those in the government and radical environmental groups that want to push them out of the foothills, it's the Salmond Ranch," the article began. "At one time, the ranch would have as many as 5,000 cattle, grazing the herds on open range, moving along the foothills from the North Fork of the Sun River north to the Blackfeet Reservation near Browning." But the government, Luke Salmond told the reporter, rushed in to crush that freedom: the 1934 Taylor Act "imposed the first limitations" on cattlemen's use of federal lands and in 1964, "when they created 'the Bob', they kicked us out again.

"It comes back to this Heritage Act," Salmond continued. "They make it a wilderness area and you're gone." His son Mark, joining the interview, lambasted the media for falling for the constant claims by supporters that "everyone had a seat at the table." He dismissed town meetings as a forum for pressuring ranchers, and the few ranchers that supported the bill as "all hat and no cattle." "No one will ask the tough questions of these people who speak for the Heritage Act. No one is asking who's footing the bills for the money being spent to push the Act."

The Salmonds' resentment of federal constraints on Western lands has many antecedents: the "sagebrush rebellions" of the 1970s, the "wise use" movement of the 1980s, the 2000 "shovel rebellion" in Jarbridge, Nevada, where a group singing "The Star-Spangled Banner" used shovels to open a road the Forest Service had closed to protect bull-trout habitat. The hostilities toward Forest Service staff in Jarbridge grew so intense that Gloria Flora, who'd been transferred to Nevada's Humboldt–Toiyabe Forest after her decision to block mineral leases on the Rocky Mountain Front, resigned in 1999 to shine a light on the ongoing threats that were condoned not only by local government but also by Idaho's representative to the U.S. Congress, Helen Chenoweth-Hage.

Since the election of Barack Obama as president, the "anti-fed, give it back" movement, as Dusty calls it, has gained steam. In May 2014, a county commissioner in southeast Utah organized an illegal all-terrain-vehicle protest ride through a BLM canyon closed to protect American Indian ruins. In August 2014, the Republican National Committee endorsed bills in Utah and neighboring states that would require transfer of vast swaths of federal lands into state control.

Though the story of a free place and people trampled by government overreach holds tremendous political sway, what's left out of this version of history is the violence and desperation that came to define that lawless frontier. Water and grass, scarce to begin with, grew scarcer with each new settler. "During the homestead era, they couldn't give this land away," says Roy. "What do you do with limber pine and piles of rock? You couldn't raise *one cow* on it. The wildlife vanished because starving homesteaders ate it all." Many abandoned their holdings; others were chased off the "open range" by cattle barons building empires of water and land. Bloody range wars erupted, the most famous of which—the 1892 Johnson County War, fought by a group of wealthy Wyoming landowners (with two dozen hired gunmen from Paris, Texas) against a group of smaller ranchers—inspired first a novel and then A. B. Guthrie Jr.'s screenplay for *Shane*.

Early grazing rules were not a bureaucrat's scheme to crush the cowboy's freedom, but a response to pleas from those small ranchers for help in securing water and grass. "The Taylor Act came about because so many people were using public lands, with no allotments or accounting of what was out there, that ranchers were running out of grass," Dusty says. "One guy'd say, 'You got a thousand head in there; I'll stick in two thousand.' Eventually, like the collaboratives today, everyone went, 'We got to do something.'"

Seeing how quickly the public's natural resources were being consumed, the U.S. began to shift from a policy of giving away land in its western territories (through the Homestead and Mining Acts, and vast deeds to railroads) to holding those lands in order to secure timber, water, grazing, mining and recreational resources for all Americans. "These lands were not public but were held *in the public trust*," says Dusty, "meaning they were open for the possibility of mining and logging but it was no longer just a free-for-all. It's like a school; if it were just public, I ought to be able to go shoot baskets anytime I want. I can't. But in Montana, boy, don't tell me I can't go there and do whatever I want. That's mine, by God."

Dusty understands his neighbors' hostility toward the federal government. "A big driver of the backlash against federal lands was Clinton's designation of all those Monuments, top-down, with little concern for their impact on residents. I know Teddy Roosevelt used it and got stuff set aside in the nick of time. But Clinton triggered lots of resentment." Dusty has also had his share of encounters with tin-eared agency types. He recalls a meeting in Choteau for a bunch of ranchers with U.S. Fish and Wildlife's "grizzly recovery czar," whom Dusty describes as "a total biology introvert." It was "a somewhat hos-

tile meeting, like they all are. This guy was saying how until they do
their big DNA hair study and see how many bears we've got, they
can't delist. The ranchers are saying, 'There's plenty, they're everywhere.' And they are thick, you know. This guy says, 'Well, we can't
defend it in court.' The ranchers say, 'Yeah, you beeracrats just want
job security.' And on and on. Finally everybody simmered down and
this biologist said he had a presentation he's wanting to do. It's kinda
the wrong crowd for this but he's got a PowerPoint, you know. The
ranchers are all sitting there with their arms crossed and he starts into
his canned speech. 'When Lewis and Clark came down the Missouri
there were 50,000 bears in the intermountain West." Pretty soon,
from the crowd: "Hold it, you're telling me you know how many bears
were here when Lewis and Clark came through, but you don't know
how many there are now?' It just blew him out of the water."

For all their mixed feelings, however, Dusty and company are also
certain of the need for regulation, again citing the lessons of history.
"They killed all the elk until you couldn't find an elk," says Stoney,
"until Fish and Wildlife come along. It's greed or machoism; so what
stops us? Regulation: name that the Rocky Mountain Front Heritage
Act and people have to live by it. We've got to regulate ourselves or
we destroy everyone else's rights." Dusty's belief in the need for regulation was reinforced by what he calls the "gangster bangster" stuff
on Wall Street. And though "no one would fight harder than me for
private property rights—which, you know, always gets thrown out
there—with rights, there's responsibilities. We have to do the right
things. We can't just say, 'That's nobody else's business,' all the time.
'Cause a lot of what we do affects a lot of people down the chain." As
Ronald Reagan said in 1988, laws "protecting environmental quality
promote liberty by securing property against the destructive trespass
of pollution."

As for the Salmonds' claims to be the true defenders of Montana's
ranching heritage, Dusty notes, with an edge in his voice, that "Mark

Salmond's family had 60,000 acres from Ear Mountain to the Sun River. Luke and his brother split it up and sold off big chunks [to Gordon Dyal, then co-chairman of the Investment Banking Division at Goldman Sachs, and to David Letterman, who donated a conservation easement on his acres to TNC]. No one pushed them off that land. They sold their heritage; sold off their legacy.

"And Mark's opposition to the Heritage Act was over a permit for sixty-five head of cattle, every other year, that none of us wanted to take away. Grazing on public land is part of this culture; the balance of working and wild is what's so great. If someone objected to the Heritage Act because they just generally don't like the Forest Service or wilderness, that's not a reason. If they could say this bad thing will happen to me, then you had to listen and try to fix it. Mark Salmond can't say, 'I'm going to lose my grazing permit,' and he certainly can't say, 'I'm going to go broke and lose my ranch.'"

Dusty seems resigned to the fact that the Salmonds "hate my guts, and think I'm a traitor to my ag roots. There's almost a code you don't violate. And if you do, you're out of the club. I kind of violate it, working on the Heritage Act, the Nature Conservancy board. If I go in the feed store it'll get quiet, sometimes." Danelle, who is rarely without her warm, open, eager smile, is more stung: "Some of our neighbors who are mad at Dusty don't know that in those meetings he defended them every step of the way. They don't want to see the complexity, that there's more layers than just that one." Some of Dusty's partners are also pained by the rifts: "You're afraid you're going to get in an argument," Chuck said in the film *Common Ground*, "and they're your neighbors and your friends and you don't want to."

"James Salmond told me I ruined his life," says Gene. "He hates me. I don't hate him."

While working to pass the Heritage Act, Dusty forged links with groups across the Continental Divide similarly committed to overcoming historic enmities to protect their livelihoods and land. All recognize the Crown as a single, interdependent landscape. Wildlife, weeds, pine beetles, fire—all move across great distances and none recognize property or jurisdictional boundaries. "It's more than about Karl and Dusty's ranches, or my little piece of the Bob Marshall," says Roy. "It's about a huge system that's still intact in this day and age."

Dusty's key partners include Ovando rancher Jim Stone, chairman of the Blackfoot Challenge, a widely emulated group of western Montana landowners working together on resource issues, and Melanie Parker, whose group Northwest Connections monitors rare carnivores for the public agencies tasked with protecting them. The three make up the leadership of the Working Lands Council, a network of landowners and community-based non-governmental organizations (NGOs) advising state and federal agencies. "The idea," says Dusty, "is to partner with agencies to get better outcomes than if the cowboys and farmers just dig in their heels and the agencies just bull their way through."

Because the west side of the Divide is wetter, more heavily populated and densely timbered, Dusty's partners face even higher tensions: more-intense development pressures and "urban–wildland" collisions, including increasingly destructive forest fires threatening new homes and more frequent human encounters with grizzlies and wolves, water conflicts amplified by the presence of a famous threatened fish (the bull trout, star of the film *A River Runs Through It*), a beetle epidemic that has killed forty million acres of forest across the West, and even more bitter hostilities, as intimate and inescapable as a family feud, left by some of the most virulent natural resource battles in the nation's history, over the fate of old-growth forests and the spotted owls, lynx, fishers and fish they shelter.

As the most established of the networked groups (originating in 1991 with community meetings at Trixi's Antler Saloon, home to legendary swing dancing), the Blackfoot Challenge has taken on the mission of distilling the principles Dusty and others have arrived at to get past such rancor. Conservation is both the Challenge's mission and a kind of case study (per its executive director, Gary Burnett), revealing a democratic process useful to any community trapped in its own divisions.

The first principle Dusty's alliance and the other groups share is to start small, with the people still willing to give talking (and even more, listening) a try. Big public meetings wind up dominated by loudmouths and can "take you five steps backwards," says Jim Stone. "Our early ones were damn near slugfests; everyone just went home pissed off. So we found the handful of people who aren't going to tear heads off, and started there."

Their second principle is to focus on the 80 percent that unites them rather than the 20 percent that divides. If they can stay with the conversation, as Dusty did, month after month, year after year— braiding in as many different perspectives and as much information as they can find, keeping extremists of any stripe from hijacking the conversation—inevitably they come to agreement on what to do.

The third principle emerges from those many hours of deliberation: respect for the contribution each partner brings. That means locals recognizing the expertise of biologists working for public agencies and NGOs, and those experts, in turn, seeing the essential value of local knowledge and shared history. Melanie's husband Tom knew wildlife behavior from years of tracking and hunting, but without a college degree he had difficulty persuading the Forest Service that the decisions they were making based on science done elsewhere ('Well, grizzlies do this in the Yukon, so . . .") were wrong. To establish credibility, he and Melanie began rigorously documenting the movement of ermine, mountain lions and wolves: not trapping or drugging or radio-collaring but following signs like paw prints and scat. They especially love snow-tracking, says Melanie: "The story is just laid out for you. You can see family groups hunting together." They set cameras and baited stations that lynx or wolverines can reach only by crawling across Velcro or barbed wire, leaving bits of their hair behind. With genetic testing Tom and Melanie have been able to tell the agencies exactly how interrelated their 765 grizzly bears are. "We can tell that a cub's mom is from the Crown and its dad from the Cabinet Mountains," says Melanie. "We're not Jarbridge," she says, referring to the Nevada group that menaced Gloria Flora and her Forest Service staff. "We're not agitating for local control. But we do want to get our local knowledge into the decision-making process, and the work on the ground. Often federal money is prescriptive: 'Go do x, y, z to restore fish habitat.' Our view is, 'We know way more about it—give us the money and performance targets, and we'll figure it out.'"

That can lead to unexpected outcomes. The Forest Service discovered, for instance, that the only people capable of helping them understand severely endangered lynx were local fur trappers, with whom they've now teamed up on protection strategies. "That requires a culture shift," Melanie continues. "The agency view has long been, 'We're the experts.' The chiefs of the Forest Service and Fish and Wildlife now get it, but need to give the field guys direction and funding to engage; they don't have the staffing, and their performance targets aren't based on involving the community."

The final and most important principle is that any path forward has to support the local human culture and economy, for the sake of those families and communities and also because, as Dusty came to understand, private lands provide enormous public value. If ranchers (and, in these heavily timbered regions, loggers) can't stay afloat, everyone loses. Dusty's circle of partners includes innovative local logging and milling enterprises like RBM Lumber, a family-owned company restoring forest health by harvesting, and finding marketable uses for, the dead, dying and scrawny trees that desperately need to be thinned. These small companies are nothing like the multinationals that came in with "the old slash and trash," as Dusty calls it. "Those big timber companies logged right to the edge of the stream," recalls Roy, "took every stick of wood and all the shade, leaving water too warm for salmon to spawn, big eroded washes and knapweed everywhere." Instead, companies like RBM log selectively, without tearing up the ground or taking out the ancient trees that anchor these ecosystems. "Of all the resource industries, probably the hippest guys are the loggers now," says Dusty. "They can do it pretty cool."

As on the Front, this steady, open engagement with people once viewed as the enemy has left everyone transformed. Melanie, who moved to Montana after graduating from UC Santa Cruz, arrived with ready-made views of her neighbors. At early meetings, she avoided the trappers, unable to imagine she could find any common ground. But she wound up respecting and loving the late, legendary trapper Bud Moore. "Bud would say, 'Where are your eyes, Melanie? They're on the front. You're a predator.' He'd explain how he laid his traps so they wouldn't take the core population. I realized these guys trap because they love these animals, love thinking like them—'Let's see, this bobcat is denning here so will walk there'—almost becoming them. And *they also* went through a transformation, getting past the idea that the agencies and academics were just out to shut them down. If you take the time to sit down and listen to how a person thinks, how they see the world . . . if you don't patronize or placate but take them seriously, people quit feeling the need to defend their lives."

Of all the issues confronting families on both sides of the Divide, none stir deeper passions than the predators they live alongside. Wolves, which were nearly eliminated in the contiguous U.S. before being reintroduced in 1995, and remain shrouded in the terror instilled in generations raised on the brothers Grimm, are especially controversial. "Grizzlies have been here all along, but the wolves were long gone," Melanie says. "Even if you're seventy you don't remember them on the landscape." With a litter of pups each spring, they've also been quick to repopulate, with just sixty-six wolves now grown to an estimated 1,700 across the northern Rockies. "They're a biological success story, but less successful in winning cultural acceptance," Melanie says. "Especially by sportsmen, who fear they'll unravel the deer and elk."

That fear is largely misplaced. As Aldo Leopold famously explained in "Thinking Like a Mountain," one of the essays in his 1949 *Sand County Almanac*, "apex predators" are critical to maintaining the health of prey species. Leopold, who like Dusty combined reverence for a fellow animal's life force with a pragmatic assessment of that animal's value, described watching the "fierce green fire dying" in the eyes of a she-wolf he'd just shot. "I thought that because fewer wolves meant more deer, that no wolves would mean hunters' paradise . . . Since then I have lived to see state after state extirpate its wolves. I have watched the face of many a newly wolfless mountain . . . wrinkle with a maze of new deer trails. I have seen every edible bush and seedling browsed . . . every edible tree defoliated. . . . In the end the starved bones of the hoped-for deer herd, dead of its own too-much, bleach with the bones of the dead sage . . ."

Wolf advocate and *Audubon* magazine columnist Ted Williams describes other public services performed by the packs. By taking out coyotes, which kill sheep and eat 80 percent of the region's mice, wolves actually reduce sheep losses and repair several food chains: restoring to foxes, weasels, badgers, owls and hawks their whiskered dietary staple.

Having seen three wolves cross his land just as the 2014 calving season began, Dusty—like most ranchers—fears them far more than bears. "They'll run the cattle to where calves don't gain weight, and heifers don't get bred up 'cause they're scattered, and the grass and fences get tore up," he says. "It's a huge hit: not dead animals but dry cows in the fall and calves a hundred pounds lighter than they should have been and sick." The ranchers' response is emotional as much as economic, says Gary Burnett. "Elk can eat a lot more money, but when elk eat grass it's pretty; when a wolf kills a calf it tugs at your heart."

Dusty is also leery of the wolves' most virulent defenders, and what he views as their misuse of federal law and courts. "When people sue over wolf delisting, they say it's about genetic connectivity, but many simply do not want a wolf killed for any reason," he says. "Rather than state their purpose they use the minutiae of federal policy. They're law firms disguised as conservation groups, and do little to actually make the world a better place." In fact, says Williams, they often make things worse. "Wolf lovers have been no less a threat to recovery than wolf haters," he wrote in a July 2013 piece for the Yale School of Forestry blog. "By 2008 the recovery goal for the northern Rockies . . . had been exceeded by at least 300 percent. But there are some environmental groups that want nothing delisted *ever*." Their lawsuit to prevent Montana and Idaho's delisting on a "technicality . . . so enraged Congress that it delisted [the two states'] wolves itself . . . It was the first congressionally mandated ESA delisting—a horrible weakening precedent for our best and strongest environmental law." Grizzlies, says Melanie, have been similarly harmed. "All the litigation to stop logging or development is based on the Endangered Species Act. So much of it is 'in the name of the bear.' That just creates bad social habitat here for the bears, which is as critical as physical habitat. If people hate 'em enough, we'll easily eradicate them."

Instead of litigating, Dusty's partners have devised collaborative strategies to keep both livestock and wild carnivores alive. The Blackfoot Challenge sends out "range riders" on four-wheelers, to ferry information about predators' whereabouts and activities back and forth between wildlife managers and landowners, and to create a human presence and scent to ward off the animals. They also collect and compost the few calves inevitably lost at birth or soon after. "Calving overlaps with grizzlies coming out of hibernation," explains biologist Sean Wilson, who designed the program. "For hundreds of years, every ranch in the West had a boneyard right near the calving area; they're not sleeping for two months so if they lost a calf they just buried them right there. We sat down with a hundred ranchers and maps of where there'd been conflicts, and realized the boneyards were drawing the predators at the calves' most vulnerable time." To remove that bait, the Challenge began carting the dead calves (and roadkill) to a composting facility and from there to reclamation projects. "The bears are back to eating their natural food instead of teaching their cubs to camp out on ranches, which has dropped the need to kill bears more than 80 percent."

Wolves remain a tougher challenge. Jim Stone believes that "ranchers need to do everything possible to prevent wolves from getting killed. If we haven't, we can't complain when they add more protections." Standing in a pasture, he shows the electrified wire he can quickly spool out when needed to protect vulnerable stock, decorated like a used car lot with "fladry," bright orange flags that ripple in the wind and have been used for centuries to spook wolves away. Jim is compact and intense, a fast talker, but falls silent to listen appreciatively when a nearby prairie pothole springs to life with the staccato, muted-brass cries of trumpeter swans. "But it can't just be me against the wolf. People don't understand how thin the profit is in ranching. The loss of a thousand-dollar calf is a big deal."

One of the most striking things about the Crown partners is their willingness to disagree with one another. They have no dogma, no party line. Dusty has adjusted his practices to be more wildlife-friendly: he uses smooth, high-tensile wire rather than barbed wire, so that antelope can slip beneath the fence and elk slide over the top. On the rare occasion that a newborn calf dies, he drags it across the road from his calving pastures so grizzlies can feast on it without disturbing the peace. And despite pressures from some hunting groups who consider wildlife a public asset and lobby for laws to prohibit "hoarding" it on private land, he lets only a few (usually young, beginning) hunters on his land. "We feel it's our responsibility to provide a safe zone for the animals," says Danelle. "They're constantly bumped and pressured, with people on the road hazing and chasing them." But Dusty has mixed feelings about fladry flagging and fenced-in calving areas. "This is not a position statement; it's a thought process. But I don't think it's necessarily the best fix for the survivability of the wolf or the bear. When you look at the lower forty-eight, the Crown and the Yellowstone are their two biggest habitats, and are still not big enough to have wildlife populations that aren't going to bump into humans. For those species to survive they have to learn to avoid conflict. Every last grizzly's going to have to change its behavior because of human activity. Do we keep trying to baby-proof the room, keep putting stuff out of reach? Or do you have to start with that bear and say, 'No, you can't go there'? By now we've killed a lot of the more aggressive bears, so we're getting a more timid bear. These are questions of mine, not, 'This is how we should do it.' And it absolutely hinges on nothing changing as far as available habitat. Habitat loss is a huge game changer."

Karl has struck something like the deal Dusty describes with the wolves on his ranch. "I lost seven calves and two three-year-old heifers in 2008 to a pack led by an alpha female tracked [by radio collar] clear over to Ovando [nearly 100 miles away]. As soon as she got here, all the elk packed up and left the ranch. She had all these pups and my cows were a hundred yards from her den, so she went to killing to feed her pups. U.S. Fish and Wildlife came out and shot one male out of that bunch, but I told 'em she was the one doing the killing. That fall, the first year of the wolf season, she was killed by a hunter in the Bob Marshall. As soon as she was gone the rest of the pack went on living with the cattle and never bothered them again; her old pack is still here. Then in 2012, we lost about six calves to a different pack that came down off the reservation where they'd been killing a 600-pound calf every other day. I had a kill permit and took about half that pack, fourteen wolves, and last year the alpha female had pups on the ranch and lived with the cattle all year. Every time I went up there, her and her pups were in amongst the calves, doing no harm. They're part of the land."

This collaborative, place-based movement continues to expand. In 2008 Jim Stone founded Partners for Conservation, pulling together stewardship-minded landowners and coalitions across the country to share knowledge and bring their collective voice to policymaking. In 2013, the Working Lands Council formalized its consultative role with the U.S. Departments of Agriculture (USDA) and the Interior. Jim is also on the board of the Conservation Leadership Council, founded by George W. Bush's Interior Secretary Gale Norton to advance conservative solutions to environmental challenges. "It's a big world," says Jim. "If we just focus on my problem, a single issue or species, we become a country of every value becomes the end-all, and nobody gets anything." Dusty has visited several of the Partners for Conservation members, including the Malpai Borderlands Group (straddling Arizona and New Mexico) and the Northern Everglades Alliance. "Like us," he says, "they're people of the land who are not going to back up completely or be steamrolled, but know that things change and other people become stakeholders whether or not you think they should, and you need to try to work with those folks." In December 2014, Dusty's willingness to work across divides paid off: the Rocky Mountain Front Heritage Act, attached as a rider to the National Defense Authorization Act with critical support from newly elected Montana Republican senator Steve Daines, abruptly rose from its moribund state and was passed into law. Speaking on the Senate floor in support of the bill, Jon Tester quoted Stoney: "There are some places on this earth that ought to be left alone even if solid gold lies underneath them."

Five months later, dozens of people, including Senator Daines, gathered under squalling skies at Chuck Blixrud's guest ranch to celebrate the victory. Dusty led off, paying tribute to the role every single person in attendance had played. "How many people out there today wrote a letter to the editor or their congressman or senator? How many attended a scoping session or public meeting?" He then moved on to pay tribute to the "real pioneers in conservation, four old guys that we need to immortalize for time to come." They'd commissioned a sculpture, he said, that's "not completed but we do have an artist's rendering." With that, he unveiled the punchline: a picture of Ear Mountain as Mount Rushmore, with the heads of wilderness advocate Bill Cunningham,* Gene, Stoney and Roy.

"You're never going to replicate the West as it was, but we've got a big enough chunk here. Hopefully years from now they'll say, 'They got it right.'"

Since passage of the Act, Dusty has had more time back home, bringing his stewardship ethic to his cattle and 2,400 acres, though he concedes that he's in the early stages of all he hopes to learn and change. Danelle, who has degrees in animal science and medical technology, is his full partner in running the ranch; both are committed to passing on good stock-handling and range-management practices to their kids.

What they're proudest of, so far, is their stockmanship, which is based on the principle that cows should be handled as minimally and gently as possible. Both have taken courses in Bud Williams's "Low-Stress Livestock Handling" methods. "He's a legend," says Dusty.

---

* Another of Dusty's heroes, Cunningham played a key role in mapping roadless areas eligible for protection under the 1964 Wilderness Act. "He's so wiry and little his pack drags on the ground," says Dusty, "but he celebrated his seventieth birthday bushwhacking the Chinese Wall."

"He herded wild caribou in Alaska." Their unhurried approach is particularly evident at branding time, which on many ranches becomes a kind of rodeo, says Dusty, "with calves drug by the heels to the fire, everyone hollering and galloping; too many people and too much beer."

"We try to work quietly," says Danelle. "We never use four-wheelers but do all our work on horseback, with our kids and the friends that come to help. Nobody's yelling or beating on cows. We don't break out of a trot, sometimes even rope at a walk. We just sneak into the herd, gently lead out that calf, do what we need to do and let 'em go right back to their mom. It's all about their body position and yours; you got to read when to pressure that cow and when to let her find her own way." For many years, the Crarys had leases on a neighbor's land at the base of Ear Mountain, where they would do "rodear" branding, meaning the cows stay together and out on the range. The whole family would ride out for several hours, in places through withers-deep grass, with one or another of the kids driving their team of Belgians pulling a wood-wheeled chuck wagon, and all camping in cowboy teepees.

"It's a more respectful way to treat them," says Dusty. "And less stressful, which means less chance they'll get sick and need antibiotics. We also breed for disposition. You can get some pretty wild animals, who just want out of here, and you're not going to change them. We've selected bulls and cows that are quiet and trust us. It's all about mutual trust and respect."

Dusty's patience extends to his horses and mules, which he counts good practice for dealing with politicians (though "mules are nicer," he says.) When a big male mule (a john) keeps jabbing a threatening hoof his way, Dusty just eases him into shoeing stocks, quietly sweet-talking and stroking his nose. Learning not to rush a mule or "get in the middle of a mule argument" is a lesson Dusty learned the hard way, when on a backcountry trip he was kicked in the kidney by a jenny getting harassed by a john. "It felt like stepping into traffic and getting hit by a truck. Your kidney's inside a membrane like a ziplock bag. I got back on my horse and rode to where I could call for help. They told me if I'd rode any longer it would have burst and been the end."

Their philosophy of letting their animals be has also reshaped the Crarys' feeding practices, again informed by their continuous pursuit of education and improvement. In this case, the inspiration came from Dave Pratt's Ranching for Profit School. "A key has been matching our cows' nutritional cycle to the availability of grass," says Danelle, "so they get the extra prenatal boost they need directly from grazing." Though they still plant and irrigate alfalfa and sainfoin (both perennial legumes), they now just turn the cows out into the hay meadows to harvest it directly. That saves significant money. "One of our biggest costs is bringing in hay," says Danelle. "You've got to cut the hay, windrow and bale it, then get a tractor and trailer to bring it in and stack it. Then, when you need it in winter, you get the tractor out again and move it back out to the field where it came from so the cows can eat it. That's a lot of diesel and labor. In the dead of a Montana winter you still need some hay. But from three ton a cow each winter, most years now we feed less than half a ton. At $200 a ton, that's a huge savings." It also benefits the land, enabling the Crarys to ditch the monocropping required by mechanical harvesters in favor of a mixture of grasses and legumes that fix nitrogen and other nutrients and shelter the soil, a mixture that a cow has no problem harvesting.

Like other Crary innovations, that change unsettled some neighbors. "In those first years," says Danelle, "the cows'd hear the tractor and line up thinking they were getting fed. They were over on the river and fine; we had this little longhorn who would lead them out into the best forage. But our neighbors would go by and see them lined up and say, 'Crarys are starving their cows.'"

"Ranchers have taken cattle that were more like a wild animal, able to take care of themselves, to breeding a cow that doesn't know how to go rustle and eat," says Dusty. "When you challenge them there's some fall-out. Those that don't handle it, we say, 'OK, I'm going to feed you but I'm going to sell you.' We're selecting for those cows that hold their condition well."

That hands-off philosophy is even at work during the most intense time of the year, when their cows give birth. While most ranchers calve in winter—to have fatter animals to sell in the fall—the Crarys again buck convention by calving in late spring, when they can count on milder nights and sunshine to warm and dry the newborns.

In that more temperate weather, they can leave the older, experienced moms out in back pastures to fend mostly for themselves. Young heifers having their first calf "need to be watched a bit more closely," says Danelle, so are moved into the front pasture between Dusty's house and Bonnie's, where all can keep an eye on them. Still, the family gets close to the heifers only when they have to. Danelle scans with binoculars from the front window, and with Dusty (and the kids when they're not in school) takes turns going out every hour, day and night, to quietly check on the laboring moms and the little ones struggling to stand and nurse. Bumping around the field in a pickup, Danelle pauses at some distance to watch a cow pacing, but moves on satisfied when a drenched, glossy black little creature slips out onto the ground, cord hanging from its belly, and its mom begins licking it and her own spilled body fluids.

Danelle does get out of the truck to check on a dwarf calf she's named Little Bit, who was so small at birth she got chilled by a cold wind and had to be thawed in a warming box Dusty built; Little Bit has now figured out how to stand directly beneath her mom so she can reach her udders. Danelle also greets Beary, a pregnant two-year-old with a deep divot on her back who earned her name at birth, too, when a grizzly got hold of her and stripped out her whole loin. "She had a very protective older mom," says Danelle. "I can just picture what must have gone on: that cow fought for that calf's life. But you could see her spine and hip socket exposed. The vet said she wouldn't make it, but I love these animals so much, and don't give up easily. Her wounds were full of maggots. Dusty said leave them on there to clean out the proud flesh, and they did, and it healed."

Danelle's pace abruptly accelerates when she spots a calf hanging half-born out of its mom, completely limp. When the cow startles, the calf falls to the ground in a lifeless heap and Danelle rushes to kneel next to it, folding its gangly limbs closed to warm it. "She just kinda quit pushing. The calf was trying to take a breath and couldn't. His eyes were starting to roll back in his head." An hour later, back in from checking, she tells Dusty, "That little one that fell out dead is running circles around his mom."

Bottle-feeding a particularly wobbly little calf, she recounts other midwifery dramas. "I had two cows last week going too long, and I thought, 'OK, I've got to intervene.' I crawled on my hands and knees so the mom couldn't see me, got hold of that calf's front feet and put just enough pressure to get her pushing a little harder. When you can see a nose and hooves sticking out right side up, that means he's coming out right. If the hoof bottoms are up, he's coming out backwards. Sometimes you can reposition them. This one [she strokes his head] we had to just pull out backwards, and he's still kind of slow."

Though they've not had a C-section in many years—"we buy bulls that throw calves of appropriate size"—she and Dusty used to do the surgery themselves. "We'd give the mom a spinal block, clip and scrub her down, make the incision through the different layers, split that uterus and get that calf out. Then you've got to be really fast: the uterus starts shrinking right away and you've got to close it back up."

As she talks, nine-year-old Carson comes out with another half-gallon baby bottle, offering it to a lonely little calf who eagerly latches on with his little snub mouth—a perfectly designed sucking machine. This baby's mom got mastitis "and was killing him," Carson says. Their young dog named Cat hangs about, licking the suck-ing calf's mouth to catch any spilled milk. Nearby, another cow is restrained in a chute so the calf she's rejected can nurse; she stamps angrily while outside the barn a deafening wind roars. Dusty comes in to rub the bowed legs of "little backwards guy," explaining that though they rarely lose calves, they're prepared if they do. "If you've got a good cow and her calf dies, gets sick or is born dead, and you have a calf like that rejected one waiting in the wings, you skin the dead calf out and drape his skin like a cape over this other one. If she's a good cow she'll take 'em and after a couple days you can take that cape off."

At summer's end, it falls to Danelle to decide which of the previ-ous year's calves will remain as replacement heifers, and which will go. "You do get attached; when you got to send those ones down the road, it's tough." She drives them to Kalispell, three hours away, to a small family-owned slaughterhouse called Lower Valley Meats. "They're really good people," says Dusty. "They don't have a bunch of goons working there; they handle the stock right. The only thing better to me would be all on-farm kills, where a big old steer's having a perfectly wonderful day in his pasture and it's lights out, no stress. That's how we all want to go: a perfectly wonderful day and 'Presto.'"

Danelle also handles the marketing of their beef, through direct sales, retailers like Choteau's local and organic Mountain Front Mar-ket, and the EatWild website. "We've drunk the Kool-Aid," Dusty says, on the health benefits of grass-finished beef, and their ads reflect that, along with their stewardship values. "Locally-owned Crary Beef is grass-fed-and-finished with no antibiotics, no hormones, no animal by-products, no grain and no confinement from start to finish. These Black Angus cattle are born and raised on our fourth-generation ranch. From birth to harvest, they happily consume omega-3-rich native grasses and forage along Montana's Rocky Mountain Front. They are never confined, are always handled in a low-stress manner and share pastures with grizzlies, elk and sandhill cranes. We are passionate about producing a nutritious and delicious product, rich in good fatty acids, conjugated linoleic acid, vitamin E and beta-carotene, while improving and maintaining native grasslands and riparian areas. Whole, half or quarter beef are seasonally available late June to October, to capture the greatest nutritional benefits of the lush grasses. Individual cuts can be shipped year-round."

As the global appetite for meat grows, especially among the half billion people in China's fast-growing middle class, the environmental costs of meat-eating have come under increasing and necessary scrutiny. The costs of feedlot, grain-finished beef seem clear: they include the land, water, fossil fuels and chemicals used to grow and transport feed grain; the pollution that runs off some feeding operations, often into the Mississippi and Gulf; the overuse, by some producers, of growth hormones and antibiotics, contributing to the rise of resistant bacteria dangerous to human health; and the brutal conditions in which some animals live.

But grass-fed beef is also complicated. Though grasslands cover 25 percent of all land on earth, that still isn't enough to provide grass-fed beef for all the world's new meat eaters, even with big dietary changes in the developed world. Grass-finished cows also take longer to fatten than grain-fed beef and therefore over their lifetime emit more methane, a powerful greenhouse gas. But their wild ruminant brethren—pronghorn, bison, elk—emit methane too. And one way or another, these grasslands *need* to be grazed. "Across the west, ungrazed rangelands are the worst places we have," says Wyoming-based USDA rangeland scientist Justin Derner. "The ground is bare or overrun with species we don't like. These grasses grew up evolutionary-wise with grazing; you take that grazing away and bad things happen." Keeping grasslands healthy is vital for the myriad species they shelter and for keeping in place their vast reservoirs of carbon, essential to soil fertility and climate stability. In the U.S. alone, grasslands lock up 158 million metric tons of carbon each year; with better management, of which good grazing is a critical part, they could sock away 70 million metric tons more. Animals that can turn indigestible cellulose into digestible protein also help keep grasslands intact, by providing a way to use them to feed humans without converting them to crops—nearly always an ecological disaster. Even the best farms are far poorer in biodiversity than the native ecology they displace. A December 2015 study, for instance, found grassland-to-cropland conversion to be the prime culprit in the decline of pollinators.

While immensely proud of his stockmanship, Dusty counts himself a beginner in managing the challenges on his land, which are both a farmer's (on his irrigated hay fields) and a cowboy's (out on the range).

"On my irrigated pastures, because we've had cheap inputs, I haven't needed to know how to read the land; we lost a lot of soil management and rotation skills. You could just put on fertilizer and make stuff grow. You were depleting the soil, but that can take a while to show. That's why we got all these Ponzi schemes: people getting a few good yields out of virgin land, then selling it just as its fertility was exhausted. You go to a lot of ground and boy, you won't find much for worms or bugs or stuff that should be there; dung beetles have about disappeared and are huge partners in soil health. And I've got nowhere to go but up on organic matter, though I'm just learning about things like water and nutrient cycling. I hope tomorrow I know a little more and a little more."

On his rangelands, though the sod was never busted, he faces the equally daunting challenge of restoring native grasslands damaged by a century of overgrazing: stripped of vegetation and a protective microbiotic crust, pounded hard, and increasingly overrun by gangster weeds. The northern prairie's native perennial grasses have deep root systems that provide many crucial services, including stabilizing the soil and capturing and transporting scarce water. But they can't compete—particularly when grazed too hard—with invasives, which are typically annuals or short-lived perennials and reproduce far more quickly. Checking on his horses, Dusty points out—amid the native Western wheatgrass, shrubby cinquefoil, needle and thread, arrow-leaf balsamroot and fringed sagewort—wet patches thick with Garrison creeping foxtail, an introduced perennial. Though an "increaser against native sedges and forbs"—meaning an indicator to alert ranchers that the land is being overworked—the foxtail is "pretty good forage," Dusty says. The real menace to him and other ranchers are the thirty noxious weed species that have invaded close to 8 million acres of Montana's prairie, outcompeting native and more desirable forages, reducing fodder for domestic and wild grazers by up to 90 percent, and exacting economic damages across the northern prairie in excess of a hundred million dollars a year. The invasives' names are as nefarious as their deeds. Spotted knapweed, a thistle-like pink pom-pom, secretes chemicals that kill native grasses. A single Dalmatian toadflax—a vamp with rubbery, heart-shaped leaves and showy orange-throated flowers with yellow snapdragon lips—can produce a half million seeds on papery wings that remain viable for ten years. Most dreaded is the yellow-green perennial leafy spurge, capable of reproducing both from its thirty-foot roots and from its pods that explode when mature, launching seeds fifteen feet from the parent plants. Most of these weeds are of Eurasian origin and many were planted in earlier reclamation efforts, says Dusty. "They didn't know a hundred years ago that it'd be a wreck."

The harm goes far beyond the cowboys and their cows, says Dusty. "Because leafy spurge is rhizomatous,* it can create a monoculture, a disaster for the pollinators, the birds, everything." Invasive weeds also accelerate loss of soil carbon to the atmosphere. "Native grasslands have about 95 percent of their biomass below ground, holding soil

---

* Sending out subterranean stems that sprout new shoots all around.

and carbon in place," Derner explains. "But invasives have tremendous above-ground growth. Weeds actively suck carbon out of the soil to grow faster; with their fine roots, they bring the carbon nearer the surface, where it can be acted on by microbes, temperature, rain—all burning it back up into the air. Many also grow early in the season, using soil nitrogen before the natives can get to it."

The solution, again, is not to quit grazing but to graze well. What that means is complicated.

For four decades, a dominant voice in grassland management has been Allan Savory, who developed his ideas on how grazing could reverse desertification in his native Rhodesia, importing them to the U.S. through his Center for Holistic Management in New Mexico. Savory's idea is that livestock can actively restore lands if managed to mimic wild herds: bunched tightly; allowed to graze intensely, stimulating plant growth and churning and fertilizing the ground with their teeth and hooves and manure; and then quickly moved on, leaving that worked-over land to rest for significant time. Long a cult figure, he gained mainstream fame with a 2013 TED talk that went viral with its (alas, overly optimistic) claim that good grazing could by itself reverse climate change.

If "mob-grazing" is not a silver bullet, Dusty's experience and teachers have persuaded him of its value. In his irrigated pastures, he has put up big perimeter fences around his watering pivots, with plans—when he finds the money and time—to fence off pie slices so he can rotate the cows from one slice to the next. "I'll die building more fence," he says. "How do you stop the squirrel cage long enough to make big changes, especially when you've got bills to pay? You never get there. You're always just trying to improve."

Out on the range, watching his horses move across the steppe, Dusty points out how their hooves "break off the litter and smash it into the ground, which shades and protects new plants, keeps the ground cooler, helps hold moisture and mineralize the soil. That's how this country evolved. Even in the Pleistocene, this land evolved with grazing." Grazing is particularly valuable, he says, as a first line of attack on noxious weeds. If he piles his cows into a weedy pasture in the spring, when invasives are the only show in town, they have no choice but to eat the undesirables. "And rather than turn 100 head in to 3,000 acres for the whole summer so they can cherry-pick the ice cream plants," says Dusty, he packs them tightly enough that they have to grab whatever they can before the next cow gets it. "If we create enough grazing competition, they'll eat everything."

But this arid landscape, with rainfall averaging just eight inches a year, presents additional challenges. "The carrying capacity on some of this land is *ten acres to the cow.*" That means herds need to be moved over a much larger expanse of land than most individual ranchers control. Ideally, says Dusty, grazers' associations will step up, combining their land and leases and using herders educated in Bud Williams's techniques, which are more flexible than fencing. "Nothing's static," says Dusty. "It constantly ebbs and flows with moisture systems and disturbance and pressure and fire. You can look at what's on the ground and say, 'I want cattle here now.' But you also have to give this native stuff a nice long rest. We might only graze it every fourteen months, so we never hit it at the same season and it has a chance to seed. The number one thing is your grass and your soil; cattle are just the vehicle to market this grass through. As one guy said, 'We're not cattle ranchers, we're used sunlight salesmen.'"

Those limits on where and when he can graze mean that Dusty needs further strategies for battling weeds. He and Danelle have tried goats, famous for eating anything, to strip the weeds' leaves, and they regularly release weed-eating insects such as hawk moths and root-boring beetles. "The bugs'll have a good year and hit it hard," says Dusty. "But then the bugs might get hit and the spurge gets ahead again." Some ranchers teach their calves to savor spurge, dousing the weed clippings in molasses and feeding them to the little ones by hand, a strategy Danelle has begun to try. When all else fails, they spray. "We never spray by the river," says Danelle, "and we're not out putting big booms and dousing everything. But we do have a hand sprayer and walk the pastures to hit individual weeds. I hate buying the stuff, having to touch it. But I feel it's my responsibility to save the native range. And if I don't do anything about those single leafy spurge plants, pretty soon it's all going to be yellow and we won't have any range left." They get help from her brother, a specialist in precision agriculture, in pinpointing the most important targets. "One spurge plant is far more of an ecological catastrophe for the native community than the pesticides I'd use to spray it," says Dusty. "But the time for spraying is our busiest, with calving, irrigating, fixing fence, moving cattle. So we get together with neighbors to have spray days. This is pretty socialistic, collectivist thinking, but we need to look at the biggest threat."

In all of this work—calving, feeding, fence-mending, branding—the Crary kids join in without (it seems) ever being asked. "Conor's a teenager," says Dusty, "but he works like a man now." When he bursts in the door one afternoon, returning from a friend's, his mom asks him to go check on the laboring heifers. He already has. "They know the signs now as well as we do," says Danelle. "How that cow's udder gets tighter and she gets restless, switching her tail." Conor and Chottie are expert ropers, regularly joining their parents at clinics and in the thick of things at branding time. Carson stays equally busy: on the coffee table is a field guide to local birds he wrote and illustrated, which along with his Lego backhoe won a prize at the Choteau county fair; when his dad comes in from shoeing he rushes to tell him about the injured magpie he's caught in a kennel and is feeding with dog food. (With AC/DC blasting from an iPod docking station, he has to shout.)

The whole family is also deep into 4-H, with each kid raising several animals every year and all spending weeks in preparation, scrubbing the steers and pigs for the show and auction. Then the painful day comes.

"It's a terminal sale," says Dusty. "That's the end of the road."

"That Sunday," says Danelle, "it's all the kids in tears."

"Yeah, the little ones, some of 'em, not too bad, they're pretty brave, it's just . . ." Dusty snuffles and sniffs.

"Remember Conor the first year he sold a steer? The night before we had to go put him on the truck, he cried and cried at home here. We had an awful night. Then the next day he was so brave. He led that steer up there himself. And last year, Carson's first, he stayed right by the truck and talked to his pig the whole time. I was like, 'I got to go to the car.' All the moms are kind of disappearing in the background with their sunglasses on."

But work, and toughing it out, is what this family does. For fourteen years after she and Dusty married in 1993, Danelle drove 140 miles five days a week to work the night shift at the Great Falls hospital, "trying to make ends meet," while Bonnie watched Charlotte and Conor. The young family lived for almost twenty years in the original 1889 ranch house, the kids learning to walk on slanted floors hammered together with square nails. "One night they're sitting on the floor eating popcorn and watching a movie," recalls Dusty, "and here come a couple bats flying back and forth. They're so used to them they just duck their heads like that [he bobs] without ever taking their eyes off the screen. And I thought, 'Yeah, this is getting pretty bad.'"

And though the kids work and roam the mountains independently, "we believe in strong parental involvement. We know what they're doing and stay connected to them at all times," says Danelle. For three years, she home-schooled the boys. Dusty banned video games, "because you get numb to all that violence and pretty soon it's normal to kill people. What is this complete erosion of conscience in our country? Is it YouTube and I want to be a hero for a day and I've got absolutely no concept of death?" Conor has a .22 and a coonhound he's training to hunt mountain lions, but abides by unbreakable rules: he can't take the gun out without asking his parents and can't ever go shooting with just a bunch of boys. The legal hunting age is twelve, but the Crary kids have to wait. "Twelve is too young," says Dusty. "They're not physically ready to control the kind of rifle you need to responsibly take game. And they're not mature enough to understand that these animals are not just targets in a shooting gallery. Now the game department wants to lower the age to *nine*, to sell more licenses. Nine-year-olds don't know what death is." Recalling the infamous advertising icon that R. J. Reynolds used to attract kids to their cigarettes, he adds, "I'm like, 'Joe Camel's got nothing on you guys.'"

It's for his children, and his neighbors' children, that Dusty never gives up the fight. Standing on a ridge with the mountains stretched out behind him and Carson fidgeting alongside, his thoughts turn to the future. "I look at Carson and, man, look at how much is changing in the world and uncertainty and all the unrest and go, 'Jeez, if there's something I can do . . . I hope there's Yeagers here a hundred years from now, ranching and kids riding horseback, and I hope there's Crarys here and their descendants. I wouldn't be much of a man if I didn't feel some obligation to keep that intact for everybody."

There's a tradition among some Choteau residents, every September 11, to watch the seven-minute, shell-shocked monologue their neighbor David Letterman delivered six days after the 2001 terrorist attack, when he went back on the air for the first time. After recognizing the heroism of New York City's mayor, police and firefighters, Letterman ended with a tribute to his Montana neighbors and the hope they represent.

"I'll tell you about a thing that happened last night. There's a town in Montana by the name of Choteau. It's about a hundred miles south of the Canadian border. And I know a little something about this town. It's 1,600 people. 1,600 people. And it's an ag-business community, which means farming and ranching. And Montana's been in the middle of a drought for . . . I don't know . . . three years? And if you've got no rain, you can't grow anything. And if you can't grow anything, you can't farm, and if you can't grow anything, you can't ranch, because the cattle don't have anything to eat, and that's the way life is in a small town. 1,600 people. Last night at the high school auditorium in Choteau, Montana, they had a rally, home of the Bulldogs, by the way . . . they had a rally for New York City. And not just a rally for New York City, but a rally to raise money . . . to raise money for New York City. And if that doesn't tell you everything you need to know about the . . . the spirit of the United States, then I can't help you. I'm sorry."

2

FARMER

IN THE desiccating heat of a late June Kansas afternoon, Justin Knopf is in the air-conditioned cab of his cherry red combine—a 13-foot-tall, 18-ton, $375,000 mobile harvesting factory—cutting a 40-foot-wide swath through a field of ripe grain. With his two brothers, he's bringing in close to a thousand acres of hard red winter wheat, enough to bake 3.6 million loaves of bread. Strong spring winds have left some of the wheat "lodged": bent to the ground, the slender heads too heavy with ripe grain to stand back up on their own. So Justin is moving slowly, giving the combine's rotating rake, or "reel," a chance to right the stalks before the sickle lops them a few feet from the ground. A conveyor funnels a steady flood of the long, spiky heads into the machine's monstrous mouth and on into its belly to be threshed and winnowed: the grains pried loose from the heads by a spinning cylinder armed with sharp serrated bars, blown clean by a fan and dropped through sieves. As the chaff and straw fall back into the field, the grain piles up in the hopper behind Justin's head, glowing like a mountain of fairy-tale gold.

Sitting happily in Justin's lap is his two-year-old son Andrew, content for hours to watch the workings of the bright shiny machines his Grandpa Jerry, raking a nearby field of alfalfa, calls "great big toys."

When his dad allows, Andrew pushes the buttons on the joystick that tip or raise the combine's 40-foot-wide header with the satisfying whoosh and zing of an arcade game. Four-year-old Charlotte sits nestled against Justin's side, absorbed in her books and princess dolls and ignoring their slow slalom through the field until Grady Proffitt, a hired hand, brings the bright green grain wagon alongside. Expertly aligning it with the combine's path, he matches speeds so that for a few minutes the two behemoths move as one. Charlotte loves to push the button that swings out the combine's 28-foot auger and pours a torrent of grain into the waiting wagon. Her dad did too, thirty years ago when he was her age, though then it was a mechanical lever he'd get to pull, and sometimes they'd take off their shoes and climb right into the wheat to play. Filled to the brim, the wagon peels away and rumbles across the stubble to unload into the white tractor trailer waiting at the edge of the field. Stenciled on the truck door is the logo of this family business, Knopf Farms, whose local roots stretch back 160 years: an alfalfa leaf and a cross.

Grady radios to say the hitch has failed on the grain wagon, so Justin pulls up and climbs down from the combine to improvise a quick repair. Though six feet three inches and broad-shouldered, he has the lanky legs, baggy dungarees, close-cropped hair, round cheeks and joyful guffaw of a sweet, earnest boy. Slower to reveal itself is his complexity and strength: a formidable firmness of conviction, a calm solidity as imperturbable as the prairie he loves.

Retrieving a set of giant washers from the shop, Justin threads them onto the hitch-pin to keep it from slipping loose again. Though he's still as resourceful and adept with a wrench as his pioneering forebears, most of Justin's tools are light-years distant from the open tractor his dad used to bounce around on, its bare metal seat exposed to the sun and wind. The GPS that steers his tractor maintains a course accurate to the inch, freeing Justin to attend closely to the field. When he spots a patch where the wheat is thin, its heads pale and scrawny where the alfalfa crop that preceded it drowned in pooling water and failed to renourish the soil, a shadow briefly clouds his handsome face and he makes a note to tell his dad which corners of the field will need leveling. Over his right shoulder, multiple data streams flow across the combine's computer screen: moisture levels in the grain, ambient temperature and humidity, pounds harvested each second—all mapped against ground speed, latitude and longitude to generate a real-time acre-by-acre map of yield. (As recently as 2010, he would know his yield only when it was weighed at the elevator, and only in 160-acre blocks.) Justin's smartphone brings still more information: the history of the crops he's grown here and the fertilizer he's applied, wheat prices on the Chicago commodities floor, near- and long-term weather forecasts, details on the soil strata beneath every acre of his land.

The ill-timed winds have left the Knopf family stretched for hands. Jerry will work most of the night with a crew in the alfalfa fields, gathering back up the scattered new-mown hay before threatening rains leach out its proteins and much of its value, and leave a soggy mess that chokes off new growth underneath. (A perennial—which they cut four times a year for sale to dairies in Ohio and Indiana—alfalfa grows back like a lawn when mowed, if it's not buried in darkness under wet hay).

Justin and his brothers Jeff and Jay will drive combines deep into the night—navigating by headlight the ghostly fields they know by heart—climbing down to catch a few hours' sleep just before dawn, when the dew makes the straw too moist and pliable to chop. The faster they harvest, the less chance they have of falling prey to a hailstorm, which can shatter the herringboned heads of grain and in three minutes destroy an entire season's crop.

Justin texts his wife Lindsey to tell her they're in the "field across from Grandma's" and she brings out sandwiches and cookies for a quick lunch, though with the wind still blowing the kids huddle in the car and their parents picnic standing up, in the lee of Lindsey's SUV.

Under the midday sun the wheat is a towhead's white-gold, brilliant against the dark green of young corn and sorghum fields and, where the landscape swells, the stark blue of the Midwestern sky. As the harvest hours stretch into the late afternoon the wheat turns caramel-colored and fragrant, like the nourishing loaves it will be baked into in Lagos or Cairo, the destination for most of the grain in the elevator Justin sells to, one of the largest in the world.

At six o'clock the extended family and hired hands gather again, this time for a harvest dinner prepared by Rachel, Jeff's wife. Jeff and Rachel's house, one of several scattered about the 4,500 acres the family farms, overlooks the wooded banks of Gypsum Creek where it meets the Smoky Hill River. It's a special place for the family. "Our grandpa grew up fishing right in this spot," says Jeff. "I have a lot of channel catfishing, flathead catfishing memories. It's a wonderful place to raise my two boys." From here, the water flows into the Kansas (or "Kaw") River, the Missouri and finally the Mississippi, where it travels 700 miles to the Gulf. "Farming along this creek, seeing it all the time, you're aware of the interconnectedness, of everyone and of the water and the land," says Justin. "You're reminded that what we do as we farm impacts folks on down the river."

Before the meal, Jeff asks Jerry to offer the prayer. "Thank you for providing for us," he begins, "for our good helpers, for this beautiful land." The conversation meanders quietly, roaming from how to solve this or that challenge in the field to funny stories about neighbors or cousins all have known all their lives. Justin thanks the farmhands for their work and they reciprocate with thanks for "letting me come help." Despite the hours of hard labor ahead, a low, steady harmony pervades the room.

By eight, the men are back at work, long shadows now falling across the amber fields, the sky turning lavender, a pheasant flushing as dusk falls. In all directions a gently swelling ocean of farmland stretches to the horizon, broken only by the hulk of 300-foot concrete

grain elevators, "the giants of the plains." Looking across this bountiful vastness, it is not hard to understand why America's anthems all
sing of these landscapes, so overwhelming in their beauty and promise. And why everyone who can has come to help with harvest: Lindsey's parents from Kansas City, Grady's dad and little sister to ride
along in the cab of the grain wagon as he ferries his golden cargo. And
why Justin, who's been running farm equipment since he was eight
years old, can't imagine any place he'd rather be. "I appreciate the long
day, the ability to see the sun rise and set, the opportunity to observe
Creation. When I can't see the horizon, I feel closed in. The prairie is
comforting to me."

This family is so tightly joined in work and faith, so deep in their
knowledge and love and gratitude for this fertile land, they might
seem out of another time.

But for all the continuity here, there are signs afoot of ominous
change. Jerry talks of dust storms blowing across farms to the west of
them so thick you can't see; Justin of the "hot periods grown hotter
and longer" and "rainfall events more intense when they come." Living in a world where a few minutes of freezing cold or pounding hail
can wipe out a season's work, this family feels the human vulnerability
to natural forces more directly than most, a vulnerability that has
grown more acute as Kansas has endured year after difficult year of
scant rain, freak storms and temperature extremes.

Three recent years give a sense of the intensifying challenges.
In 2012, Justin and his neighbors faced the double burden of a "hot
drought": the hottest year to date in American history, combined with
the most severe drought on the Great Plains since 1895. The cloud-
less, torrid days took a toll both on Justin's ripening wheat and on his
summer crops. "Past a certain temperature, a plant has to cool itself
by respiring more water," he explains. "That takes energy away from
photosynthesis and growth, so we start losing yield. And if we're in
drought and the plant can't get the water it needs to cool itself, you

see heat scald, necrosis: browned leaves that don't ever come back. That's lost factory. We're an energy factory: we capture energy from the sun and transform it into energy humans can use. And when you diminish the size and efficiency of that factory and lose yield, you've wasted resources."

By fall, Justin was planting his winter wheat into the driest conditions he'd ever seen; the following month, Ken Burns's documentary on the Dust Bowl premiered. "It's unusual for farmers to be into PBS documentaries, but it was surprising how much it came into the conversation," Justin says. "We knew that if we'd been managing things the way we were back then, we would have had it all over again."

Spring 2013 brought more worries, when just as the young wheat hit the most fragile part of its growth cycle the weather took another extreme swing. Farmers are "on edge across Kansas tonight," Justin wrote on April 23, "as our low temperature is forecast to be around 22–23 degrees. This is very late to have temperatures so low. Our wheat crop is vulnerable to severe freeze damage if the temperature is at or below 24 for two or more hours. It could reduce yields by 80–90%, or even be a complete loss."

A few weeks later, the Kansas Hard Winter Wheat Tour (an annual statewide field survey by the Wheat Quality Council) recorded every imaginable kind of damage. Dryland farms to the west of Justin's were in "abysmal" condition. One had seen just six inches of rain in two years, less than 15 percent of normal; others that averaged forty to sixty bushels an acre would "yield in the single digits, or not even be harvested." In the north, they found wheat buried to death under the first recorded May snowstorm since 1907; in the east, fields drowned by torrential rains—three times normal precipitation that spring in much of the Mississippi River basin, concentrated in intense, wildly destructive downpours.

Those sudden rains flooded a few of Justin's fields and ruined some soybeans but also raised his hopes for the coming wheat harvest.

Then the rain stopped, as abruptly as it began, and the heat returned. By fall planting time his fields were parched again, a bad start for the wheat made worse by an early cold snap that killed many of his seedlings. An exceptionally bitter and windy winter kept his fields nearly barren of snow, leaving the surviving young plants without their usual insulating blanket. Then spring 2014 delivered another shock, with an abrupt heat wave hitting just as the crop broke dormancy—April temperatures reaching the high nineties—pushing the wheat straight into early maturity. Wheat typically waist-high on Justin "headed out" at just eight inches tall. By June, two-thirds of the state's wheat was declared in poor or very poor condition. Tens of thousands of acres of dryland wheat, as a *National Geographic* story on the "New Dust Bowl" put it, "died beneath blankets of silt as fine as sifted flour."

That continuing volatility took a toll on Justin. "When weather is extreme, when we're in a very dry or very hot or very cold or very wet period, there's no doubt it weighs on a person." The recurrent crop failures were felt across the nation: total U.S. crop insurance payouts for those three years totaled nearly $30 billion, most of which is paid by taxpayers.

With America the world's largest wheat exporter and Kansas the nation's largest producer, poor yields in Justin's community reverberate around the world. Global trade in wheat is greater than in all other crops combined. Wheat supplies 20 percent of all calories humans consume and is the top source of vegetable protein; that central role in the human diet, dating back at least 10,000 years, has often thrust it into the center of history. A lack of bread helped spark the French Revolution and many subsequent upheavals. The Soviet Union and nearly all its Communist satellites had as their national emblem a wreath of wheat, a tribute to the peasantry but also to the sovereignty secured by this versatile, storable grain. Today, it remains the primary staple in the world's most volatile regions, including the Middle East.

Wheat helped seed the Arab Spring, when simultaneous droughts withered both U.S. and Russian wheat production, ruining yields and prompting Russian president Vladimir Putin to halt exports, causing bread prices to spike in North Africa and Egypt. In 2014, the Pentagon warned with unprecedented urgency that drought and crop failures pose an immediate threat to national security, creating recruitment opportunities among the starving and displaced for ISIS and other extremists.

Living on the eastern edge of the historic Dust Bowl, where in the 1930s "black blizzards" blew away hundreds of billions of tons of top-soil, ruining and exiling 2.5 million people, the greatest threat Justin faces is the ongoing loss of soil. Though the way soil just lies there, you might think it was dead, it is in fact one of the world's largest reservoirs of biodiversity, containing a third of the planet's organisms. If you were to sift the living microbes out of a single hectare of land (2.5 acres), they would weigh as much as twenty-five horses. In just one "unprepossessing lump," writes biologist E. O. Wilson, you would find "more order and richness of structure and particularity of history than on the entire surfaces of all the other lifeless planets." The visible inhabitants—springtails, mites, millipedes, roundworms and their "fanged predators," including spiders and ants—are just the beginning, dwarfed in number by ever smaller creatures: single-celled slime molds, Kickxellale fungi, Zoopagales preying on amoebae and finally bacteria, any one of which, writes Wilson, "if allowed to expand without restriction for a few weeks would multiply until it weighed more than the Earth."

These bacteria and fungi are as beautiful as the most exotic creatures of the sea: lacy, feathery, spiky, like vivid corals or luminescent worms. Some look like aliens and do things only aliens should, including breathing toxic iron or methane gas. Nearly all remain creatures of great mystery—of the thousands of species in the soil, scientists have identified, and gleaned the functions of, very few. "We know more about the depths of the ocean and outer space," says Justin, "than we do about the relationships in the soil beneath our feet."

Yet these microbes are quite literally the foundation of all terrestrial life. They are vital to photosynthesis, from which all food and all oxygen comes. They clean our water and air, filtering out pollutants and breaking down the organic wastes that would otherwise swamp the planet; turning "distemper'd corpses," as Walt Whitman wrote, into "the grass of spring," the bean, onion, apple buds, calf, colt and "the resurrection of the wheat."

Just as our gut microbes are essential to metabolism and health, so the microbes in the soil are critical to Justin's crops: holding the soil together in their sticky, silvery fungal webs; keeping it pliable and permeable to both roots and water; warding off diseases and pests that afflict both plants and humans; cycling nutrients through the system. "There's a whole community working together, roots and organisms sending each other biological signals, communicating in ways we don't understand, even asking for help," says Justin. "Say a plant detects an insect feeding on its leaves, and needs more phosphorus to repair that damage. It may send an enzyme signal to mycorrhizal fungi saying, 'I need more phosphorus.' And the fungi may spread out and bring phosphorus to the root, maybe getting some nutrient in return."

Soil microbes are also directly vital to human health, the source of more than 90 percent of antibiotics and 60 percent of anticancer drugs. Scientists are now assaying soil microbes from far-flung places in search of new antibiotics able to kill pathogens that have evolved resistance to the old. In January 2015, *Nature* reported that a breakthrough method for extracting drugs from soil bacteria had yielded a powerful new antibiotic that cures severe infections, including resistant staph. Until now, such efforts were stymied by scientists' inability to grow the relevant bacteria in lab conditions; the breakthrough came in culturing the microbes in a beaker of the soil they came from— stunning evidence of the need for the whole system, the ecology, to remain intact. As an article published the same month in the *Proceedings of the National Academy of Sciences* put it, we are "Animals in a Bacterial World," utterly interdependent in our evolution, genomes, development and survival.

But soil is also one of our most vulnerable and endangered resources. It is effectively nonrenewable, taking up to 500 years to build up a single inch. And it is being lost at a precipitous rate. Geomorphologist David Montgomery, author of *Dirt: The Erosion of Civilization,* calculates that since World War II soil loss worldwide has caused farmers to abandon an area of land equal to one-third of all present cropland. Annual erosion now exceeds soil production by 23 billion tons, 1 percent of the world's total agricultural soil inventory; at that rate it will all be gone in 100 years.

A prime reason for that loss is the most traditional of all farming practices, a staple of bucolic scenes from Currier and Ives through Grant Wood: plowing the soil. As Dwayne Beck, manager of the Dakota Lakes Research Farm and a formative force in Justin's life, puts it: "In natural systems, tillage is a catastrophic event (associated with glaciers, volcanoes, avalanches) that occurs only rarely." Done year after year, stripping and churning the soil can do deep damage, especially to grassland soils. Ecologists like to call the prairie an "upside-down forest," because the vast bulk of its biomass is not in the visible flora or fauna but in the invisible world underground. Plows hurt that upside-down life the way chainsaws and bulldozers hurt right-side-up forests, and have done so since long before the invention of big mechanized farms and agrochemicals.

Standing in a field that was part of his great-great-grandparents' original farmstead and has been farmed by his family ever since, Justin explains why his ancestors had to break out the sod. "There was nothing they could eat from that prairie grass. They could feed animals on it, but they also needed some grain, with more protein in it than native prairie seeds. And they didn't have a way to plant seeds in that tall grass, which would have just outcompeted their crops anyway. So they had to cultivate the soil."

The results, at first, were stunning. "Those tillage passes, inverting the grasses into the soil, brought a huge sudden flush of food and oxygen into the soil, setting off a tremendous burst of microbial digestion and respiration," Justin says. Fertility that had been built up in the soil over millennia by glaciers and wind and the prairie's biology "was released in a giant flood. The native fertility in these soils was so high, it sustained the first three generations of my family.

"But that mixing and mixing over decades made it difficult for soil life to thrive." Like a tornado spinning through a small town, the churning steel of the plow scrambles microbial communities, separating and disorganizing symbiotic partners, chasing out worms and other creatures big and mobile enough to flee. Opportunists like fast-growing bacteria capitalize on the disaster, overwhelming more-stabilizing fungi and rapidly eating their way through the soil's "organic matter"—the dead remains of everything that ever lived in or atop the soil—losing much of it into the air. That organic matter, made up mostly of carbon (since life on earth is carbon-based), is the most important determinant of soil health. "By the 1950s," says Justin, "my grandparents were planting a monoculture of wheat after wheat, each time on a beautiful, cleanly tilled seed bed, discing whenever they needed to uproot weeds. And the native fertility was getting used up. After years of tillage and oxidizing the carbon, we had gone from 5 percent organic material in the unbroken prairie down to less than 2 percent."

That depletion of native fertility, dovetailing with a need to repurpose World War II munitions plants, ushered in the postwar age of chemical fertilizers, especially anhydrous ammonia nitrogen. In a few decades' time, that fertilizer revolution led to a *doubling* of the amount of nitrogen on earth available for plants and other organisms to use, and a quadrupling of the amount entering the oceans. The resulting increase in global food production freed a billion people from hunger but has also done severe and ongoing damage. More than half of all nitrogen fertilizer (whether synthetic or from manure) is lost into the water or air, making agriculture the biggest polluter of lakes and rivers. In the Gulf of Mexico, it causes algal blooms that deplete oxygen, snuffing out marine life. In drinking water, fertilizer can elevate nitrates to levels unsafe for humans or cause spikes of toxic bacteria like those that shut down the Toledo water supply. The turn to chemical fertilizers also reinforced a destructive idea: that soil, like a factory, could be swept clean and then fed the handful of inputs humans understood plants to need—a dire oversimplification that has led to the degradation of much of the nation's best agricultural lands.

Beyond disrupting microbial communities and dissipating organic matter, the plow does sheer physical damage to the structure of the soil. Half the volume in healthy soil is water and air, held in a fretwork of caverns excavated by roots, rambling worms and Whoville-sized beasts. Like an earthquake leveling a city, the plow collapses those structures, leaving behind a dry, compacted hardpan, both less permeable and more vulnerable to water erosion. "Each raindrop hits like a little explosion," explains Justin. "The soil seals over and won't take water." Instead, water runs along the surface, carrying precious topsoil with it. "You'll often see the best crops at the bottom of hills, where eroding soil lands before leaving the field entirely."

Justin leads the way into a leased field he farms to show what he means. Late rains during the 2014 wheat harvest had confronted him with a difficult decision. Though this field and a few others were too soggy to harvest without leaving deep tracks, a wet forecast with possible hail made it too risky to wait. More rain would decrease the grain quality and value; hail would destroy it entirely, losing all that had gone into the crop not only for his own family but also for the land's owner. So Justin took a combine into the mud, knowing that "in the intensity of harvest, we'd create problems we'd have to fix." The storm never came, so his rush to harvest, he soon realized, had been a mistake. Once the ground dried a bit, he went back in with tillage equipment to smooth the ruts. But no sooner had he tilled than an intense downpour hit, shredding the bare top layer of soil into small bits easily carried down the slope by the pummeling water. "As small streams move soil downhill, they get bigger and stronger, moving more and more soil down the field," he says. The shallow gullies left behind visibly pain him. "That's the complications of working in a live system. You have to make the best choice in the face of uncertainty and things beyond your control."

Of even greater danger to soils laid bare and turned "wrong side up" are the winds that blow across central Kansas nearly all the time. "Between February and April we'll have 20-mile-an-hour winds just steady for many days in a row, and gusts to 40 miles an hour; it's rarely quiet." On soils left brittle and crumbly by decades of tillage, the wind sets particles of soil rolling, gathering them up like a cartoon snowball. "If it's dry and windy for days on end, this field would just start blowing away. You see your soil being lifted into the air and leaving in a big cloud of dust. It's not a good feeling." Such dust storms are increasingly common, especially west of Justin, where the drought has not loosed its grip since 2012. "Sometimes a farmer has tried to plant a crop, but the seeds never had enough moisture to even start. When it's that dry for such a long time, the environment becomes so fragile you literally can't grow any biomass to protect the soil. Then the only way to stop the blowing is to go out in the field and make passes every hundred yards or so with a harrow, to make ridges to physically try to catch some of the soil particles rolling along the ground."

Justin has watched some neighbors resort to that desperate measure. But if the winds keep blowing and the rains don't come, much of that irreplaceable soil will be lost forever. That's what happened in the Dust Bowl, when parts of Kansas and neighboring states lost more than 75 percent of their topsoil, an average of five inches across ten million acres boiling across the prairie in clouds "ten thousand feet or more in the sky," as Tim Egan wrote in his beautiful and terrifying history of the event, hitting Kansas on Black Sunday with a murderous wall two hundred miles wide. Even in areas relatively untouched by that disaster, the cumulative soil loss from tillage has been vast, and continues: 15 to 18 inches in parts of Nebraska; a current rate of loss in Iowa estimated by Iowa State University agronomist Richard Cruse to be sixteen times the rate of natural regeneration. All told, in the U.S. alone, 1.7 billion tons of topsoil is blown or washed off of croplands annually, a loss valued at up to $44 billion a year. As president Franklin D. Roosevelt warned in the aftermath of the Dust Bowl, "A nation that destroys its soils destroys itself."

Justin describes the sense of failure and culpability, the "helpless hopeless feeling when it's dry and under the relentless prairie wind the soil begins to blow. It's just terrible to see it piling up inches thick in ditches, lost for good. And it keeps cutting away; the sand and static electricity kill the plants. God gives us a conscience inside us, and it's like you've gone against that conscience . . ."

That's a challenge Justin is rising to, by revolutionizing industrial-scale farming.

Justin's roots in central Kansas reach back to 1868, when his maternal great-great-great-grandfather James Thorstenberg left Sweden in search of religious freedom and economic opportunity. Lured by a steamboat company brochure promising a "land of gold," he and his wife Hannah (who in pictures appears twice James's size) spent two months in steerage crossing the ocean with five small children, including a newborn baby girl. They settled first in Galesburg, Illinois, on land secured through the Homestead Act signed by President Lincoln in 1862. But Illinois soon became too crowded for their taste, so James and his brothers founded a church "colonization committee" and set out to find new land on the western frontier. Their scouting brought them to Assaria, Kansas, just south of Salina, where they found a small community of Swedes already settled and "rich black soil that went deep into the earth and on it green grass as tall as they were," according to a homemade family history. "How different it was from the shallow, sandy soil on top of rocks back in Sweden!" Pooling funds, the committee secured 14,800 acres from the Kansas Pacific Railroad. Each man drew from a jar to learn exactly which acres would be his, paying $1.50 an acre for upland and $5 an acre for bottomland, in five payments at 10 percent interest. When the deal was done they joined a worship service under a large cottonwood tree. "Pastor Dahlsten recalled a wild turkey cock performing music from the treetop during the service," notes the family history. "When the last hymn was sung the turkey flew away."

By 1870 James and his brothers had settled their families in Assaria and begun the crushingly hard labor required of homesteaders to "improve" their claim. With mules and a two-bottom plow, they "broke out" sod held fast by the prairie's densely woven mat of roots, a mat that sometimes ran more than ten feet deep. Kansas had been a state for just nine years, and was still reeling from the violence that had convulsed it in the years leading up to the Civil War. "Bleeding Kansas," Horace Greeley called it, as dozens died in battles between John Brown's abolitionists and proslavery factions over whether Kansas would be slave or free. (It did finally join as a free state, a fact Justin invokes with great pride. The state's motto, "Ad Astra per Aspera," means "to the stars through difficulty.") One hundred and fifty miles southwest of Salina, Bat Masterson was trying to rein in the wild streets of Dodge City. Two hundred miles to the southeast, in Independence, Kansas, Laura Ingalls Wilder was in her little prairie house, contending with Osage Indians.

In Assaria, Hannah had two more babies: Justin's great-great-grandmother Emma and then Olaf, who lived just four years; his small grave was one of the first to be dug in the cemetery where all Justin's maternal ancestors now rest. In 1875, the Thorstenbergs helped found the Assaria Congregation, raising $2,685 to build the church that Justin and his brother grew up in and his grandmother still attends. James's brother O. H. milled the lumber, always keeping a cone of hard-pressed sugar and a hammer on his desk at the mill to break off small treats for the children.

In 1890, great-great-grandmother Emma married A. J. Anderson, whose family had also come to America in 1868. A. J. was a frontier dandy. His granddaughter Jean (Justin's grandma) remembers him in laced leather leggings with a big cigar. Pictures show him in a woolen three-piece suit, watch fob and bow tie, his hair pomaded; Emma stands behind him in a jacket with a high lace collar and long dark skirt. By 1907, A. J. had built them a beautiful white farmhouse, with crabapples blooming in the front yard and trumpet vines climbing the rails of the big wraparound porch. The house, which is still standing, had a music room, tornado room, root cellar, blacksmith shop, barn, granary and chicken house. Set in the best soils in the county, it was surrounded by beautifully tended fields, plowed into long, straight, warm, fragrant furrows. It was the industry and rectitude of this generation that inspired Willa Cather's 1913 novel *O Pioneers!*, about a community of Swedish settlers wresting God's fruitful order out of the useless wildness of this High Plains prairie. Hers may be the most lovely and wrongheaded of all paeans to plowing: "There are few scenes more gratifying than a spring plowing in that country . . . the brown earth, with such a strong, clean smell, and such a power of growth and fertility in it, yields itself eagerly to the plow . . . with a soft, deep sigh of happiness."

Emma and A. J. had four daughters, but only one of them, Esther, raised a family, marrying Joel Isaacson, who was the youngest of eleven children. Joel, who died when Justin was in third grade, was "kind of a sport," says his daughter Jean, always sharp in a clean white shirt, fedora and tie. Esther—who lived to be 101 years old, dying when Justin was twenty-eight—was just as fashionable, in her fringed dresses and marcel-waved hair. The couple had three children. Jean, Justin's grandma, was born in 1926, the same year that Kansas "wheat queen" Ida Watkins claimed profits of $75,000, "bigger than the salary of any baseball player but Babe Ruth."

At age eighty-eight, Jean can still recall the chores given her when she turned six: gathering eggs, fetching in the eight cows from the field to milk, helping harness up the team of six draft horses to drag the plow or pull the grain wagon out to the wheat threshers. Though just two generations have passed, wheat harvest then was a universe apart from Justin's. In fields kept small enough to be able to plow and plant and harvest in good time, the men would cut the wheat by hand, stack it in teepee-shaped shocks and await the threshing crew (pronounced here as "thrashing") with their mule-driven machine. The threshers moved through the community one farm at a time, with all the neighbors coming together to help feed the machine and then to share a meal cooked by the daughters and wives. Justin now farms some of the Isaacson fields. "They probably laid pretty similar," he says. "It's meaningful to me that my great-grandfather and great-great-grandfather spent time working this land."

Jean helped skim the cream from the milk, churn butter, and deliver both to neighbors' porches. During the holidays, she helped serve the potatis korv (potato sausage), sill (imported herring) and smor bakelse, butter cakes she pressed into little curved tins to bake, making fluted cups to fill with lingonberries and whipped cream. Jean loved the little dogcart her dad rigged for her and riding her pony the two miles to school. She got in trouble the day she brought a tiny goat kid home in a gunny sack slung over the saddle horn, and again a year later the day Rudy the goat climbed atop her dad's Model T.

Jean was still in elementary school when the Depression and Dust

Bowl hit. She remembers her mother putting wet tea towels in the windows to catch the dust, but not much economic hardship; their land was paid for and they could fairly well feed themselves. Some neighbors did not fare so well; several families had to live in the church after losing their land. And life got a bit harder for Jean's grandma Emma in 1934 when A. J. died. She took in and cooked for a "lady schoolteacher" boarder, making gravy to serve on boiled rabbit or baking whole-wheat "graham bread," an early health food, for chicken salad sandwiches. Jean remembers Emma making rag rugs and always entertaining on her birthday, inviting everyone in town but asking each guest to put a dime in the collection dish for the church.

Grandma Jean married Carl Erickson, whose parents Carl and Esther had arrived just two decades earlier and still spoke mostly Swedish. They hosted all the relatives for Christmas Eve, serving lutefisk with mustard sauce and ostkaka, a kind of cheesecake made from curdled milk and almonds with kram (a fruit sauce). Jean and Carl had four children; the second oldest was Justin's mom Karlene.

The farming history on Justin's dad's side is more broken. Jerry's grandfather was an itinerant Baptist preacher, moving his family from Canada to Oklahoma and several places in between. In the 1920s, when Justin's grandpa Sam was in middle school, the family settled briefly in a little church outside of Hope, Kansas, where Sam found work carrying water to the threshing crews. Sitting down for a harvest meal at the Shirack farm, he met twelve-year-old Bernice, whom he didn't forget even when the Knopf family picked up and moved again, this time to a farm in Alva, Oklahoma. After high school, Sam went back to Kansas to persuade Bernice to marry him. They bought a little farm, just fifteen acres, in the flat, dry, sandy soils of Alva, just as the Dust Bowl hit. Justin remembers their stories of the doors and windows seeping sand, a relentless tide they could never keep ahead of. Many families abandoned their land: by the time Jerry was born in 1951, there were so few left in the area that he had just three kids in his first-grade class (then two, after one flunked out). Even with the memories of the Dust Bowl still fresh, Jerry remembers his dad's eagerness to plow the "virgin dirt"; few were yet thinking about what contribution humans might have made to that calamity. When Justin was in grade school, Jerry took him and his brothers to see the Alva homestead; it was small and simple, says Justin, and by then "abandoned and bleak."

Sam and Bernice moved the family back to Kansas in time for Jerry to attend Southeast-of-Saline High, where he met Karlene, an ethereal beauty with white-blonde hair and glacial blue eyes; they married in 1972. The *Salina Journal* wedding announcement noted that Karlene wore sheerganza and venise lace in a demi-bell silhouette gown, with a Juliet caplet of lace daisies trimmed with pearls holding a veil of "imported silk illusion." Karlene studied special education at Marymount College and became a teacher; Jerry graduated in air conditioning and refrigeration from the Salina Area Vocational-Technical School, and worked as a service technician for Montgomery Ward.

By the time Justin was born, in 1978, Jerry had returned to farming; at age eight, Justin joined him, "going directly from the school bus to find dad in the field." At sowing time, they had to fill the seeder by hand with five-gallon buckets, each weighing fifty pounds. Justin always envied the beautiful splash of seed his father's strong swing of the bucket would make. He was soon working on his own, "driving a tractor to pull a hay wagon. Dad'd set the seat up high and steering wheel low so I could see over it, but I couldn't reach the pedals. It had a hand throttle and you didn't have to clutch it to get into gear. But if I needed to stop suddenly, I had to hop down off the seat and stomp down on the brake."

By age ten, Justin had stepped up to tilling the soil, making multiple long, slow passes through the fields. The first and slowest step was plowing: cutting and turning over the soil to incorporate any straw, roots and leaves left from the previous crop. "It takes a lot of horsepower to rip through everything, so you can't go very fast, and you're pulling a narrow implement so you have to go back and forth many times." Plowing left a rough "cloddy" surface, so Justin would have to pass through, twice more, with a disc to break up clods and churn the soil. Then came two more passes with a spring-tooth or field cultivator—a big, shallow rake—to smooth a firm seed bed and kill emerging weeds: five trips in all. He sometimes wished he had the freedom and leisure of some of his friends. In summer, he'd be out from breakfast to supper, stopping only long enough for a lunch of fried hamburgers and Pepsi at his grandma Bernice's.

When Justin turned fourteen, Jerry entrusted him with still more responsibility: giving him the means to rent land and buy seed and fertilizer, and the authority to make all decisions on what to grow and where to turn for advice. "I got two-thirds of the crop, and decided when to sell it into the market." Justin rented eighty acres of terraced cropland and forty acres of pasture from a neighbor, on a beautiful hill with views in all directions across the swelling prairie, including down to the red and white Knopf family barns. He was keen to farm just the way his dad did. "He's a good farmer. And at that age, whatever your dad's doing, he's your hero. He has all the answers." Finding the fields weedy, full of goatgrass and downy brome, Justin plowed as often as he could to bury the weed seeds deep, where they couldn't germinate. When he couldn't manage the time to plow he would burn the fields instead, down to bare dirt and ash.

Still, his fields failed to thrive. No matter how much he plowed or burned, some areas remained so choked with goatgrass that he could barely produce any wheat. His soil was also washing away whenever a heavy summer rain would hit, digging gullies at just the depth of his last plowing. And he kept running out of water, getting terrible yields when he'd try to rotate from winter wheat to a summer annual like soybeans—yields a third or less of what he needed to be profitable.

"Some of those years we got as much of our income from USDA disaster payments and crop insurance as we earned from the crop."

Then Justin had a life-altering experience: he went to college. Kansas State was only an hour east of home but "until I went there my world view was Saline County, and everybody was farming like we were." He still vividly remembers his first class on his first day, a course on crop science taught by Dr. Stan Ehler. "He was talking about hogs, and I thought, 'Why is he talking about hogs?' But I soon understood: he was showing us how much you can learn by watching how they behave, that living things speak to us if we just pay close attention to what they're trying to say. I took that lesson in observation back to my wheat: if the plant is yellow, what is it saying to me?"

Kansas State was itself in transition in the 1990s. The college had recently hired its first soil microbiologist, Chuck Rice, reflecting the recognition that the "chemical century," as Rice puts it, was giving way to the century of biology. Rice had done his PhD in Kentucky, gaining exposure there to some of the earliest research, led by Shirley Phillips and Harry Young, on soil microbes' influence on plant health. "Dr. Rice introduced me to these foundational concepts about the importance of microbiology, what soil diversity does. If that doesn't push a future farmer to ask questions about how to enable that life and diversity, you've missed something." Now one of the world's foremost experts on soil carbon, and as of January 2016 holding a key post at the National Academy of Sciences, Rice is still a mentor to Justin, whom he calls "my star."

Justin's classmates also played a role in his changing thinking, especially Andy Holzwarth of Gettysburg, South Dakota. Andy had lived through the pain of seeing his dad struggle and have to give up the family farm, a tragedy repeated many times across the grain belt in the 1980s and an anguish Justin would come to recognize in the three years he traveled the region as a full-time agronomist for a seed company. "During times of drought, I'd go with farmers to see their corn or soybeans and they'd be just completely dried up. When a farmer loses a crop . . . he's put so much of himself, his time and effort and his family's time and money, and his own judgment. To be there with him and know he understands that it's all headed in the wrong direction, and that's partly because of choices he's made, that's not a hopeful feeling."

Andy had heard that a guy named Dwayne Beck was doing a new kind of farming not far from his hometown, designed to prevent the soil and water losses that were devastating so many farmers. So junior year, he and Justin organized two vanloads of students to drive the 500 miles through the Sand Hills of Nebraska to visit Beck's farm. Beck and neighboring farmers spent a day showing the young agronomists how they rotated crops to cycle nutrients, control pests and build soil, and how they'd modified their machinery to allow them to plant *without* tilling the soil. "No-tilling," they called it, its very name expressive of the degree to which tilling had been the default across thousands of years of agricultural history. Seeing the health of their soils and the abundance of their yields was a "dawning moment" for Justin; "I thought, 'My lands, there's really something to this!'"

Justin soon brought that new knowledge—and habit of experimentation—back to his own land. His first move was to plant alfalfa on thirty-five acres where the goatgrass was worst, leaving it there several years to kill the weeds. A winter annual like Justin's wheat, the goatgrass simply couldn't compete with alfalfa, a perennial with roots that can reach more than twenty feet and in their first year grow as deep as Justin is tall. The alfalfa's robust roots also did "deep nutrient cycling: bringing nutrients from six feet down up to where crops with shallower roots could reach them." When Justin took the field back out of alfalfa, it was transformed: clear of downy brome and goatgrass, delivering wheat yields one and half times what they'd been.

Jerry had long understood and taught his sons the benefits of planting alfalfa for a few years between wheat crops. But understanding the underlying science and putting it to work on his own land made a deeper impression on Justin. Now, he wondered, how could he start rotating on the rest of the field, even where it was too steep for the alfalfa machinery? He'd tried putting on summer annuals like milo (another name for sorghum), but they always dried up and produced dismal yields. What if he could store moisture the way Dwayne Beck was, by forgoing tilling? By 1998, he was no-tilling his own fields, on the worst of which his yields tripled. "I went from 30 bushel milo [per acre] and 10 bushel soybeans in my first tries to 90 bushel milo and 30 bushel soybeans."

By 2000 the Knopfs were no-tilling more than half their 4,500 acres. "Dad was hesitant to make such a dramatic change too quickly and asked significant questions, but we're fortunate to have a father who has always given us a lot of trust and freedom and responsibility, and respected our opinions. He just wanted to see if it worked before investing in hundreds of thousands of dollars' worth of new equipment. We didn't go right out and buy an air seeder and sprayer, but tinkered with the machinery we had.

"A defining moment, for Dad leastways, came in 2004, when after a spring thunderstorm he stood at the bottom of the farm driveway watching the water in the roadside ditch come off our fields and, right across the road, off of a tilled field. That water was all muddy, but ours was so clean you could see to the ditch's bottom." Riding motorcycles with Karlene also brought it home for Jerry: when they'd pass through fields shaded by the stubble left in a no-till field, he says, they'd get a delicious breath of cool air, but "where someone had turned the soil or burned off the stubble, it was like getting in a hot car." With that, the Knopf family bought state-of-the-art equipment and committed the whole farm to this revolution.

Justin's fields now look nothing like his grandfather's tidy furrows. Though his acres of wheat and alfalfa and milo are still marked off like Willa Cather's "vast checkerboard . . . light and dark, dark and light," they are messy: the knee-high stubble standing like Bart Simpson's hair, with a thick, tangled mat of straw or frozen-out leaves covering the ground.

Where his predecessors set out to clear away or tame nature, Justin strives to emulate the native prairie. "We have large, sometimes very violent dangerous thunderstorms in the spring, with heavy rainfall, heavy winds. And then we get dry, hot and windy, and then very cold with blizzards. The prairie can weather all that; it is steady, resilient, across long periods of time. Our farms on the prairie have to learn to do the same, to be resilient to these dramatic, harsh forces."

Farming like the prairie requires, he says, "learning how to come alongside that biology, to partner with it." It also requires "resisting the temptation to think we can control it. A living system is complex, always changing and adapting; that's part of its mystery. And that mystery, the need to deal with lots of uncontrolled things, is where my faith plays into my self as a farmer. Science is part of it. Technology is part of it. But humility is part of it as well." (The word *humility*, fittingly, derives from the same root as *humus*, the dark, decomposed organic matter Justin is building in his soil.)

A first lesson Justin learned from the prairie is the importance of physical protection: keeping a roof of living and dead plants over the soil and its inhabitants. Walking into the tall, stout, crunchy stubble of a cornfield, Justin lifts an edge of the thick, dense carpet of "residues" he left behind when he harvested here a few weeks earlier, a tatami mat of yellowing leaves, stripped cobs and year-old wheat straw. "If we have heavy rain, that raindrop is going to land on this residue instead of on bare soil. It could blow or rain like mad and nothing will leave this field." The mat also protects everything living in the field, beginning with the crop itself. Justin doesn't grow much corn, which pollinates and needs lots of water and cool temperatures in just those months when Kansas is blistering under the hot sun. But when he does, having this "natural mulch of wheat residue between rows really helps reduce evaporation and cool the soils." It also suppresses weeds, giving the crop a head start.

Aboveground, the stubble provides shelter for grassland birds, a highly vulnerable group in fast decline as ever more of their habitat is converted to farms. Justin often hears the low, fussy clucking of ring-necked pheasants in his fields, startles up a family of bobbling quail, or pauses to watch a coyote skulk about in search of fledglings or eggs. In Illinois, researchers have found that no-till fields host significantly more and diverse nesting birds, including threatened upland sandpipers, Eastern meadowlarks and field sparrows.

Leaving the dead remnants of the previous crop atop the soil also protects the life underground, not only keeping the microbial hordes cool and moist but also—the second lesson of the prairie—feeding them at a more temperate pace. "See how everything is gradually crumpling as the fungi eat it," asks Justin, "from the bottom up? We'll find pieces of milo stalk three years later. My grandfather was taught that by plowing under the residue they were feeding it back to the soil. They didn't realize the soil could feed it back to itself, at a slower pace that's much better for the soil." Turned under, Rice explains, all that leftover plant material overstimulates the microbes, "like giving sugar to a kid." Fast-growing bacteria explode, eating their way through the organic material and exhaling much of it into the air. A slow-food diet, by contrast, nurtures more efficient fungi that store most of what they eat in their own growing bodies, keeping it there (and thus in the soil) even after they die. Their long, weblike hyphae also perform services for Justin's crops: bringing the plants nutrients they can't reach on their own and binding the soil in the big, soft lumps roots love.

Over time, the proper care and feeding of soil fauna rebuilds the organic matter that is essential to crop health. Most soils are made up largely of minerals—weathered from rock and deposited by the wind (loess), by flood (alluvium), or at the base of hills and mountains (colluvium). Some of those minerals are like tiny electric snowflakes, grabbing with their negatively charged points and edges onto nutrients—like calcium and potassium—which are mostly positively charged, so "a root can get in like a big straw and suck them up," as UC Berkeley biogeochemist Whendee Silver explains. That's good, but organic matter is better: not crystals but amorphous, rotting goo, the "bits of slime" Justin loves to find in his soils. That goo coats soil particles, says Silver, "jamming in lots of charges" and therefore grabbing and holding onto lots more nutrients, making the soil both more fertile and less "leaky." Its stickiness also helps glue together the desirable "aggregates" that hold water and resist erosion. "Drop a chunk of healthy soil with good structure into water," says Justin, "and it will hold together and float around. Try the same thing with overworked soil, and the water will turn cloudy as the soil dissolves."

Farming in concert with the prairie requires constant, close, almost paternal attention. Justin knows the land intimately through soil testing and data analysis but even more through his own senses and hands. In an essay she wrote for an online writing class, Lindsey describes watching Justin "studying, seeing things I can't see . . . turn[ing] the wheat over in his hands, examining every square inch with the eye of a scientist and the touch of a father. His nail scrapes away at something, his nose almost touching the wheat."

Justin is as often striding across a field as he is cocooned in a tractor: opening a furry soybean pod to bite into one of the tiny egg-yolk-yellow seeds to see if it's ready to harvest, stopping to dig with the small shovel he carries everywhere, or crouching to sink his big hands deep into the black, fragrant soil. He is often joined by his young employee Garrett Kennedy, a bright young K-State graduate who reminds Justin of himself fresh out of college—though Garrett is so skinny the shovel barely notices him when he jumps on. Sometimes

Andrew comes along too, with his own pint-sized shovel, diligently digging alongside his dad. It would be hard to say who is more excited by their discoveries, especially the fat earthworms whose excavations and excretions inspire in Justin a thrill and awe so fervent as to verge on comical. Even a pile of their castings brings him joy. "Oh this is wonderful. This shiny material is their poop, or 'frass,' very rich for plants. They burrow down five feet or more, eating all the time, filling their bellies with nutrients that they carry from deep in the soil up to the root zone, and carrying other things back down. In deep drought they go deeper still, tie themselves in a knot and go dormant. We can't see fungi or bacteria, so earthworm activity is my best measure of biological activity. What a gift this is."

Sometimes Justin follows the worms, climbing right down into the ground to investigate the subterranean layers. One autumn day, he and his dad bring a small excavator out into a recently harvested alfalfa field into which Justin will soon plant wheat. Jerry, his handsome, lined face set in its usual bemused smile, happily digs the rectangular pit, teasingly reassuring onlookers that he'll go only five, not six, feet under. As soon as he's done, Justin slides in and with a spade begins peeling away layers of the soil wall. Finding a fungal growth resembling tiny pine needles, he hands it up to Garrett to take a picture of it with his phone, to identify later. At the deep clay level two feet down it gets hard for Justin to spade, but he shows how the thick alfalfa taproots had no trouble breaking through. "The alfalfa roots release a big flush of nitrogen the wheat can use. Then when they decompose they leave behind a nice, fat channel" through which both the wheat's weaker roots and water can penetrate deep underground. Had he been regularly tilling this field, Justin says, he would have collapsed those channels and left a "plow pan" at the depth reached by the steel, so hard and impervious it would have turned even the powerful alfalfa roots sideways. Shoveling into a field tilled for generations, he shows the aftermath: no worms, no passages for roots or water, no moisture, no rich black color, no chunks of decaying life. Instead of squishing juicily and ribboning out like Justin's soils, it crumbles to dust in his hands.

Jerry's soil pit reveals the benefits of another practice central to Justin's farming: long and complex crop rotations, in patterns designed to mimic natural succession. Again, his model is the prairie, which derives its self-sufficiency and resilience in large measure from its diversity of species and the complex barter economy they share.

By switching back and forth between grasses such as wheat, with fibrous roots, and broad-leaf plants with deep taproots, Justin's rotations build soil structure. They also provide the microbes with what he calls a "balanced diet": high-carbon residues like wheat straw that last a long time, but also nitrogen providers like alfalfa and soy. Tromping into a soybean field ripening from dull gold to ash gray, Garrett uproots a plant to show how legumes "fix" nitrogen. He points out the small nodules stippling the roots; though gray now, they were pink earlier in the season, he says, engorged with leghemoglobin. The nodules house bacteria able to do what the beans can't do for themselves: grab nitrogen out of air pockets in the soil. The bacteria give that nitrogen to the bean plants in exchange for plant sugars. They keep giving even after they die, their decomposing bodies releasing the nitrogen slowly into the field, nourishing whatever crop follows in sync with that crop's needs, so little is lost into the watershed or air.

By using plants to fix nitrogen, Justin has significantly reduced the amount he adds per bushel of wheat. He's also become far more precise in its use, aided by the yield maps generated by his combine's computer. "My grandpa probably put a set amount of nitrogen across all his fields. Today we try to get the exact right amount on every field." Using an algorithm developed by Kansas State, Justin plugs in his yield target, previous fertilizer applications and, most importantly, the "credits" he's earned by persuading "biology to work with us and for us."

Those biological credits include the nitrogen that will be released by his soybean and alfalfa fields for at least two years after harvest. "That's difficult to measure because it happens in an ongoing cycle as the next crop grows," says Justin. "And the nitrogen cycle is complex. It's mobile in the soil and a lot can affect it: rainfall, temperature, biological activity. It can also be immobilized by microbes using it as they break down residue. So we rely on K-State estimates: 'OK, you had a 30-bushel soybean crop here so we calculate that's going to release this amount of nitrogen.'"

His second source of biological nitrogen credits is the organic matter Justin has been building in his fields over the past twenty years. "We take soil samples to quantify it. The fields we've no-tilled for a longer time have higher levels and will mineralize, release, more nitrogen." Combined, the biological credits can reduce his need for nitrogen fertilizer as much as 80 percent, sometimes entirely. "Two wheat fields we planted this year had been in alfalfa and had credits of 60 pounds an acre." At the current price of 55 cents per pound applied, that saved the family $33 an acre.

Getting nitrogen levels right not only saves money but also maximizes productivity and minimizes pollution. "We don't want to underfertilize and leave yield potential on the table. We want to get the most bushels we can per inch of water and acre of land. And if underuse costs us two bushels an acre, and wheat's at five dollars, we lose ten dollars an acre. But we also don't want to overfertilize. If we put ten pounds more nitrogen on than we need, we've overpaid $5.50 an acre. We also risk having it leach into groundwater or run off the field or volatize into the air as nitrous oxide, which can happen if it's applied to the soil surface and we get the exact wrong conditions. Our next step will be getting precise not only at field level but on part of a field, then to a particular plant telling us it needs more nitrogen."

That the wheat prefers the biological nitrogen is evident in the dark, dewy green of young wheat planted after a legume, compared to the more wan hue of fields planted in continuous wheat and reliant on synthetic fertilizer. Jerry is also glad to see the anhydrous go: "It's expensive, dangerous to you, kills the earthworms and makes your ground hard."

When Justin sees his residues breaking down too quickly to protect the soil until he plants his next cash crop, he renews those residues, keeps the ground covered and continues building soil, feeding his microbes, storing nitrogen and managing weeds by planting a non-commercial "cover crop," which he won't harvest but leave to die where it stands. In one field he has planted a cover mix of eight winter annuals. Like his no-till residues, the mustard, canola and cabbage plants provide a protective canopy: spreading like umbrellas so that raindrops don't blow soil particles to bits but "fall on the leaves to absorb that kinetic energy." He has mixed in vetch and cowpeas, both legumes with root nodules, to scavenge nitrogen from the soil and bank it for the next crop, and fast-growing oats and barley to outcompete "diligent little weeds" like red root pigweed and tall waterhemp. His radishes, one of which Justin pulls up to show a fat root that will be a foot long by the time the crop freezes, "provide a nice macropore for water to go into," and a varied menu for his microbes: "Some might like feeding near a turnip root; some might like grass." Most importantly, keeping something growing on every acre supports life's habit of begetting more life. "A living root is constantly exchanging chemical and enzyme communicators with surrounding microorganisms," he explains. "There's a flurry of life around that rhizosphere. If we can keep something growing there, life never slows down but builds and builds and builds."

Cover crops cost Justin about $25 an acre to plant and earn him no income or immediate return. Though a 2014 USDA survey of farmers found corn and soybean yields following cover crops about 4 percent higher than in adjacent fields, and yield improvements in the drought conditions of 2012 of up to 10 percent, Justin hasn't yet seen those gains. "In four years of on-farm research, we've never seen a direct yield benefit," he says. "We're doing it with the hope that long-term in our system we'll have production increases by having healthier soil. But with so many variables, that's hard to measure." Cover crops do have immediate benefits for aquatic life and fishermen downstream. Keeping living roots in the soil in all four seasons—roots that can take up nutrients and hold them in place—significantly reduces run-off of nitrogen and phosphorus into the Mississippi River. Iowa State researchers estimate that with a tenfold increase in cover cropping, to 25 percent, their state could achieve its full share of reductions needed to shrink the Gulf of Mexico dead zone by two-thirds, the target set by the EPA and the twelve big agricultural states in the watershed.

Justin will count on the hard Kansas winter to kill off the autumn cover crop so that he can plant into this field in the spring. The rest of the year, however, he does the killing himself. For that, he uses chemical herbicides. In fact, without recourse to plowing to "burn down" cover crops or bury weeds, no-till farmers use more herbicides—infamous chemicals like Monsanto's Roundup—than conventional farmers, leading some to denounce no-till as "a short-term solution, requiring that we poison the soil to save it."

And it is not just herbicides Justin uses, but also synthetic insecticides and genetically modified seeds. He knows that to some his use of those technologies automatically casts him among the villains of Big Ag, destroying nature and human health with their "industrial" (or "corporate," or "factory") farms. Popular culture abounds with such dystopian portraits, from the documentary *Food, Inc.* to Chipotle's viral YouTube video of a sad little scarecrow traveling through a wasteland to his job on a ghoulish food assembly line. Sustainable farming—in this influential view—is necessarily organic, small, local: everything, in short, that Justin is not.

It is true beyond dispute that chemicals of all types—in everything from cookware and cleaning products to clothing and toys—have been dangerously underregulated in the United States, and that many have done incalculable harm. Concern about pesticides, in particular, was the root from which modern environmentalism grew: Rachel Carson's *Silent Spring* brought to light the devastation such long-lasting toxicants as DDT were inflicting on birds, humans and other living things, and inspired the small band of scientists that would become Environmental Defense Fund to bring an unprecedented series of legal actions that in 1972 got the poison banned nationwide. That same year, Congress passed a law giving EPA far more effective means to regulate pesticides.

For Justin, the fear of chemicals has hit close to home. Lindsey gardens organically, diapered their babies in cloth and in another of her essays juxtaposed herself in those chemical-free settings against Justin finding orange streaks of leaf rust, a fungal disease that can cut wheat yields 35 percent, and rolling out the sprayer she regards with suspicion: a "futuristic superbug 13 feet tall with a 90-foot wingspan" that she only rarely and reluctantly lets their kids go near.

The more she's learned about the chemicals Justin uses, the less afraid Lindsey has become. She has also come to see, with Justin, the complex trade-offs inherent in every one of the choices they make. For instance, most organic farmers avoid using herbicides by tilling. But in Justin's view, and that of many soil microbiologists, that causes *greater* ecological harm. Justin always turns to biology first: keeping weeds in check with rotations, residues and cover crops. When he does use chemicals, he works to ensure that, like his soil and water, they rarely leave his fields. But going to zero would exact, he and his mentors believe, too high a cost. "If you till to avoid herbicides," says Rice, "You do massive damage to soil microbiology. If herbicide allows you to leave the soil intact, it is a net environmental positive. We have fields that have been in continuous no-till for twenty-two years, using herbicides, with ever more microbial diversity and life."

Skipping synthetic insecticides can also be the more damaging choice, Justin believes, if it means losing a crop and with it all the precious water and other resources that went into growing it. "In late fall, adult weevil beetles come into the alfalfa and lay their eggs into the stems. If the larvae hatch in the spring, they can turn a hayfield from beautiful lush green to gray, tattered shreds in forty-eight hours. We spray a pyrethroid insecticide that's far safer than the old organophosphate chemistries. If we didn't, we could lose half of our annual production."

Opting for pesticides "certified for organic use" would not resolve his dilemma. Some pesticides that are naturally occurring are classified by the World Health Organization as more hazardous than some synthetics. And since many organic pesticides are applied in granular form, says Justin, they are "far more likely to 'dust-off' into the air and water" than the liquids or seed treatments he uses. Non-pesticide strategies can also bear high ecological costs: organic fruit and vegetable producers frequently rely on so-called plastic mulch to control weeds and insects, covering entire fields with a kind of Saran Wrap that can overheat soils, killing off microbes, and typically after harvest, winds up in landfills.

One chemical has weighed especially heavily on Justin's mind lately, a class of insecticides his family has used since the 1990s that has lately become a prime suspect, along with habitat loss and the varroa mite, in the steep decline of honeybees and other pollinators: neonicotinoids, or "neonics."

When applied to seeds before planting, Justin believes neonics provide protection at "an absolutely critical time. A seedling is like a baby: those first thirty days are when it's most fragile, most vulnerable to sickness and death. The neonicotinoids prevent not only the physical damage of a biting wire worm or aphid, but also the diseases those pests vector, the wounds other pathogens can enter." Three years of on-farm testing persuaded him that they were worth the significant

cost: "our bean yields improved two of three years and we had significant stand improvements [the ratio of seeds planted to mature plants]."
He's found them to be of even greater value in sorghum, which is vulnerable to chinch bug invasions worthy of a horror film. Having spent
their nymph stage in Justin's winter wheat fields, the adults move by
the millions at summer harvest into neighboring green fields. Hundreds surround the base of each young sorghum plant, "sucking their
juice, killing as much as a quarter of the field." If the sorghum seeds
have been treated, however, the bugs are stopped at the field's edge,
reducing losses to just a few percent.

Still, says Justin, the evidence is clear "that if a bee has contact
with neonics, it's going to be harmed. So the question is: what's their
opportunity for exposure? Early concerns came from some cornfields
in southern Germany, where a bunch of things went wrong at once:
the corn seeds were treated with unusually large doses but without the
glue that keeps the neonic on the seed; then farmers planted in high
winds, in late spring when nearby canola fields were already blooming. Talk about a place where pollinators are abundant. The dust blew
and settled in the canola field; the bees carried it back and suffered a
huge colony loss.

"We apply much lower doses, 0.8 ounces per 140,000 seeds,
and with a polymer that keeps it tightly attached. Pollinators have
no access to it when it's underground. Then it's systemic—inside
the plant tissue—which they have no piercing mouth parts to penetrate. It's been hard for me to envision how that tiny amount could
be diluted through the whole plant and still be expressed in the pollen
grain, especially because we see in our own fields that after forty-five
days insects can do damage; at that point the neonic has no noticeable
activity. But lately I've been seeing research indicating that the active
ingredient may be expressed in pollen, and may last longer in the environment than we thought."

Operating within that uncertainty has been immensely challenging for Justin. "I know that to many people it's not complicated: the neonics are either black this way or white that way. But for those of us struggling through perspectives, it's really complicated." Justin has been trying out a new alternative on some of his corn, which he hopes may replace neonics. He's also been looking at wax-based replacements for the lubricant he uses to help the seed slide through the planter, a talc that he worries might leak out and carry neonic dust to flowering plants on the perimeter of his field. In every case, he weighs the environmental costs (to the degree they're known) of using the pesticide versus not using it. In a soybean field, for instance, "how many years will I have bean-leaf beetles and have to spray a pyrethroid insecticide?" He has closely watched the experience of European farmers as they've struggled to find substitutes in the wake of the 2013 continent-wide ban on the entire class of neonicotinoid insecticides.

When in January 2016 EPA released a 291-page report on imidacloprid, the neonic Justin uses, he spent hours digging in. "It's the first really in-depth assessment and research review I've seen on different crops and timings and application methods and how that translates into interactions, or not, with honeybees." The report found the highest risks in cotton fields, where the insecticide is broadcast on the soil in granular form, and in citrus, where it's sprayed on trees in bud; the risks were low in corn and wheat, which produce no nectar. "For corn, they have the research showing detectable levels in the pollen but at concentrations below the threshold of risk to colonies. They don't have that kind of data yet for soybeans, so they classified the risks there as uncertain. Given that uncertainty, this coming spring I plan to use less in my beans, and only on the highest producing fields where our risk for yield loss would be the highest. I'll also do another on-farm study reassessing its value."

Justin challenged himself all over again at the 2016 No-till on the Plains conference, attending several talks by one of the most high-profile critics of neonics, entomologist Jonathan Lundgren. "His view is that the chemical may become inert enough to not affect our target organism but still have low-level suppressive effects on the whole insect community. He urged us to keep moving toward greater diversity. It really drove home for me again the importance of diverse rotations and especially perennials, which you don't replant for five years or more so the field gets a long rest from any seed treatment."

Justin continues to dedicate many of his solitary hours in the tractor to thinking this through. "Pollinators are important to farmers. Even soybeans, which self-pollinate, get a boost if a bee goes in the flower; it triggers the little biological spring that drops the pollen. And industry assurances aren't enough. I'm absolutely obligated to keep abreast of the science and, where I become convinced they're a danger, to find an alternative."

As he does with weeds, Justin tackles insects first with biology. Boosting the health of his soil fortifies his crops' ability to fight off infestations. And his rotations make it harder for the pests to track down their targets, though insects are harder to outsmart than weeds. "If there's any alfalfa within miles, we can plant a brand new alfalfa field and it will soon have just as much weevil pressure as the older fields," he says. Insects possess an astounding ability to figure out farmers' rotational patterns, according to Beck. Following a long history in the Corn Belt of the standard corn–soy–corn–soy rotation, corn rootworm beetles have adapted. Some now fly to soy fields to lay their eggs, so the larvae hatch into the corn that will be there by spring; the eggs of others now take two years to hatch.

Justin has heeded Beck's guidance to be unpredictable in the interval and sequence of his crops. He has also nurtured populations of beneficial insects, using parasitic wasps as his first line of defense, though he has "not succeeded so far in boosting that population high enough to overcome the pressure." He next plans to add livestock to his land, contracting with ranchers to bring their cows or sheep into his fields to trample insect larvae and eggs and to eat his cover crops, supplanting both insecticides and herbicides. (A friend of Justin's with a small organic farm uses chickens to eat larvae, but Justin laughs that the two million birds he'd need might be hard to manage.) Grazing cows in his fields will benefit ranchers and their omnivorous customers: beef finished on a mix of cowpeas, clovers, oats, barley and tubers will be more nutritious and delicious than feedlot beef, according to Jim Gerrish, who for twenty-two years led the University of Missouri's Forage Systems Research Center. And at the going rate of $3 per animal per day, they will get fat as cheaply as on feedlots. (Applied broadly enough, Gerrish says, custom grazing could replace every feedlot in the nation. That's a patent waiting to happen, jokes Justin: "farm-finished beef.") Justin's soils, in turn, will benefit from "inoculation" with the nutrients and gut microbes abundant in manure and cud, as well as from the benefits of hoof action and chewing that Dusty Crary enjoys. Justin has been greatly inspired by Gabe Brown's 5,400-acre farm near Bismarck, North Dakota: by integrating animals, Brown says, he has eliminated synthetic fertilizers and cut pesticide use by 75 percent.

Just as the use of chemicals is not black-and-white, neither is scale a reliable indicator of sustainability. Many small farms are terrific, but not all. Rice grimaces as he recalls a visit to an organic turkey farm where poultry waste ran right into a stream. And big farmers, says Justin, are leading the adoption of no-till and cover crops. "There's a perception out there that's tempting to buy into, that to take care of the environment you have to farm at a scale like my grandfather would have farmed on: a couple hundred acres, with smaller machinery, very limited technology. And I think that's not quite accurate. As I think about the farmers in our community and probably agriculture as a whole in much of the Midwest, I would argue that many of the larger-scale farms are the ones on the cutting edge of environmentalism." That's in part, he says, because of the expense of the technology, "so you need to be at a certain scale to implement some of these management practices." And a conscientious farmer like Justin on 5,000 acres can have many times the impact of a farmer managing a hundred acres.

As for locally grown: also not always best. In *The Third Plate*, renowned farm-to-table chef Dan Barber challenged himself and his acolytes to reconsider a religion of eating that pushes farmers to grow everything (heirloom tomatoes, microgreens, asparagus) everywhere, even in climates and soils ill-suited to support them. Far better, Barber argued, to design a "whole pattern of eating—a cuisine," around what the land is suited to provide.

As Lindsey has discovered in her own garden, trying to make hot, dry, windy, freezing, brutal Kansas grow vegetables usually ends in heartache. (So far, only her jalapeños and green beans have thrived.) Locavores notwithstanding, it would make little ecological sense for Justin to shift from crops that thrive in Kansas's aridity and temperature extremes, producing strong yields of high-quality grains and legumes with few inputs and *no* added water, to crops that require irrigation or greenhouses warmed with fossil fuels. "Kansas is so effective for winter wheat because it fits our climate, producing in the cooler time of year and when we get most of our precipitation," says Justin. Wheat is also close enough kin to the region's indigenous prairie grasses that it can support, rather than disrupt, the native ecology. It is because Justin grows grasses in a grassland that he can—*through* that intensive production—rebuild his soils and nurture several thousand species of co-inhabitants. In a wetter place, says Justin, he would choose crops akin to the native neighbors that love moisture; in a grayer place, like Minnesota, plants efficient at capturing and using every bit of sunlight.

There are not, in other words (cartoons aside), simple, perfect, universal answers. Agriculture can't be formulaic or dogmatic because, as Justin says, "diverse ecosystems require diverse practices."

He sees, for instance, "crops and geographies and family circumstances for which no-till is not the right solution," including in very cool and poorly drained soils where under thick residues the soil can never dry or warm enough to germinate seeds. Cover crops, too, may not work everywhere. One of Justin's closest friends, Kevin Wiltse, has experienced significant losses planting cover crops on his farm in western Kansas, where the drought has been far deeper and more relentless. "We get 29 inches of annual rainfall; they're below 20 inches. Many farmers out there will grow a wheat crop one year then skip the next, just to store water in the soil. Kevin hoped the cover crops would reduce evaporation and improve water capture enough to compensate for the water they used. But he wound up with a moisture deficit and a tremendous yield hit in the commercial crops he followed with."

Every region, even every farm within a region, differs in history and soil conditions, topography and climate—demanding the same close attention and responsiveness Justin brings to his own land. "It's never simple and it's always complex. We have to think through how things interact in the local environment, and what's right here may not be right even down the road." That's why Justin never criticizes other farmers' choices, assuming they are based like his on hard-won knowledge and experience of their land. Nor is he ever defensive about his own choices, but remains open to any question or challenge to his thinking, eager to see any research that might change his mind. "I believe we're created to be in a community and impact and learn from each other," he says. "I really like John 1: 'In the beginning was the Word: and the Word was God . . . Through him all things are made.'" That sacred presence throughout time and in every person means that all deserve a respectful hearing.

Fortunately, Justin's antidogmatic approach looks to be the wave of the future. A growing number of farmers, and some of the most influential voices in farming and food, increasingly reject inflexible ideologies, understanding that purist positions often result in bad trade-offs. Like Justin, most see themselves as perennial works in progress, inching along a continuum from chemistry to biology, monoculture to diversity, high inputs to relying on nature to cycle nutrients and manage pests.

Author and UC Berkeley professor Michael Pollan was among the first to complicate the conversation. Writing in 2001 in the *New York Times Magazine*, he challenged the idea that buying organic necessarily means "casting a vote for a more environmentally friendly kind of agriculture," noting that the label provides only the narrowest information about a farm's practices. Though Pollan's insights are still just sinking into the broader culture, many celebrated producers increasingly reject what New York farmer Amy Hepworth calls the "two-party" system that divides farmers into good (organic) and bad. "The most sustainable, responsible system is a hybrid system," she told Tamar Haspel, a Cape Cod oyster farmer who also writes for the *Washington Post*. "Organic was the greatest movement in my lifetime, except for the one I'm doing now." *Mother Jones*'s Tom Philpott has also complicated the Big Ag story; in one series, he unwrapped the lumpy, seemingly sinister category of "corporate farms," documenting not only that American farms remain overwhelmingly family-owned and run, but also that agribusinesses like Monsanto want nothing to do with the unpredictable, low-margin work of actually growing food. Former *New York Times* food writer Mark Bittman is increasingly aligned with this cutting edge. "I'm not talking about imposing some utopian vision of small organic farms on the world," he wrote in a blog about Iowa State's Marsden Farm and its use of rotations to cut herbicide and fertilizer use and groundwater pollution. "Debates about how we grow food are usually presented in a simplistic, black-and-white way, conventional versus organic . . . loads of chemicals and disastrous environmental impact against an orthodox, even dogmatic method that is difficult to carry out on a large scale. . . . The Marsden Farm study points to a third path . . . using chemicals to fine-tune rather than drive the system."

That more nuanced conversation has opened up even around the most inflammatory of topics: GMOs. The use of genetically modified organisms in agriculture has stirred intense opposition not only because people fear the technology itself but also because it is so closely identified with its leading proponent and patent holder, Monsanto. With its roots in chemical manufacturing, including such poisons as DDT and PCBs, Monsanto could not have done a worse job of introducing genetically engineered crops to the world. It laid the groundwork for a backlash with its narrow insistence on applying the technology for the sole purpose, as one scientist put it, of "making the world safe for more pesticides." Then it compounded the damage with arrogance. When, in a meeting in his office in the mid-1990s, U.S. Agriculture Secretary Dan Glickman challenged Monsanto CEO Robert Shapiro to address mounting public concern about GMOs, Shapiro snapped, "They should be thanking us. We're doing God's work." Over the next two decades, the company's pattern of attacking critics and dismissing valid concerns was so egregious that Shapiro ultimately felt the need to publicly apologize. In a 2015 Harris Poll assessing corporate reputations, Monsanto ranked 97th out of 100, behind Halliburton and BP. And that was before the September 2015 front-page story in the *New York Times* on how industry "enlisted academics in [the] GMO lobbying war."

Yet as hard as Monsanto has made it for anyone to see past its dark shadow, in recent years many even of its most fervent critics have come to see genetic modification of organisms as a tool that can be used destructively but also for good. The most startling turnabout came from Mark Lynas, who in a blistering speech at Oxford University apologized for his leadership of Greenpeace's anti-GMO campaign: "as an environmentalist . . . I could not have chosen a more

counterproductive path." Since then, deeply reported series by Amy
Harmon in the *New York Times*, Nathanael Johnson in *Grist* and
others have explored both the growing body of science (which to
date has found no evidence that eating GMOs poses risks to human
health) and the public backlash and its costs. In one piece, Harmon traced the environmental harm done by the opposition to a
genetically modified orange resistant to the greening blight that is
threatening millions of citrus trees in the U.S. and around the world,
reporting that without that alternative, farmers have applied ever
larger doses of pesticides. (Those include neonicotinoids, which are
known to harm pollinators when sprayed.) In the *Post*, Haspel has
championed DuPont's efforts to engineer yeast to produce omega-3
fatty acids, potentially relieving pressures on marine resources.

Like most scientists, Justin sees the risk of GMOs not in the technology but in the misuse of the two dominant genetically modified
crops: corn and soy engineered to resist Roundup so that farmers can
spray the herbicide on a growing field. Combined with policies that
have kept corn prices high, the two GMOs have enabled and encouraged the spread of destructive monoculture—as Beck says, "Roundup
Ready corn followed by Roundup Ready soy is not a rotation." With
that monoculture has come rampant overuse of the herbicide, spawning a plague of resistant weeds. "As with any technology in an ecosystem," says Justin, "overuse will cause that ecosystem to shift and
adapt." The rise of superweeds has driven farmers to keep ratcheting up their herbicide use: the "pesticide treadmill," ecologists call it.
Justin also shares critics' concerns that a few companies control the
research and proprietary seeds.

The solution, he believes, is not banning the technology, "because
if we aren't open to responsibly using science and technology to produce more with fewer resources, if we take away management tools
because of fears that are not scientifically sound but emotional, then
impacts on fragile ecosystems will increase." Rather, he advocates

increased support for both good farming practices and public research. Public funding for seed development is half what it was thirty years ago. In 1980, 70 percent of the acres planted in soybeans used seeds developed by the public sector; by 1997, the same percentage used private-sector seeds. "If we don't want corporations to be the primary innovators of biological products," says Justin, "we need to support applied research at public institutions and a regulatory process that a university can navigate."

Walking through the greenhouses at Kansas State, past breeding experiments focused on the shape of wheat heads (big and long versus thick and square), tables full of the cockleburs and pigweeds he had to identify to pass his exams, and photos of severely eroded tilled fields, a third of their soils washed away ("I can't hardly look at it"), Justin reveals an advocate's fervor for America's land grant colleges. "There's no other institution affecting production ag more. They're training the next generation of people who'll be managing the land; they have a high level of trust with producers and a network in place to disseminate research. But as it gets harder for them to get funding, especially for the long-term studies needed to understand the impacts of cover crops and rotations, we aren't getting the foundational science we need to answer critical questions. There's a dearth of research, research that is so important to me as I try to understand this vast unknown. My extension agent* says, 'I get questions day in and day out about things like cover crops. But I don't have research to share."

Alongside many of his peers, Justin has stepped up to fill that void: turning his farm into a laboratory, with rigorous scientific protocols. An article about Justin in the *Salina Journal* (headlined "Old

---

* State extension offices, housed at land grant universities and supported by the USDA, dispatch local agents to educate and share new research with farmers and rural communities.

McDonald Has a Laptop") described him replicating tests across diverse fields, using "spatial data recording . . . to find out if a certain micro-nutrient . . . was improving yield." He partnered on a similar experiment with Andy Holzwarth, again doing replicated tests, this time of "precision" (sensor-based) nitrogen application. In November 2015, Justin hosted U.S. Senator Pat Roberts (chairman of the powerful Agriculture Committee), several dozen growers and industry leaders for a demonstration of cutting-edge technologies. Great Plains Precision Ag flew a drone over one of his dad's young alfalfa fields, using infrared imaging to pinpoint where a heavy rain had washed out seeds and they would need to replant; dark greens alerted Justin to weed problems he hadn't spotted from his pickup truck, which would have taken hours, he says, on a four-wheeler or walking to find.

In experimenting with his farm and his family's livelihood, Justin assumes great personal risk. He often invests years in no-tilling before depleted soils recover, facing possible losses in the meantime. And he has made costly mistakes. "I tried to move to intense rotations* in a dry year and had crop failures when other guys didn't. I've let cover crops go too long, which cost me yield." One of his riskiest experiments began in fall 2014, when for the first time ever he seeded a rye cover crop, hoping to capitalize on its robust production of roots and residues to carpet a balding field and ready it for soybeans. "It's a stretch for me and particularly for dad to plant rye, because it can be an aggressive invasive in wheat. Growing up, we'd have to go out in the fields a month before harvest and rogue out any rye plants before they went to seed, or pretty soon five plants would be hundreds. As a child, out pulling rye by hand on a hot summer day in a Kansas wheat field, I would have never thought I'd be planting the stuff." Beyond the risk of invasion, spring 2015—again, abnormally warm and dry—

* Planting one crop after another in close succession.

brought another worry. "If the rye uses up the water in the soils and rains don't come to replenish it, our soybean yields could wind up ten bushels lower. We can't afford that kind of economic hit. So we wade through in real time. If we stay dry we'll kill the rye earlier. We won't know whether that was the right answer till the soybean harvest, but that's how we learn, observing and adapting to go forward."

Making the wrong choice can land hard in an industry that—popular conceptions of fat-cat corporate farmers notwithstanding—operates on slim margins. A June 2014 article in the *Kansas City Star* (on the drought's "harvest of hard luck") laid out "the math . . . for the Kansas grower, based on a simplified rule that each wheat bushel makes 67 loaves of bread: Last year, an average 1,000-acre field equaled about 3.3 million loaves. If a nickel of each loaf went to the farmer, that's $170,000. But if this year's wheat yields are half what they were in 2013, it would mean half as many loaves and $85,000 less produced from the same farmer's field."

The consequences of life on those margins are visible on a drive through nearby Gypsum, under a sky that has gone from vivid blue to a wash of unbroken gray. Though a few storefronts flicker with resurgent life—a new hair salon, a gun shop, a diner where Justin and family sometimes meet at midday for the meat loaf special, tater tots and banana cream pie—Main Street is mostly shuttered. Many of the town's houses and trailers stand dilapidated or boarded up; the school where Jerry went to high school and Justin to preschool has been abandoned, its windows broken and its roof caved in. "Dad won't hardly drive by," says Justin. "It makes him too sad."

Still, Justin keeps experimenting, never satisfied but always "searching for the next thing we ought to be doing." And so far, most of his bets have paid off. Through all the crazy weather, and even as average yields in Saline County slumped from 52.9 bushels an acre in 2013 to 29.5 in 2014, Justin's yields stayed high. In 2013, "the field across from Grandma's" produced close to 80 bushels an acre—without extra nitrogen. Those high yields and low input costs have in turn translated to a good family income, as evidenced by Justin's 2012 books. Accounting for both variable costs (seed, fertilizer, pesticide) and fixed costs (land, labor, machinery, fuel), Justin saw a $4,000 loss on 62 acres of corn: so dry during its critical reproductive phase, when it drops its pollen onto the silks of the ear, that it yielded just six bushels per acre (170 bushels is the national average). But his 520 acres of soybeans cleared $67 an acre; his 580 acres of wheat earned $63 each, and his 102 acres of alfalfa earned $145 apiece.

Justin's performance on ecological measures has been stronger still, with his fields increasingly (like the prairie) self-contained: holding soil, water and nutrients in place and building organic matter, fertility, and biodiversity where it counts the most, underground. Among those who regularly measure Justin's progress is Dr. Ray Ward, a legendary soil scientist who runs a private testing lab in Nebraska. (Ward calls his drying facility "the Kansas room" because it is always 120 degrees with a wind blowing like mad.) Ward has invented tools to measure fatty acids, a marker for living microbial biomass. Combined with additional tests of the soil biology (diversity of microbes, their metabolic activity) and chemistry (including the balance between organic and inorganic nitrogen), the lab arrives at an overall grade for each field called the Haney score. Lance Gunderson, director of soil health at Ward Lab, deems Justin well along in the right direction. While a tilled, monocropped, chemically fallowed field typically scores about a 1 on the Haney scale, and Gabe Brown's full-system approach (no-till, diverse cash and cover crops, livestock) scores a 25, Justin's fields come in between 6.5 and 13.5.

Justin's successes are not atypical. A 2014 survey by Kansas State found the majority of no-till farmers reporting increased wheat and sorghum yields, higher net returns, increased fertility, and decreased weed pressure and erosion. Those using complex (at least three-year and three-crop) rotations also reported higher yields, as did those using cover crops, with an additional benefit of reduced insect disease. (However, all also reported significant increases in "management intensity.")

These heartland farmers are, in other words, exemplars of essential aspects of "agroecology," the framework increasingly guiding the world's most influential thinkers on how to secure food for people everywhere over the long term. Though few agroecology champions might think to look for heroes out on these big, high-tech, commodity-producing Midwest farms, Justin and his peers are working through many of the core principles: using nature, not factories, as their model; focusing not just on market crops but on the whole living system they inhabit; promoting symbioses among a vast diversity of organisms; using biology to recycle nutrients, water and energy; and relying on knowledge gathered with their own hands and experience.

Most importantly, Justin is building resilience to climate change, even as the conversation around it remains complicated.

The projections are dire, particularly for the Great Plains, which have been identified by scientists as one of the planet's most vulnerable places for water availability. Weeds like Palmer amaranth, which can yield a million seeds per plant, are already expanding their range north as the Plains warm, alongside fast-adapting insects and pathogenic microbes. In 2013, Kansas State forecast that every one degree Celsius increase in mean temperature would cut yields of dryland hard red winter wheat, the kind Justin grows, by 20 percent. Even the most optimistic scenarios project three degrees of warming in the grain belt within Andrew and Charlotte's lifetimes.

Still, the topic is not easy to broach. Dr. Gary Pierzynski, head of the Kansas State agronomy department and another of Justin's mentors, describes how he and his colleagues navigate that tricky terrain. "We have no doubt that climate change is happening. But we recognize that talking directly about it raises issues with some of our elected officials, who remain unconvinced and don't support investing state resources to study it. So we emphasize our focus on challenges like extending the life of the Ogallala aquifer; we don't disguise it but take away the climate change message."

Justin generally takes a similar path. He does allow himself a mordant joke, passing an experimental greenhouse that looks like the Sonoran desert, full of cactus and distinctly out of place amidst the prairie grasses and grains: "Maybe this is the climate change lab. Maybe this is my farm in thirty years." But he quickly grows more tempered. "I don't know if our more severe weather events are part of a long-term trend or not. But it's not important for me to know. What's important is this: if it's happening, what steps am I taking to make my little bit of land a more buffered system against an extreme climate? And if it's not happening, well, everything I'm doing makes sense to do anyway."

He does find the conversation growing somewhat less tense, as the "mainstream ag press moves toward a more straightforward treatment of climate change." Among the examples he shares is a fall 2014 article in *Successful Farming* magazine headlined "How to Cope with Climate Change." The article begins with author Gil Gullickson, the magazine's crops technology editor, acknowledging his readers' likely response: "I know what you're thinking. Climate change is just some figment of Al Gore's imagination. A communist–socialist–liberal plot hatched by . . . eco-Nazis aiming to run the U.S. economy into the ground."

Gullickson then prods his farmer readers to "ask yourself the following questions": are springtimes getting wetter, rainstorms more intense, droughts more severe? The answer, he continues, is yes: in Iowa, more frequent and intense spring rains are keeping farmers out of waterlogged fields and washing away nitrogen. In Oklahoma, farmers endure wheat failure after wheat failure as rains that historically helped germinate seeds in the fall and bust the wheat out of dormancy in spring have become unreliable, freezes hit both early and late, and droughts increasingly end "with a bang," with a year's rainfall sometimes falling in a few weeks' time.

Gullickson goes on to cite the scientific consensus that climate change is not only happening but is driven by human activity, then quotes Kentucky farmer Don Halcomb complaining "that we are poorly served by farm groups on this issue." Focused on the short-term costs of addressing climate change, Halcomb says, they ignore the fact that "not addressing it also has a cost," like the $5 billion in ag production lost in the 2012 Texas drought. Ag's "suits," Gullickson adds, show no such qualms: Cargill's chairman Gregory Page was one of those who joined Democratic climate activist Tom Steyer in issuing a report quantifying the costs to the U.S. economy of inaction, including declining yields. Monsanto's scientists, finding that even a one to two degree Fahrenheit change in a region will bring new infestations of insects and disease, have shifted resources into developing new hybrids, seed treatments and microbes protective against those emergent stresses.

Finally, Gullickson urges his readers to look to no-till and cover crops as their best strategies for boosting climate resiliency, pointing to findings that those practices reduce evaporation up to 80 percent and save as much as two inches of water, translating to twelve more bushels of wheat per acre.

No-till does more than help farmers adapt to climate change; it also helps reverse its underlying cause, by putting back into the soils carbon lost to the atmosphere through cultivation. Like all plants, crops take carbon dioxide (the primary greenhouse gas) out of the air through photosynthesis, turning it into carbohydrates. If those plants are plowed under to be digested by hyperactive bacteria, much of that carbon is lost back into the air. But when they are left in place, to be eaten by slow-growing microbes who lock the carbon into their bodies or wrap it in their threadlike hyphae, that atmospheric CO_2 becomes valuable soil carbon, providing myriad services to Justin. Soil carbon is the staple of his microbes' diet, helps to make other nutrients accessible to plants, and keeps soil both glued together and porous; it is the major ingredient in his vital organic matter. "Carbon is the currency of the biological system, passed from one organism to the next, with value at every step along the way. So the CO_2 in the air is something I'm interested in capturing and holding with my plants and soils; it's not in my interest to have it leave my fields." Having been nearly depleted of carbon by decades of cultivation, his soils are now more than halfway to the 4 percent carbon levels in native prairie soils. With another decade or two of no-till, says Rice, Justin will close that gap.

The benefits to the atmosphere, even if collateral, are immense. Soil is the planet's second largest carbon repository, storing nearly twice as much carbon as the earth's atmosphere and every plant combined. Land cultivated for centuries has lost half or more of its stored carbon. But if applied globally, restorative agriculture on cropland and pasture soils could achieve up to 15 percent of the total carbon reduction needed to stabilize the climate. "So when I think about carbon and greenhouse gases," says Justin, "it's from a sense of, 'OK, if climate change is a problem, I have some solutions to bring to the table.'"

Beyond sequestering carbon, Justin's approach to farming helps slow climate change in two additional ways. His reduced reliance on synthetic nitrogen avoids both consumption of the fossil fuels used to make that fertilizer and the release into the atmosphere of nitrous oxide, a greenhouse gas 300 times more potent over a hundred-year timespan than CO_2. (Overfertilization is the largest source of greenhouse gases from U.S. agricultural production.) His diesel fuel consumption per bushel is also way down, because he makes far fewer tractor passes and needs far less horsepower when he does. With his no-till equipment, he can work as much as 100 acres an hour; as a kid, dragging steel through the field, he could cover just 5–10 acres an hour, using ten to twenty times as much diesel per acre.

Justin is part of a still-small vanguard, with less than 20 percent of corn, soybean and wheat farmers across the nation practicing continuous no-till. The slow uptake, even in geographies well suited to it, is in part cultural. "Ag's technology adoption curves, even going from horses to tractors, tend to be slow," Justin says. "Those old farmers were sentimental about their horses, reluctant to replace them with hard cold steel. And all farmers, including me, love a nice, dark, warm seedbed, the smell of freshly tilled soil. It feels orderly and productive. So now, after five or seven generations, to say, 'That's wrong; what we need is a big mess out there,' can be hard to accept.

"Change is also expensive. The machines you had aren't useful anymore. You've got to have the right kind of seeder—ours cost $175,000—and a sprayer, which cost another $225,000. If I only farmed my own ground I wouldn't be able to afford it." On top of those capital costs come, typically, difficult transition years, with profits dipping before the nitrogen savings and yield gains kick in. As Beck says, "This complex web of life, soil structure and organic matter does not reappear quickly on soil that has been tilled for many years."

No-till is also a more demanding way to farm. Justin's younger brother Jeff provides an example: the challenge of planting tiny alfalfa seeds into thick wheat residues. A baby alfalfa plant forms a crown, where it stores carbohydrates to sustain the green tissue through winter and initiate growth in the spring. That crown needs to stay below the soil surface all winter to be protected from cold. But if planted into deep no-till residues, the confused little plant will push up into the darkness, thinking itself still underground, and wind up "winter kill." For now, the Knopf brothers use a hybrid solution: once or twice before seeding, they run a vertical tiller through the field to chop and mix the residues into the top couple inches of soil, so it degrades a bit faster while still leaving enough to prevent erosion.

"Switching to no-till is not like buying a GPS," says Justin. "It's a whole complex system that takes learning, analyzing, experimenting. In tillage-based farming, you're managing visible things. If you see weeds or clods you go work the ground. In no-till you're also managing the invisible, a vast hidden complexity of biological interactions among multiple crops and thousands of microorganisms. You're going to have failures, question marks. OK, something changed in my field, but what caused that change? It doesn't happen overnight and you only have a short time in this life to see what's happening. It's really an investment in the future."

Yet for all the obstacles, Justin can increasingly rely on a growing network. His closest partnership is with three like-minded Kansas farmers. Kevin Wiltse is from Great Bend, eighty miles southwest of Knopf Farms, in the epicenter of the 1930s' Dust Bowl. Doug Palen and Chad Simmielink farm about the same distance to the northwest, close to Willa Cather's Nebraska haunts. The four talk by phone weekly, regularly visit each other's farms to brainstorm over management challenges and often do joint on-farm research. "Having such a far-flung network is a big change from thirty years ago," Justin says. "Back then, it was about having your next-door neighbors' extra hands to do all the hard physical work; people would share equipment, do their threshing or work cattle together. We still work together with neighbors, but there's no longer that survival necessity, to rely on them that way. As in everything else, the value now is in information and knowledge, and for that, you need bigger networks." The four men also share "personal stuff," he says. "We are all fathers of young families and talk about how not to let the farm take over our lives."

The four are in turn part of a larger network, a producers' association called No-Till on the Plains dedicated to examining "the sym-biotic relationships between the soil, microbial life, plants, insects, nutrients and wildlife," recognizing that "nature provides answers for agriculture to follow." Their annual conference is held every win-ter in Salina. In 2014, it drew 1,500 large-scale farmers of all ages from Nebraska, Iowa, Oklahoma, Texas, Missouri, the Dakotas, Colorado and Kansas: third-, fourth- and fifth-generation commod-ity growers trying not to go under in the face of diminishing water and collapsing soils. They spent three days in a stunningly radical discussion: on agriculture's devolution into an "extractive industry," the damage done by over-reliance on chemical fertilizers and pesti-cides, how to reduce their contribution to atmospheric CO_2. Justin attends every year. So do many of the stars of this new wave, includ-ing (sometimes) Dwayne Beck, who in front of the rapt audience is half oracle, half stand-up comedian. When a farmer asks him how to fix a "saline seep" at the bottom of one field, Beck advises him to grow hay to take up the salt, then cut and carry the hay to the top of the hill to feed his livestock. "That's what buffalo do. They feed in the bottomlands, then sleep up top where it's safer, ferrying the salt in their manure." Fellow speaker Ray Ward jumps in to tease him: "I wish I had time to sit around and watch buffalo." Beck's droll reply: "Become an academic."

But the PhDs are not the main attraction. This is a grassroots, col-laborative affair, focused on farmers standing up to tell their stories, share what they've learned, fortify one another's courage. "I'm not an expert," they say. "I'm just trying to figure it out like you." Like Justin, all are trying to farm in sync with the particular history and ecology of their land, and all are rigorous empirical scientists. "Don't decide in the coffee shop whether to spray insecticide," says Dan Forgey, man-ager of an 8,500-acre South Dakota farm. "Get your sweep net out; see what's there." Many have applied their improvisatory skills with a cutting torch, welder and pliers to customize equipment; in an exhibit hall set up for the purpose, they showcase their inventions alongside startups showing off custom cover crop seed mixes or precision fertilizer technology. The U.S. Natural Resource Conservation Service (NRCS) has brought its precipitation simulator, a portable shower that rains hard onto two-inch-thick squares of differently managed lands, with buckets to capture the runoff. The tilled soils rapidly fill the bucket with muddy water; flipped over, two inches down, they're dry as a bone. The native grasses and no-till soils are wet all the way through, their runoff buckets empty.

The stories the farmers tell often begin in sorrow, with falling yields, a family going broke, a farm on the brink of foreclosure or sale. But then, the critical turning point: they switch to no-till and their land and yields recover. Nearly all end with a slide showing the ultimate reward: a closeup of worn, dirty hands cupping black, chunky, moist, wormy, root-laced soil. Justin is particularly moved by the story of Texas farmer Jonathan Cobb, and seeks him out afterward. Cobb's family nearly lost their farm in the Blackland Prairie near Temple, Texas, in the historic heat wave of 2011. "It was like living on the face of the sun," said Cobb, "a hundred days over 100 degrees. After plowing we would measure our soils at 155 degrees; by harvest it'd look like we'd had a blow-dryer on the crop. I thought that after three generations, dating back a century to my grandfather's sharecropping days, our family farm would end with my dad. It took me three months to work up the courage to tell my dad his hopes were gone."

Cobb put the family home on the market and was looking for work in Austin when he heard a talk by NRCS agronomist Ray Archuleta about regenerative farming and "felt called to stay." He began walking his fields, "because going fifteen miles an hour on a tractor you don't notice much," going out even in punishing rains to see where soil was washing away and needed protection. He quit tilling, sold off his equipment and slashed inputs—he jokes that he put his land in rehab after years of "chemical crutches." His route to profitability, he believed, lay in reducing costs: "It's expensive to fight nature."

Cobb now plants a dizzying diversity of cover crops: winter peas, wooly pod, white ladino, crimson and Persian clover, sainfoin, Elbon cereal rye, black oats, nitro and Graza fodder radish, Winfred hybrid and purple-top turnip, forage collards, kale, plantain and chicory. Some are grown for seed to sell to other farmers. He also raises sesame, mung beans, and barley for local craft-beer makers, and has added cattle and laying hens to graze off his cover crops and devour bugs. He now has millions of worms, he says, on the land he renamed Redemption Farm, including a "famous" arm's-length monster which he keeps a picture of, along with 1,500 soil pictures, on his iPhone.

These no-till farmers are part of a still larger network of innovators, focused on such cutting-edge opportunities as drones, agro-bots and "biologics" or plant probiotics, the intentional introduction of beneficial microbes into the seed or soil. Justin inoculates the soybeans he plants and sells with rhizobium bacteria (the ones that fix nitrogen), a strategy farmers have used for years, but new molecular techniques have broken the field open. "Even Monsanto," says Rice, "is changing from a chemical company into a biological company"; it has partnered with biotech giant Novozymes to search the microbiomes at thousands of sites for novel enzymes.

Justin is also part of a network of market innovators who are striving to give consumers and retailers more meaningful insight into the farms behind their food. In 2014 Whole Foods announced that by 2018 they will classify produce as good, better or best based on farm metrics including soil health, farmworker health, water conservation, biodiversity and greenhouse gas emissions. Perhaps more surprisingly, Walmart—the world's biggest grocer—found that the single most effective step it could take to reduce its enterprise-wide climate impact was to ask suppliers of the food products it sells to use grains grown (like Justin's) with minimal nitrous oxide emissions. Walmart's partner in the effort is Environmental Defense Fund (EDF), whose Sustainable Sourcing Initiative also includes General Mills and United Suppliers, a wholesale provider of seed, fertilizer and other ag products cooperatively owned by its 600 member retailers. Though the initiative initially focused on the corn supply chain, in June 2015 Campbell Soup Company and its subsidiary Pepperidge Farm joined the partnership and became the first to ask its wheat suppliers (initially in Nebraska and Ohio) to incorporate conservation practices. As the online magazine *Grist* noted, "Farmers often resist top-down regulations, because what makes sense politically may not make sense in the context of every unique farm. Farmers are often more open to changes when the proposals come from trusted partners [including] food makers and farm suppliers . . . Now, more than policy and governmental agencies, the private sector is proving to be the biggest driver of farm conservation."

The latest network Justin has joined is the world of farm policy and those who shape it. In 2015, indicative of the respect he's earned among his peers, he was asked to serve on a new environmental regulations committee at the American Farm Bureau Federation, the nation's largest and most powerful farm organization. The first meeting he attended was focused on the Endangered Species Act and the controversial Waters of the United States rule. Justin was pleased to find that, rather than simply vowing to fight those regulations, his committee was focused on finding alternatives, including voluntary, farmer-led solutions that would make regulation unnecessary. "It's rare that I think the expansion of federal regulatory authority at the local level is a good thing. Their implementation is usually poor, with lots of unintended consequences. And how many guys at EPA have managed a farm? Maybe we could do some of this water quality work through NRCS, an agency more accustomed to working with ag. Or put more resources into their Conservation Stewardship Program,* to fund practices like cover crops that protect clean water."

Justin and Chuck Rice both see many opportunities for policies to better advance conservation practices. Crop insurance, for instance, could allow farmers more flexibility, says Rice. "The best way to salvage a wheat crop hailed out in spring is to treat it as residues and plant a summer crop right into it. But you have to let it go through its whole cycle"—right through its meager harvest—"to get insurance." Incentives could also help farmers, including Justin, plant "buffer strips" of native grasses along waterways and at the low edges of his fields. "Even as good a job as he's doing, Justin's not going to have zero losses," says Rice. "He'll have some runoff of nutrients. Grass buffers could capture any sediments and nutrients," keeping them out of waterways, while also moving carbon deep into the soil, recycling phosphorus, providing habitat for birds and bees and reducing downstream flooding. But "buffers add costs without direct returns," says Justin. "So farmers have to have proof or at least faith—as with cover crops—that we'll recover that cost. We need the science that quantifies the benefits to us: the beneficial insects it will promote, the nitrogen captured. Or we have to know that we'll get some kind of support for providing benefits beyond our farm—reducing nitrogen loading and algal blooms downstream, supporting insects that provide services to the broader ecosystem—via ecosystem service payments, or nitrogen credits, or the Farm Bill cost-share." The financial risks, otherwise, are too high. As Beck says, "a farmer that pursues conservation at the expense of profit will wake up to someone else mining the soil he's conserved."

Justin was excited to learn in early 2016 of one such emergent opportunity, aimed primarily at rescuing monarch butterflies, whose

---

* Farmers earn CSP payments for conservation performance; the higher the performance, the higher the payment.

numbers have plunged by 95 percent as the milkweed they depend on for nesting and caterpillar food has been killed off by herbicides. EDF announced the first steps toward a butterfly "habitat exchange": farmers will be able to grow habitat like any other crop and be paid for it by public or private "conservation buyers," from US Fish and Wildlife to pesticide manufacturers. "We could plant a mix of milkweed and other insect-friendly plants at field edges," Justin says. "And on stream banks, where crops have to compete with trees for water anyway so yields aren't as high; we wouldn't be giving up as much production." Either place, the flowering strips would do double duty as buffers.

The largest network Justin works within is the global food system, a system under mounting strain as population growth and growing affluence drive up demand for food, while land degradation, water and temperature stress and opportunistic pests put downward pressure on supply. Feeding the world is not just a problem of production. Many have noted that reducing meat consumption and the waste that now claims a third of global food stores would make existing supplies go much further. But even with progress on those fronts, farmers will by most estimates have to double production by 2050.

And they will have to do so in the face of multiple downward spirals. The first is the accelerating loss and degradation of the world's most fertile lands. The U.S. loses fifty acres of agricultural land every hour, a total of one million acres between 2007 and 2010. Worldwide, 20 percent of all cultivated lands are degraded: turned to desert or rendered useless by toxic contamination, salt damage, erosion or depleted fertility. The second stress is water, insufficient already on a quarter of the world's croplands—including nearly half of all wheat production—and fast declining. Though irrigated agriculture produces 40 percent of the global food supply, nearly every one of the world's major aquifers is overdrafted. The third accelerating challenge is pests: weeds, insects and pathogens vastly expanding their range;

these include *Blumeria graminis* fungus, which causes powdery mildew on wheat.

Combined, those stresses will drive up food prices (doubling by 2030, according to Oxfam), hunger and instability. They will also drive the clearing of still more land. Humans have already expanded agriculture into every grassland and temperate forest on earth, converting more land to agriculture since World War II than in the eighteenth and nineteenth centuries combined. They are now colonizing the sole remaining biome: tropical forests. Between 2000 and 2012, 230 million hectares of rainforest, an area equal to the U.S. east of the Mississippi, were destroyed worldwide, nearly all to make way for farming. "Never before," according to data from the United Nations Food and Agriculture Organization, "has such a large swath of the earth been tilled."

Turning tropical forests to farms is nearly always a bad deal. Yields are lower in the tropics than in the world's temperate regions and the ecological trade-offs far higher: not only destruction of the most biodiverse ecosystems on earth but also release into the atmosphere of the huge stocks of carbon locked up in those tropical peat bogs and massive trees. That carbon loss sets off a devastating feedback loop: accelerating climate change, which further suppresses global yields, which leads to more agricultural expansion through deforestation, releasing still more carbon, and round and round.

As producers for export of an essential global staple, "we spend a lot of time thinking about this," says Justin. "How will farmers feed the world without degrading our own lands or destroying more rainforests?" The answer almost certainly lies in replicating globally the kind of "restorative intensification" Justin practices on his own land: maximizing the nutrition produced per acre in ways that don't exhaust but rather rebuild the natural wealth needed to sustain those high yields over the long term. (Rice calls it "diversified intensification";

Beck, "regenerative agriculture.") Focusing those efforts on *existing* farmlands currently producing yields far short of their potential will be crucial; the University of Nebraska is leading work on a "Global Yield Gap Atlas" to guide such efforts. Crucial too will be concentrating production in regions like the American Midwest, where the biodiversity and carbon losses from giving land over to agriculture are relatively low and the yields high.

As trustee and science advisor to the International Center for Tropical Agriculture, Chuck Rice has become an important bridge between Great Plains innovators and their sometimes more advanced southern peers. He points to Brazil as his "poster child." Not only do 80 percent of farmers on the Brazilian and Argentine savanna practice continuous no-till, Rice says—essential in their high temperatures and rainfall, where soils are destroyed within three years of plowing—but the Brazilians also lead on intensified cropping, using seamless rotations to achieve year-round productivity. "As they're harvesting their winter wheat the soybean planter follows directly behind; they won't harvest what they can't immediately replant. Then they do the same with a cover crop going in the moment the soybeans are harvested, then another round, then corn, rotating through ten plant species" to foil pathogens and—with multiple crops to market—reduce economic risk. "They're managing not a field but a landscape."

Justin, too, is increasingly part of this international exchange of knowledge, traveling on several occasions to meet farmers in South America and Africa. He and Lindsey went together to Malawi to advise a friend working with women farmers (many widowed by AIDS) struggling to grow enough white corn to feed their families. (They grind the corn to make *nsima*, a porridge Justin says tastes like Malt-O-Meal.) "They have rain and good soils, but even a good crop can be ruined by red ants or grain borers," says Justin. "We helped get them better seed treated against the ants and a little bit of fertilizer, and they're working to improve storage methods. But it was difficult to see malnourishment alongside natural resources that could sustainably produce abundant food."*

He heard a more promising story at the 2016 No-till on the Plains conference from Dr. Kofi Boa, a Ghanaian cocoa farmer and agronomist who founded Africa's first Center for No-till Agriculture with support from the Howard Buffett Foundation. Buffett, a no-till farmer himself, is working globally to advance, as he says, not another green revolution (more chemical inputs) but a "brown revolution"— soil health.

Justin's sense of responsibility, his faith in forces beyond his knowledge and control, even his embrace of science: all are rooted in his deep Christian faith. "I believe in Creation and intelligent design. That we evolved from primates, I don't believe. But can species change, develop over time? Absolutely. We can use the intelligence we've been given, the science we've been able to understand, to better utilize resources. To me the science validates what I believe, that things were created for a purpose. Here's this huge energy source of the sun, these beautiful green leaves filled with chlorophyll, this brilliant mechanism of photosynthesis to capture that energy in ways we humans can use to sustain ourselves. It's a beautiful, remarkable design, but we need the air and water and soil to effectively carry it out. My job as a farmer is to maximize photosynthesis as effectively as I can, to capture and transform that solar energy into life. I think we were given a certain amount of dominion, but with our ability to shape nature and use resources comes tremendous responsibility."

He doesn't expect it ever to get easy. Though Walt Whitman called this landscape "A NEWER garden of creation," Justin cites the harsh judgment in Genesis (turning to Lindsey for the exact passage) that God

---

* Malawi is now home to the Soils, Food and Healthy Communities project, a farmer-led organization using agroecological methods to improve food security and nutrition.

delivers to Adam: "Cursed is the ground because of you; in toil you will eat of it all the days of your life. Both thorns and thistles it shall grow for you." In that judgment, Justin finds his purpose. "We had the Fall; things became broken and then Jesus's ministry was about restoration. We can't restore things like He did, but we can think of our lives as a ministry of restoration. We can't restore soils to the way they might have been before the Fall, but it gives us leastways a direction and goal."

Justin loves the centrality in the Bible of wheat and bread, the "staff of life." He loves the symbolism in John, that the grain of wheat must die to be resurrected as the young plant. "That's the seed's purpose: to die for life to carry on. Which I think means we need to look beyond ourselves; add life to the people and community and environment around us; produce things that outlast our brief lifetime. If you're not working on a project that will be completed beyond your lifetime, then you're not thinking right. People that leave a legacy have a perspective bigger than themselves."

Wheat has never, for Justin, lost its place at the center of the human story. It's there on Sundays, when the Knopfs always rest, even though they've sometimes lost an alfalfa field to rains. During a service in their New Community Church, held in a repurposed basketball court filled with folding chairs, Andrew climbs down from Justin's arms to steer a toy combine, red like his dad's, back and forth across the chairs. Men with close-cropped hair raise big arms to the sky and sing along with the electric praise band: "In this time of harvest when His providence and bounty flow . . . the seeds I receive I will sow."

Wheat has been central, too, in Justin's story with Lindsey, particularly their romantic, and nearly disastrous, courtship. Mutual friends introduced them late in 2001, when Justin was traveling for a seed company and Lindsey was an accountant in Kansas City. A former University of Missouri triple jumper, Lindsey has an athlete's muscle and grace, dark-rimmed, gray-green eyes and a habit of self-deprecation (even as she lays out a beautiful table and dinner of farmers' market steak, grilled vegetables and homemade ice cream and apple pie). "She clearly stood out to me," recalls Justin. "I made sure she got invited along to things. Then I saw her at a wedding in a black dress, and it was all done for me."

The two dated for a year. But in 2003, as Justin was going into debt to buy land and facing a period of financial uncertainty, he broke up with her. "I felt I had to get the transition under my belt and knew we were at a time in our relationship that I had to fish or cut bait." Lindsey, upended, quit her job and signed up with Americorps, moving to Cleveland to be a house mom for highly motivated but at-risk kids.

She later wrote an essay recalling that year's confusion, anger, anguish and lingering hope. The story begins on an early May morning with a curly-haired girl bouncing into her backyard and stopping short when she sees a four-foot square of wheat suddenly tall enough to reveal itself at the center of her garden. "The head has emerged from the flag leaf . . . still a light green tint . . . [with] at least two dozen small kernels woven together like a zipper," she writes. "The wheat not only has survived the harsh Cleveland climate, but it will produce a harvest." Calling the "only person who can share in this triumph," she leaves a wary, excited message. "The wheat headed out! Call me if you want. Bye."

The story then jumps back in time. "Eight months ago, in late September, she stood in this same spot with a shovel and a bucket of Kansas wheat kernels, a birthday present from the farmer, who thought she could grow a memory of Kansas right here in Ohio." A letter in his "scratchy print" had accompanied the wheat: "There seems to be so much truth in seed . . . it dies to produce new life and grow." But she'd hesitated, fearing more heartache. "A field of golden wheat, heads bowed, had witnessed their first kiss. They'd dreamed together, hour after hour, with the steady hum of golden wheat filling the combine . . . But then those words were whispered on the bench behind her apartment, 'I just can't give you what you deserve right now.' . . . He'd moved back home to farm fulltime, and there wasn't room for the girl." Finally, she "jumped on the shovel, slicing through the thick sod with a satisfying crunch." (Justin will later tease her for being a sodbuster. "She had to kill the grass somehow so went the organic route and tilled it.")

What Lindsey didn't know, on that morning the following May, was that her call to Justin had gone unanswered because he was standing beside his blue Chevy pickup, filling it up with gas, headed her way. "By that spring I felt ready," he says, settling into a story he clearly loves to tell. "I knew she'd complete a lot. But in the meantime she'd become a lot more independent. I decided I'm going to try to get this girl back." Discovering that Billy Joe Shaver, a country singer they both "appreciated," would be performing in a Cleveland bar, Justin picked some heads of wheat and drove all night to get there, nearly a thousand miles. "I didn't know if she'd even show up—I just wanted to surprise her and ask her to be with me again." He brought along their favorite beer, an unfiltered local wheat beer available only in Kansas. When he got to Cleveland he found the bar, knocked on its closed door and "told the guy who answered, who turned out to be the owner, about the girl I'd come to see." He arranged for them to serve her a bottle of the beer with a wheat head sticking out of it, passed the hours till show time washing his truck and cleaning up, then headed into the bar hidden under a ball cap.

And she didn't show up. "I knew she'd been reconnecting with an old friend, a guy who'd played Division One football when she was on the University of Missouri track team. Here I was, this skinny farm kid, thinking I blew my shot." Finally, she walked in. When the waiter took her the wheat beer, "she smiled and looked square over where I am." A friend had scouted a place for them to go talk after the show. "I worried it might be hard for her to transition from the city, to the seasonal intensity. But I shared with her the ways I'd grown in my heart, and asked if she'd like to continue this journey together."

A year later, in June, two weeks before harvest, Justin asked Lindsey to go out on the four-wheeler with him to check the wheat. He took her to those first fields he'd farmed at age fourteen, where he'd hidden a ring among the golden stalks, using a tall patch of invasive rye as a natural flag so he wouldn't lose it. He got down on one knee in the warm, heavy wheat. "After she said yes, we sat and listened to the swishing music of the wheat rustling in the wind; when heading and ripening it makes a beautiful, calming sound." (Truman Capote called it "the whisper of wind voices in the wind-bent wheat.")

For a winter wheat farmer, the year reverses the usual storybook rhythms. Here, early summer is the time for harvest, culminations, celebrations, and autumn the time for new beginnings and all the reflections those beginnings bring.

Autumn has its harvests, too. The small, hard berries of the grain sorghum have turned a rich rust color; when their glumes turn black and itchy, they will be ready to bring in. Rolled bales of "prairie hay"—native smooth brome grass cut for livestock—glow in the warm, slanted cinnamon light. At dusk, the lilting whistle of meadowlarks in the ripening soy gives way to the hoot of a barn owl, the whine of coyotes, a steady heartbeat of clicking insects.

But sowing the wheat, says Justin: "that's the most critical thing." So out comes the seeder, bright yellow and green, a big wagon attached to a line of steel octopuses, each sitting atop a sharp wheel. As Justin's tractor pulls the machine through the field, each wheel makes a small slice through the residue into the soil below, a seed is blown in through one of the plastic octopus arms, a second canted wheel sweeps soil back over the seed, and a third gently tamps down the soil.

Over the next ten months, Justin will shepherd this wheat from infancy to maturity, through a lifetime's ups and downs. "That wheat germ is the embryo, and like any embryo is delicate, fragile. So we take great care to protect it, to get it placed and fed and sheltered just right. It's a special, hopeful time. When I'm sowing those seeds I envision that crop being beautiful—lush, dark green, weed-free— throughout its life." In little more than a week the seeds will sprout and push tiny green blades above ground. "I never get tired of seeing new life emerge from the soil. But there are so many things that can happen to this crop through the year. We deal with so many forces of nature we can't control."

He suggests a walk on Gillum's Peak, one of the few nearby remnants of native unbroken prairie. Though now scattered with Osage orange trees, planted by settlers terrified by the treeless landscape, it remains easy to see why these dense grasses so often remind observers of an animal's pelt. (The charming *Kansas Bestiary*, for instance, describes bison as "long and shaggy at one end moving to short and smooth at the other, like the Kansas prairie itself, shifting over 400 miles from eastern tallgrass to western shortgrass.") When stripped, these grasslands suggest an animal flayed: in his 1907 address to Congress, Teddy Roosevelt excoriated those who would "skin" the land.

For Justin, Gillum's Peak is a reminder of "how brief a time we're here from the perspective of the land. The Gillums owned lots of this land in the twenties. They're not here anymore but the land remains, and you can still see their impacts on it. I'm living with their legacy and it keeps me in mind of the legacy I will leave. These fertile soils are a resource for the world that we have to be wise with, for my son and daughter and their children. If they're lost, I can't replace them. I believe we're building soil—but in my lifetime, will I build even an inch? I don't think so."

That sense of the past and the future layered on the land is just as strong on the site Justin and Lindsey have chosen to build their home, next to those fields Justin first farmed and later knelt in to propose.

They bought the land from the great-grandson of Peter and Marie Nelson, Danish immigrants who homesteaded it in the 1870s. As in so many families, the settlers' descendants chose to leave this hard life behind. Lindsey likes trying "to imagine Marie rhythmically pinning her soaked linens to dry, the children hiding out in the tree house, Peter heading to the barn to milk the cows." Justin loves the remnants of several generations scattered about the land. Climbing a wooden ladder up to the barn hayloft, he rubs the rungs "worn by all those hands going up and down." He admires an abandoned old moldboard plow: "This is what turned the prairie over." Most wonderful of all is a cylindrical tin grain bin just a bit taller than Justin wearing a conical cap, like the Tin Man who rusted in place not far from here. A faded stamp shows it was bought at L. H. Banks hardware store in nearby Gypsum City; inside the little curved door are tally marks from fall 1923, counting bushels sold to Ed Ott for sowing in his fields. All that's in it now is a beehive thriving between the tin and siding. Justin can't reach the honey without disturbing it but will try to move the bees, once he and Lindsey have settled here, into a regular hive.

Other plans include an experimental wheat plot, some small equipment and a hand mill so Lindsey and the kids can plant and reap and grind and bake and eat the wheat they grow. She wants a rain barrel to water her organic vegetable garden, and pear and apple trees. The kids already love galloping across the hills with their graying chocolate lab Miles, and playing on abandoned tractor tires three times their size. Tromping all together through the pasture at sunset on an unseasonably warm January afternoon, Justin points to a cold front approaching from the north. A battalion of glowing metal-gray clouds is rapidly assembling, driving ahead of them an eerie, turquoise sky and a gathering roar—heard before felt—of the formidable winds blowing this young family's way.

3

RIVERMAN

"BUT the basin of the Mississippi is the
BODY OF THE NATION."
—MARK TWAIN

Down the Missouri three thousand miles from the Rockies;
Down the Ohio a thousand miles from the Alleghenies;
Down the Arkansas fifteen hundred miles from the Great Divide;
Down the Red, a thousand miles from Texas;
Down the great Valley, twenty-five hundred miles from Minnesota
Carrying every rivulet and brook, creek and rill . . .
Down the Yellowstone, the Milk, the White and Cheyenne;
The Cannonball, the Musselshell, the James and the Sioux;
Down the Judith, the Grand, the Osage, and the Platte,
The Skunk, the Salt, the Black and Minnesota;
Down the Rock, the Illinois, and the Kankakee,
The Allegheny, the Monongahela, Kanawha, and Muskingum;
Down the Miami, the Wabash, the Licking, and the Green
The Cumberland, the Kentucky, and the Tennessee

Down from all those mountains and prairies and forests, the Mississippi River, the "Big Muddy," built the American nation.

It began by building the continent: literally, physically forming the land on which the United States stands. Carrying more than a million tons of silt and clay each day, the river spent about 100 million years filling in the whole middle of the country, turning water into land from Cape Girardeau, Missouri, all the way to Baton Rouge. For the past seven thousand years, it has devoted itself to building Louisiana and the nation's most expansive and productive wetlands. It also spread its bounty east and west, overtopping its banks in spring flood to lay down the heartland's rich soils.

Having built the stage, the river moved on to building the nation's history and commercial might. It played a decisive role in the Civil War: as supply line, battlefield, weapon (Union forces sabotaged levees to drown Confederate redoubts) and finally, with Vicksburg's surrender to General Ulysses S. Grant on July 4, 1863, the fissure through the Confederacy that sealed its fate. It was equally important to America's growth as a global economic power: first as a developing nation exporting its natural resources, then as a leader in industrialization and urbanization. Filmmaker Pare Lorentz captured that evolution in *The River*, the glorious documentary from which the passage above comes. (Scored by Virgil Thomson, the script was called by James Joyce "the most beautiful prose I have heard in ten years.")

We rolled a million bales down the river;
Rolled them off Alabama, rolled them off Mississippi, rolled them off
*   Louisiana . . .*
Ten million bales down to the Gulf—Cotton for the spools of
*   England and France.*
Fifteen million bales down to the Gulf—Cotton for the spools of
*   Italy and Germany . . .*
Black spruce and Norway pine, Douglas fir and red cedar,
Scarlet oak and shagbark hickory, hemlock and aspen—
There was lumber in the North. . . . Lumber enough to cover all Europe.

Down from Minnesota and Wisconsin
Down to St. Paul; Down to St. Louis and St. Joe—
Lumber for the new continent of the West.
Lumber for the new mills.
There was lumber in the North and coal in the hills.
Iron and coal down the Monongahela.
Iron and coal down the Ohio.
Down to Pittsburgh, Down to Wheeling,
Iron and coal for the steel mills
For the railroad driving West and South
For the new cities of the Great Valley
Corn and oats . . . tobacco and whiskey . . . hemp and potatoes,
Pork and flour, we sent our commerce to the sea.

This muddy river even gave rise to the most essential work of American literature. "Out of the turbulent, bank-caving Missouri," Mark Twain wrote in *Life on the Mississippi*, his 1883 memoir which included the first draft of a chapter that would grow into *The Adventures of Huckleberry Finn*, "every tumblerful of it holds nearly an acre of land." A drink of it sates both hunger and thirst for the "natives . . . when they find an inch of mud in the bottom of a glass, they stir it up and then take the draught as they would gruel." A character he calls the Child of Calamity praises its "nutritiousness . . . a man that drunk Mississippi water could grow corn in his stomach if he wanted to."

Finally, it is here in this great river valley that America's grandest efforts to tame nature with engineering have played out their decidedly mixed course: making possible agricultural harvests, industrial advances and global trade unlike any seen in history, but also setting in motion disasters the magnitude of which have only recently been understood.

This river and its delta are the vast and complex landscape that

Merritt Lane, CEO of America's premier shipping company, is laboring to preserve. Like Dusty Crary and Justin Knopf, Merritt has deep
family roots here: from his great-great-grandfather Dr. Joseph Jones,
who erected an 1880 quarantine against yellow fever around the port
of New Orleans, to his grandfather Joe Jones, who founded the Canal
Barge Company (CBC) in 1933, seeing even in the depths of the
Depression the potential of this unmatched network of inland waterways to drive America to the forefront of world economic power. As
a business and civic leader, Merritt marks with his story several turns
in our journey: from land to water; from dirty boots to the C-suite;
from landscapes measured in thousands or millions of acres to one
that spans half the country; from heroes driven by their reverence for a
place and all the life it holds to heroes with no particular love of nature
but only the certain knowledge that nature provides the best and only
means to protect what they do care about—people.

In Merritt's case, that protection is needed against two gathering dangers. First, the Mississippi River itself, roiled by the same
volatile weather that has repeatedly upended Justin Knopf's life, is
growing ever more dangerous to his mariners. And, dwarfing that,
the Mississippi Delta, home to the country's most important energy
infrastructure, fisheries and ports, plus two million people including
Merritt's family, is disappearing into the sea. That vanishing can only
be reversed, Merritt has come to see, by reestablishing the river's native
function. So he has stepped up to be a part of the most ambitious
environmental restoration in history.

The Mississippi has always been terrifying. Twain quotes Captain Marryat of the Royal Navy writing in 1837 of its "bloodstained"
waters: not a lovely stream on which "the eye loves to dwell . . . nor
can you wander upon its bank," but a place for the "fetid alligator,
while the panther basks at the edge in its cane-breaks." This "furious,
rapid, desolating torrent" sweeps down whole forests, reported the

Captain, often blocking its own path; then "as if in anger at its being opposed, inundates and devastates the whole country round." In *Rising Tide*, his account of the great 1927 flood that put eleven states under as much as thirty feet of water, killing hundreds and forcing nearly a million from their homes, author John Barry calls the Mississippi the most complex river in the world: "an uncoiling rope made up of a multitude of discrete fibers, each one following an independent and unpredictable path, each one separately and together capable of snapping like a whip."

For those who navigate it, the Mississippi's murkiness and changeability and feral force present great risks and unrelenting demands, the kinds of intimate confrontations with nature Teddy Roosevelt celebrated as forging true character in a man. Twain devotes many chapters to his time as a cub pilot struggling to learn the name and shape of every island, sandbar, point and bend he had to navigate through over thousands of miles. He finally masters, he thinks, every inch going upstream, but struts only a moment before the pilot reminds him that it's an entirely different river going downstream. And brand new again in high water, and again in the phantom shadows of starlight, and again in pitch darkness or the "grisly, drizzly gray mists" that erase everything. And then brand new again as it caves a bank here, builds a new one there, bounces sand along its bottom into new bars and shoals, sinks ships and uses their wreckage to snag new victims. In Twain's hands, the river becomes life itself, continually tumbling men into a "blind and tangled condition," torturing them "with the exquisite misery of uncertainty" but requiring that they forge on nonetheless.

For Merritt's captains, technological advances since Twain's day have made life on the Mississippi both easier and harder, giving them more visibility into its murky doings but also growing everything to gargantuan proportions. As captain of one of CBC's biggest vessels—

the M/V* *Lainey Jones*, a $14 million, 6,000-horsepower, snub-nosed "towboat" (confusingly named, since it *pushes* cargo)—Donnie Williams must remain extraordinarily patient. Pushing 20,000 tons of cargo upriver against what he calls a "wall of water," he often makes just five miles an hour. And that's when he's not waiting—below narrows for a downstream tow to pass, or at a dam for a creaky lock to fill. Yet through every one of those languid moments he must also stay unflaggingly, acutely alert, lest a sudden change in the river's mood upend his world. "It's amazing," says Captain Williams, "how fast things happen out here."

The stakes for Canal Barge are especially high, given the nature of their vessels and the cargo they carry. CBC has hopper barges that move millions of tons of coal each year for the Tennessee Valley Authority, and deep-draft† deck barges that take the world's largest offshore drilling rigs into the Gulf and Caribbean (rigs so huge that the captain isn't on the boat but navigates from a helipad suspended from the rig's steel scaffolding). But their core business is the most valuable and dangerous on the water: moving "red-flag" cargo—oil and petrochemicals, in loads worth tens of millions of dollars—in shallow-draft tank barges. These hazardous liquids require all kinds of special care. Styrene, a feedstock for plastics, hardens if exposed to oxygen; benzene will freeze unless kept warm in heated barges; anhydrous ammonia must stay at 28 degrees below zero Fahrenheit to remain in liquid form. Their customers are the world's most demanding. "We work for some of the largest companies in the world, with the highest standards of safety and expectations of quality," says Merritt. "To keep their trust and their business, we have to continuously meet those standards."

---

* Motor Vessel.
† Draft is the distance from a vessel's waterline to its bottom.

CBC transports these treacherous cargos—with a far better safety record than trucks or trains—across waterways connecting thirty-one states. Through the Gulf Intracoastal Waterway, they carry jet fuel from Houston to Alabama naval airfields, or gasoline and (for a time) the space shuttle's rockets to Florida. More often, they turn north at New Orleans, pausing in Baton Rouge to bundle barges into massive tows to send up the broad reaches of the lower Mississippi. That's the journey the *Lainey Jones* is making: she'll hand off a few of her barges in Cairo to a smaller boat to push up the Ohio River to Louisville, Cincinnati and Pittsburgh, then continue on to St. Louis to hand the rest to a towboat that'll push them on up the upper Mississippi all the way to St. Paul. Other CBC vessels work the Cumberland River to the TVA power plant, the Tennessee River to Chattanooga and the Illinois River to Chicago. Their first-rate performance has won Merritt the deep respect of his industry colleagues. Having grown the company to thirty-seven towboats, 856 barges and revenues of $405 million a year, Merritt frequently receives purchase offers from rivals and private equity firms; "We've turned into one of the pretty girls at the dance," he told *Fortune*. In 2014, his peers elected him chairman of the Waterways Council, the industry's voice in Washington.

From the wheelhouse of the *Lainey Jones*, perched three stories high so he can see across the full 1,100-foot length of his cargo, Captain Williams has to navigate well ahead of himself. "All that tonnage, ain't like you can put on the brake and stop. Going southbound, with the current, you got to drive a mile ahead of you at least. And I can't see a sandbar. If all of a sudden them sounders start risin' up it tells me I'm fixing to run aground; I can stop my engines but I can't stop. One time I had eighteen empties, and the boat hit ground and stopped short and broke all my wires, and the barges just went on by themselves, and I thought, 'Oh, crap.' I just backed up, went around the hump, chased after and caught 'em. I've really only had one major crash. I went into the top of Old Town Bend with thirty barges and come out the bottom with twenty. It was midnight on Christmas Eve, raining, the river rising a foot and a half a day. My loose barges hit a northbound tow and knocked seven barges out of *his* tow. 'Boy,' I said, 'Santa Claus is gonna be pissed at me tonight.' I've only ever sunk one barge. It was rock; it wasn't chemicals or anything like that. No, I don't take any chances with these chemical barges; I don't get out of the channel for nothing."

The crew's job is nearly as demanding, akin to caring for a 20,000-ton newborn: six-hour shifts round the clock, up and sleeping at all hours. Highly disciplined and vigilant, they make clear what it means to "run a tight ship." At port, the deckhands build (or break) tow: tying the barges together with sets of rigging, each weighing close to 100 pounds. "If you was laying a thirty-barge tow," says Captain Williams, "it'd take 110 of them sets, and maybe thirty-six hours." Once in motion, while the engineer tends to the whale-sized engines, the deckhands and mate go out every few hours. In darkness, pelting rain or wailing winds, on decks that can be sheer ice in St. Paul or burning hell on an asphalt barge in Houston, they walk every inch to make sure the buoy tanks—the voids that make these tons of steel float—aren't taking on water, that the cargo is secure, that everything inside and out is spick-and-span. It's monotonous work, until it's not. A man that goes overboard has a 10 percent chance of surviving, before being sucked under and chopped up by the rudders. Here a man (or more rarely, woman) can see interdependence at its most bald: eight people utterly reliant, for life and limb, on one another.

And risky as this life has always been, it's getting riskier. "In my thirty-five years, the weather has changed," says John Belcher, a second captain along on this trip to train an apprentice pilot, or steersman. "It's gotten hotter than hell. When I first come to work out here there was no central air. You absolutely today could not survive on a boat without AC; you would literally die. We also get less rain nowadays, but it seems like when it rains, it rains. For the river to come up nice you need a good long steady rain. If you get a pounding rainstorm that saturates the ground it just starts washing the soil; instead of high stage, you get a flood."

The Mississippi in flood takes every bit of a captain's wits and skill; writer John McPhee likened it to "shooting rapids in an aircraft carrier." When "the river gets up," says Captain Williams, "the current's a whole lot stronger and faster, and the set—how that current hits you and slides you over—is that much greater. If the river's up fifty foot, you got a wide open river, but the bridge piers don't get no wider. You still got to put all that tow down between 'em."

Chris Harvey, the young steersman, describes the terror of water "forty foot taller than it is now, the river midway up the trees and what seems like four miles wide and super muddy, houses floating, dumpsters going by, things barreling down the river that can tear up vessels, a current as fast as sixteen mile an hour. You try to keep the boat moving faster than the current to maintain steerage, but that river'll do whatever it wants with you. So many hit the Vicksburg Bridge they shut it down at night. And in some narrows like Victoria Bend, where the river's wide above and then pinches down, it'd be so swift, you'd get to shoving up through and start into your turn and the river'd just spit you back out. It would take two boats this size to push a ten-barge tow through that little hole."

From his bird's-eye perch at CBC's headquarters in New Orleans, Merritt watches as extreme weather cascades across the nation, the river a kind of amplifier gathering into itself the sum of every flood or drought. In 1993, the year before Merritt became CEO, the Mississippi had gathered itself into a five-hundred-year flood; lasting six months and cresting at the highest level ever recorded at Hannibal, Missouri, it killed fifty people and caused $15 billion in property damage across nine states. In 2008, the river mustered a two-hundred-year flood and in 2011, another five-hundred-year flood—making clear how in need of revision such designations had become. In the 2011 flood, the river threatened so many cities that the Army Corps of Engineers, for the first time since 1937, had to dynamite open the levee at the New Madrid Floodway and inundate more than 100,000 acres of Missouri farmland.

As these floods rise, Merritt and his team have to keep pace, revising plans in real time. In high water, Merritt says, "you need more horsepower per ton to control the barges, so you start by shrinking the size of your tows; then you may get into daylight hours only, and at some point, when barges begin flipping and sinking and running into bridges, you have to stop entirely." That's what happened during the unprecedented Christmas floods of 2015, when unseasonable warmth turned snow to rain from Oklahoma to the Ohio Valley, pushing the river to its highest-ever January levels and bringing CBC's operations to a standstill.

The challenges are greater still when the river swings to its opposite extreme, as it did just fifteen months after the 2011 flood, falling 55 feet to a level 13 feet below normal. By September 2012, it was down to the near-record low of 9.86 feet at Memphis, barely deep enough to navigate, and the Corps had reversed from blasting levees to relieve flooding to blasting rocks on the river bottom to keep barges from running aground. "Low water is even more treacherous than flood," says Merritt, "because you still get swift currents where it funnels." Steering can be next to impossible, especially where the river doubles back on itself. As another captain, Daryl Wheeler, explains: "If you're in a tight turn, you got to do what's called 'flanking': if the river makes a horseshoe and you have a thousand foot of tow, about a mile or so upstream you kick it out of gear and let it float on down till it gets to the same current speed as the river. Right before you hit the bank the current will start swinging the head around and then you come back on ahead with the engines, but you got to know when to do that." Twain found this maneuver, essentially turning your boat into driftwood to let the low river turn it, particularly nerve-racking. "This drifting was the dismalest work; it held one's heart still."

Drought also requires lighter loads. "My barges are nine foot deep fully loaded," says Captain Williams, "so I need ten foot of water; anything under that I'm coming to a sudden stop. In extreme low water we're loading barges to seven foot. We get paid by ton-miles so that hurts the company." That CBC is long-haul complicates it further, says Merritt. "The draft when you load in Texas is the same as when you arrive in Pittsburgh weeks later. If you go to a place where all of a sudden there's no water, you're stuck: you have to offload or wait. For six months in 2012, we fought a situation where we were not able to load as deeply, had more delays and at times came to a dead stop." Eventually, he told *Time* magazine, stretches of the river cease to be "economically passable . . . You could move so little cargo, you just can't go," all of which means less revenue, more expense and sharply reduced profits. As these events grow more commonplace, CBC has begun adding "disruption clauses" to its contracts, requiring customers to share the risks of more volatile weather.

The January 2014 polar vortex added yet another set of difficulties, choking the Ohio, Illinois and upper Mississippi rivers with ice several feet thick. Other barge companies looked to CBC to decide whether to keep moving or come to a halt, and convoyed behind its well-maintained ships as they broke the ice. "As it gets thicker and harder, you again have to put more horsepower behind fewer barges," says Merritt, "and you see more damage, which at some point doesn't make economic sense." CBC's *Lamont Schrader* wound up idled in the shipyard for a week for repair of an ice-torn hole in its hull. The *Joe Jones* limped in too, its tail-shaft broken. "That's a cool $100,000 right there," Merritt says. The spring thaw was also worrying: if too quick it can cause "gorging," stacking ice floes up and refreezing them in larger blocks at the locks, making them difficult to open and close.

Canal Barge took on the burden of breaking the ice—"we're not like a Soviet icebreaker in the Northwest Passage, but our steel hulls can crush through"—because they recognize their interdependence even with competitors. "Think about a shared resource like the river," says Merritt. "If the other guy has an accident, that hurts me." That interdependency, heightened by the swing between drought and flood, has spurred conflicts, particularly with the Missouri River basin. "The Missouri is increasingly closed off from the rest of the system," says Merritt. When in 2012 it appeared that the Mississippi might fall below the nine-foot depth needed for navigation and shut down traffic between St. Louis and Cairo, "people in our industry were saying, 'Release the dams in the Missouri.' But farmers in South Dakota and Nebraska were saying, 'We need that water.' There's more and more legislative fights and litigation over irrigation versus navigation. A lot of management only looks at this or that. But this is all one big system."

Like the mountains and prairies it drains, the river is expected to grow more volatile: a 2013 report commissioned by the Federal Emergency Management Agency (FEMA) forecast that the Mississippi north of Cairo will see the amount of land inundated by hundred-year floods increase on average 45 percent by 2100, thanks to increasingly concentrated rains across the basin. The ongoing loss of upstream wetlands and floodplains to absorb these torrents will make their impacts far worse. Risks are also mounting on the man-made waterways built to extend the Mississippi's reach, particularly the Gulf Intracoastal Waterway, where Canal Barge got its start. "The Intracoastal is a whole different world," says Chris. "It's just 300 foot wide; you meet another doubled-up tow, you might have just 20 feet between you." As sea levels rise and storms intensify, "you got tides comin' in and out of the sounds, and storms can rock the boat pretty good. If you got empties and 30-mile-an-hour winds you about have to be pointed straight into the wind and going sideways to get anywhere. And once you're out there, there's nowhere to tie off, nothing but open water. Two-foot swells can fill up your wing buoys or an open hopper barge and sink it pretty quick."

All of those changes jeopardize commerce as essential as it was in the nation's early days. As the *Lainey Jones* churns its way from Natchez to Vicksburg, it passes a steady parade of even bigger barges loaded with Tennessee coal, Iowa cement and Kansas wheat. Every year, shippers move more than $200 billion worth of goods across the nation's 12,000 miles of inland waterways: "the heavy and still vital materials of our economy," as *Fortune* put it, building-block commodities "measured by the thousand ton, bushel and barrel," including 60 percent of all U.S. grain exports. The $11 per ton-mile shippers save over truck or rail transportation provides the critical competitive margin for many U.S. products. "Each one of my fifteen barges carries as much as 144 semi trucks," says Captain Williams. "If you had to take all them trucks to move just this tow, and then all them other tows we've been passin', think about the pollution. And you want to talk about traffic jams; you'd never be able to git nowhere with all the trucks on the road. If you shut the Mississippi River down for a month, our economy would be in a critical state."

Even more than that growing unpredictability of nature, however, it is human engineering that has brought Merritt's world to the brink of collapse.

Until well into the nineteenth century, the Mississippi River essentially ran the show: braiding this way and that and periodically terrorizing the mid-continent. But as cities and industries grew up along its banks, those habits became increasingly difficult to endure. By the latter half of the century, engineers emboldened by the "heroic age" of bridge-building and railways, power stations and skyscrapers, set out to tame the beast, undertaking some of the most intense efforts in human history to dominate nature. In 1874, James Eads—a brilliant engineer who had invented a submersible that allowed him to walk the river's bottom—pronounced man now "capable of curbing, controlling and directing the Mississippi, according to his pleasure." In 1879, Congress established the Mississippi River Commission to do just that, investing authority in the U.S. Army Corps of Engineers.

Twain ridiculed the hubris of it all, mocking those who imagined they could say, "Go here," or "Go there," and make the river obey. Those "little European rivers" with their "hard bottom and clear water" would be a "holiday job" to tame, he wrote, but "ten thousand River Commissions, with all the mines of the world at their back, cannot tame that lawless stream . . . [they] might as well bully the comets in their courses and undertake to make them behave as try to bully the Mississippi into right and reasonable conduct."

Twain was wrong: the Corps did subdue the river, though in rendering it docile set in motion a spiral toward far worse destruction. After much venomous debate and political jockeying, the engineers settled on a strategy of complete containment. By locking the river between levees and dikes and sealing off its distributaries, their plan was to concentrate the Mississippi's force enough that it would deepen its own channel and keep itself continuously dredged, lowering its own flood. The idea of levees wasn't new; the earliest European settlers had protected their fields and homes with mud walls. But the Corps brought vast new muscle and effectively challenged the river to a fateful race: as they built the levees higher, the penned-in river, unable to spread out in flood, got higher too. So they'd raise the levees again, in places four stories high, and the river would get higher still. By the 1920s, the levees were not only longer than the Great Wall of China, but higher and ten times wider.

Still, the levees were no match for the river when in 1927 it rose to unprecedented heights. In fact, they made the flood far worse. Had the walls not been there, the river would have spread across its floodplain in a slow rise, dissipating its force over tens of thousands of square miles. Instead, it piled all that force against the levees until they finally burst like a dam, abruptly drowning whole communities. With multiple flood crests, the largest of them reported to be 20 miles wide and 40 feet high, the river brought down every bridge from Cairo to the Gulf. Decorated generals deemed it more terrible than war: "snapping trees with the great cracking sound of heavy artillery," floating bloated bodies or abandoning them on slivers of dry ground to be devoured by hogs.

In *Rising Tide*, Barry describes the fundamental ways in which the flood changed America. Shattering "the myth of a quasi-feudal bond between Delta blacks and the Southern aristocracy," it helped to spur the Great Migration to the north. It also marked "a watershed, when the nation first demanded that the federal government assume a new kind of responsibility for its citizens," including protection and recovery from large-scale environmental calamities. As one of the first times in U.S. history when radical disruption of a natural system wound up radically disrupting America, the 1927 flood is often invoked by Merritt and others as a preface to the present crisis.

The Army Corps, briefly humbled, tempered their "levees only" policy with construction in 1931 of the Bonnet Carré Spillway: 350 floodgates stretching more than a mile along the river's east bank that can be opened to draw off high water into Lake Pontchartrain when the river menaces New Orleans. (The Morganza Spillway, completed in 1954, similarly protects Baton Rouge by diverting water into the Atchafalaya basin.) But Congress had warned the Corps, "never again," and in the Flood Control Act of 1928 directed the agency to make their levees still thicker and higher. About the same time, the Corps and others began building dams: erecting massive hydroelectric power plants on the Missouri and adding locks to turn the steep upper reaches of the Mississippi into a staircase of lakes. They also introduced "wing dikes": angled walls set into the edges of the river to funnel its flow to the center during times of low water. These were the triumphs celebrated in the Lorentz film and a parade of others. McPhee quotes a Corps film vowing to vanquish the "large and powerful adversary . . . that could cause the U.S. to lose her standing as first among trading nations: Mother Nature."

Though many of those interventions did indeed help secure navigation, they have also, over the decades, complicated life for Canal Barge. The locks, now aging and decrepit, regularly leak and fail; in a good week a towboat will spend four hours getting through each of the twenty-nine on the upper Mississippi; in a bad week, she might be stopped outright. They also make floods more dangerous, sucking boats toward the river's middle and bottom. Captain Williams's scariest hours, he says, were at Smithland Lock and Dam on the Ohio River north of Paducah, Kentucky, where he came upon a boat owned by another company that had crashed, leaving six barges teetering atop the dam with the floodgates wide open. "I had a 6,000-horsepower boat and so did the other guy. We wired ourselves solid together, tow knee to tow knee,* and I backed down to within feet of that edge. I'd catch lines off the back of my boat to a barge and that other towboat'd pull and I'd push. I've got the life of everyone on the line; if anything went wrong over the dam we'd go, sink our boat. We got all but one. The current got that last barge and started pulling me backwards. I come full ahead, snapped the lines and let that barge go."

The dikes built to concentrate shallow water also worsen floods. "When Huck Finn came down it was wide open and shallow," says Captain Belcher. "Your old paddlewheel and packet boats—those ones you see in old movies where the black guy's standing on a flat

---

* The flat part the boat pushes with.

boat with a stick, more or less a raft with no propelleration—only had a four- or five-foot draft. But as they've built these dikes higher and higher, the river wound up here and the land down there; there's plenty of places the bottom of the boat is up above the land. And now they can't use the places the government set aside to take the flood. Way back when, a farmer maybe got 5,000 prime acres for two dollars a year, knowing if they have to blow that dam they will. But they've kept this water under check so long that his kids and their kids have built homes, so now they won't flood so that farmland's wore out and the river's a big mess."

All those upstream complications pale, however, compared to the damage the reengineering has done to the river's delta—one of the largest in the world and as essential to America's prosperity as the Tigris–Euphrates, Nile, Po, Ganges and Yangtze deltas were to the emergence of Mesopotamian, Egyptian, Italian, Indian and Chinese civilizations. Deltas have been crucial cradles of human progress because they link a continent to the rest of the world, opening trade routes like those Merritt and Canal Barge now ply. But even more, because these borderlands between land and sea, freshwater and salt, floods and tides—enriched by the great loads of minerals and organic matter delivered by their rivers—rival anything on earth for their abundance and diversity of life.

The Mississippi Delta is a classic example of how these fan-shaped marvels come to life—and how they can die. Roaring across the continent, gathering up mud and silt and clay (the "tribute" paid continuously by its tributaries), the Big Muddy keeps everything stirred and moving right up to the moment it can't: because the land flattens out and it runs into the big still wall of the Gulf. Losing all its energy, like a teenager arriving home, it dumps everything it's holding in a big muddy heap. All those sediments become a layer of land, and then another layer and another layer as the conveyor keeps rolling in. Even as the Gulf tries to sweep them away and the sodden sediments compact and subside—shrinking and settling like a drying sponge or a fallen cake—the river arrives with more.

In time, as Twain's Captain Marryat noted, the river manages to block its own way, forcing it to nose up over its newly-made land, or split into "distributaries" and go around. As it ambles this way and that, it drops out finer silts, creating ridges of dry land and vast expanses of wetland, wound all through with its slender fingers, the fertile bayous. After a thousand years or so, when it tires of the circuitous journey it's made for itself, it finds a shorter path to the sea and switches tracks entirely. In the last seven thousand years the Mississippi has made that big switch seven times, each time leaving behind a "lobe" of land thousands of feet thick; the more recently abandoned are home to Terrebonne, Lafourche and St. Bernard parishes.

In its present-day delta, however (home to Plaquemines Parish), just a century of engineering has brought those geologic processes to a halt. First, upstream dams began impounding sediments, trapping out half the river's load. Then the levees guaranteed that the rest would go to waste: funneling them past the starving wetlands and off the continental shelf, to be lost forever in the deep abyss. By "imprisoning the snake," as Army Engineer D. O. Elliott put it, the Corps had destroyed the Mississippi's land-building powers.

The discovery of oil and gas in Louisiana's wetlands compounded the damage. Beginning around 1930 and peaking in the 1970s, prospectors drilled some 50,000 wells and cut more than 10,000 miles of canals and pipelines through the marsh, to bring in rigs and pump out millions of barrels of oil and trillions of cubic feet of gas. There were few constraints: 80 percent of the wetlands were (and remain) privately owned. And like the unbroken prairie when Justin Knopf's family arrived there, these mosquito-infested swamps were seen as wastelands, awaiting redemption by human enterprise.

Pumping out all that oil and gas hastened natural subsidence, leaving underground voids into which the marsh could collapse. More damage was done by the "spoil banks," where canal builders dumped their dredged material; encircling and isolating portions of the wetlands, those banks interrupt sheet flows. Worst of all, the canals gave saltwater a way to penetrate deep into the wetlands, where it has killed off vast reaches of freshwater marsh and cypress swamp. As dead roots loosen their grip, wave action widens the narrow channels into broad boulevards. When salinity and inundation reach a certain threshold, the marsh collapses.

Together, the dams, levees and canals have added up to disaster. In the past century, Louisiana has experienced the greatest land loss on the planet: 2,000 square miles that once sheltered coastal communities from tropical storms, vanished into the Gulf. Sitting in his modest office, Merritt describes a visit down to Venice, not far from the Mississippi's mouth, where the river would still be adding land to Louisiana if its umbilical connection to its wetlands hadn't been cut. "A guy showed me where they used to raise cattle, and I said, 'Where? There's nothing but water.' There's no question that all of it—the levees and dams, pulling minerals out of the ground, cutting canals, allowing saltwater to intrude—exacerbated the loss." He points to an 1879 map on the wall made by his great-great-grandfather Dr. Joseph Jones. "There was a lot more Louisiana then."

Southern Louisiana is now a land of used-to-bes. It's hard to go anywhere without locals showing off this or that watery spot that used to be high and wide enough for livestock, fields of sugar cane or sweet-smelling citrus groves. Windell Curole, whose Cajun family has lived along Bayou Lafourche for seven generations, will tell you how when a hurricane destroyed Cheniere Caminada in 1893, his great-grandparents retreated inland to Leeville, where little wooden houses nestled along the banks of the bayou and sheep bedded down in deep grass amidst tidy fields of rice, cotton and corn. But then, Windell will say, his grandparents had to retreat too, when Leeville sank into the Gulf, leaving little more than a two-lane road and a listing dock lapped all around by water. Even then, Windell says, his grandfather would move the whole family down to a trapping camp, where they'd spend their days paddling out into ditches just wide enough for a pirogue (a traditional Cajun marsh canoe) to trap mink, otter and muskrat. Windell has vivid memories of those trips: "If it rained we had to hang the skins inside. And we stored meat and ducks in a big barrel of pig lard; by the time you'd get to the bottom of the barrel it was pretty strong. Life didn't smell too good in those days." That's all gone too: "Those pirogue ditches are now broad canals."

Only from the air, however, does the full extent of the loss become clear. Flying his seaplane south from New Orleans, pilot Lyle Panepinto points to the strange straight channels running far out into the Gulf, water highways seemingly built (nonsensically) across open water. In fact, he explains, those are the remnant edges of the thousands of miles of canals carved through what was solid marsh just a few decades ago. Turning north along the river's west bank, he pulls up a navigation map, which shows the plane flying across land. Then he toggles to a satellite picture and all the land disappears, a vanishing act confirmed by a look out the window to empty water below. Mapmakers simply can't keep up. Between 2011 and 2014, the National Oceanic and Atmospheric Administration (NOAA) removed thirty-five names from nautical maps of the Louisiana coast because the marshes, swamps and cheniers* that once defined their boundaries had ceased to exist. The names hint at the human history that went with them: Bay Cheri, Grand Bayou Carrion Crow, Bob Taylors Pond,

---

* A chenier is a ridge built by waves and currents piling up shells; its name comes from the French word for oak, which in Louisiana are often found on these ridges.

Yellow Cotton Bay, Tom Loar Pass. Those trying to wake the world to this catastrophe search in vain for adequate metaphors, likening the torn and ragged wetlands to a moth-eaten tapestry, an Etch-a-Sketch shaken by a giant, a maple leaf devoured by insects to its veins, an organism eaten by a cancer.

Most striking of all, as you move from the rare regions of the delta still fed by the Mississippi to those starved of its nourishment, is how quickly this landscape's Amazonian vibrancy bleeds out to lifeless gray. Take a boat into a place like Mardi Gras Pass, about thirty-five miles south of New Orleans, where the river breached its east bank levee in February 2012, and life bursts all around. Pink lilies, floating ferns, white lotus, bright green duckweed and delta duck potato ornament the water; sawgrass, sedges and reeds carpet the banks; the air itself is spun green and silver with wild muscadine grapevines and nets of Spanish moss. All is ripe with a smell so fertile it verges on rotten, and thrums with the jungle click and chirp and buzz of myriad insects. Half-seen things swim by: alligator, muskrat, a river otter hunting for blue crab. Songbirds feed on bright purple American beautyberries. Black coots (here called by their Cajun name, pouldeau) take flight in a mob. A silken white ibis takes slow, thief-in-the-night steps through the shallows; a roseate spoonbill spreads its watercolor wings.

Nowhere is the picture of what the river can do more vivid than in its west branch, the Atchafalaya. Splitting off eighty miles north of Baton Rouge, this is the channel the river would have jumped into entirely had the Army Corps not held it back in 1963, granting it just 30 percent of the river's volume. Even so, left to roam free across its floodplain, the Atchafalaya has remained the largest river swamp in North America, so lush that the Tarzan movies were filmed here. It has also built, for Louisiana, eighteen new square miles.

Fisherman Ryan Lambert and coastal scientist John Lopez are frequent guides into these various bayous, turning a boat into a "time machine," as John says, to show visitors how this world looked a hundred years ago. Ryan likes to reach down through the shallow waters and pull up whatever he finds underneath. At Fort St. Philip, seventy miles south of New Orleans on the Mississippi's east bank—where levees were left to decay after hurricanes Betsy (1965) and Camille (1969) drove everyone out, and the river has been free to roam for almost fifty years—he pulls up a handful of "beautiful, thick, mineral mud" and can't contain his happiness. "This was water two years ago. Now it's brand spanking new Louisiana. Look at this: that's North Dakota, Minnesota, Ohio, Kansas. Isn't that a wonderful thing? From every state in the Mississippi Valley, just dumping it right here."

Crossing the river, Ryan finds an entirely different story. First, he has to pull the boat out of the river and up over the levee even to access what's called West Bay. And once there, he motors mournfully about in empty water. Just thirty years ago, he says, families would come down to their "camps" here to fish and hunt and sometimes live: running traplines and hunting ducks in the marsh out their back door; crabbing in the bayou out front or paddling to friends' camps for crawfish or shrimp boils. But walled off from the river, those bayous and swamps melted away, a few sorrowful pilings in the wide open saltwater the only trace of the life once lived here.

Reaching to the bottom, Ryan gets a handful of what looks like crumbling coffee grounds: the rotting remains of submerged marsh, a mat seventy feet thick but useless for building land without new sediment infusions to offset subsidence and rising seas.

Unless the Mississippi can be freed to build land and life again, Louisiana will lose another 1,750 square miles by century's end. Already, says Phillip Turnipseed, director of the National Wetlands Research Center, the loss of these estuaries is "the greatest environmental, economic and cultural tragedy on the North American conti-

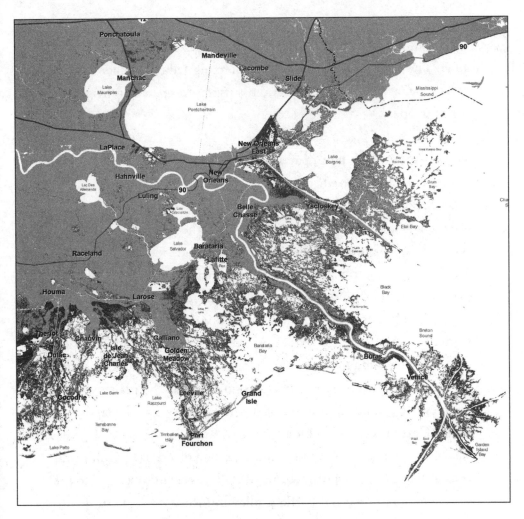

Projected land loss over the next 50 years,
absent restoration

nent." But much more stands condemned, as a map created by the state projecting a "Future without Action" (above) makes plain. The river's distinctive bird's-foot-shaped delta, including the shrimp docks of Venice and lower Plaquemines Parish: gone. South Lafourche Parish, from Windell's home in Golden Meadow to Port Fourchon, the staging area for nearly all the Gulf's offshore oil and gas: gone. Cocodrie, Dulac, all the tiny crabbing and fishing villages of Terrebonne Parish: gone. The twelve-mile land bridge protecting Lake Pontchartrain on New Orleans's northeast flank from surging to menacing heights in storms: gone.

With that land will go unparalleled mineral, biological and cultural riches. Louisiana overwhelms like no place else in America. First, in sheer scale: from Lyle's seaplane this water-world appears flat as the ocean and nearly as vast—40 percent of all the marsh in the nation. Its Chinese-scale works of human engineering are just as astonishing: a $14.5 billion storm protection system includes the world's largest pumping station and a two-mile-long concrete wall reinforced with eight Eiffel Towers worth of steel. Most mind-boggling of all are the seemingly impossible juxtapositions: a magnificent city and centuries-old traditional fishing communities alongside huge fossil fuel resources, titanic industry, food production as significant as the prairie's and a wildness that can be as impenetrable and dangerous as the landscape Dusty Crary roams—all of it at risk.

Merritt's industry is on the front lines of danger as the Gulf Intracoastal Waterway, designed for inland vessels, becomes exposed to the sea. "The loss of natural barriers, allowing these stronger storms to come in and bring significant surge,* is a threat to our people and facilities all along the coast," he says. "We wind up with vessels high and dry. After Hurricane Rita we had equipment thrown deep into the marsh, which was difficult and expensive to salvage. And multiple times a year now we have to go through hurricane planning, moving people and equipment around or stopping navigation." The river itself, below New Orleans, is also vulnerable, the shipping channel flanked

---

* Storm surge is the huge mound of water piled at the front of a hurricane by winds. Each cubic yard of seawater weighs a ton.

by thinning ribbons of land ever more likely to be breached. Even more exposed are the five deep-draft Louisiana ports Canal Barge uses when it heads into the Gulf or to Latin America. The largest in the Western Hemisphere, the port complex handles more than $130 billion of commodities each year and provides the crucial link for heartland farmers and manufacturers to global markets.

Equally threatened by this new exposure to the sea are the facilities belonging to Merritt's primary customers. Louisiana is the nation's top producer of offshore oil and produces 25 percent of America's natural gas and petrochemicals. Ninety percent of the country's offshore production and 30 percent of its total (domestic and imported) oil and gas supply moves through this coast, through a tangle of pipelines so dense it looks on a map like a fishing net thrown across the state. The eighty-mile stretch of river between Baton Rouge and New Orleans is one of the largest industrial corridors in the world, with 150 petrochemical plants and refineries belonging to Dow, DuPont, Exxon, Monsanto and dozens more lighting the riverbank at night with the smoke and flames of a thousand chimneys. Those billions of dollars' worth of facilities grow more vulnerable by the day. Oil and gas structures built on land are now surrounded by water, their pipes rusting in the salt and waves and regularly crashed into by passing boats, causing spills and fires. Once-sheltered refineries are hammered and flooded by hurricanes: the damage done by Katrina's storm surge to the Meraux refinery contaminated St. Bernard's water supply and inundated over 1,700 homes with an oily sludge. Altogether, Katrina ruptured at least ten storage tanks and terminals, spilling eight million gallons.

Most imperiled is the matchless biological productivity nurtured in this estuary. Louisiana's commercial fishery is the most productive in the lower forty-eight, providing nearly a third of all U.S. seafood, including half of the nation's wild shrimp, 35 percent of its blue crabs and 40 percent of its oysters, as well as crawfish, alligator and finfish. Louisiana also has some of the best sport fishing in the world. Windell shows off a favorite fishing hole where he sometimes spends his lunch hour, frequently joined by neighbors and pelicans, all pulling out redfish, drum and speckled trout. He has other prized spots for gar and pogies, largemouth bass, catfish and perch. "If you're hungry in Louisiana," he says, "it's because you're on a diet. This place throws food at you." The hunting here is unsurpassed, with nearly half of all the waterfowl that traverse the U.S.—black-bellied whistling ducks, gadwall, canvasbacks, pintails, green-winged teal, cackling goose—coming down the Mississippi flyway, some of them to winter here. They arrive in such vast numbers, hunters say, that they block the sun. Millions of other birds, including hundreds of species of neo-tropical migrants that breed in North America and winter in South America, also rely on Louisiana's wetlands for R&R on their transit between the continents, though as the grass they rest in and feed on vanishes, their numbers are declining rapidly. Less lovely creatures flourish as well, including water moccasins, swarms of mosquitoes and biting flies.

What makes this region so rich is the lacework the river built all along the coast. Louisiana's 400 miles of shoreline, stretched out, is actually 7,700 miles long. Its wiggles create the kind of edge conditions (like the Rocky Mountain Front) in which life flourishes best—realms of transition from fresh to salt, land to water, warm to cool, shallow to deep. On the highest ground—the "natural levees" built by the river when it was still allowed to overtop its banks—stand grand live oaks, their massive trunks and bushy heads of dark green looking like giant stalks of broccoli. Next come bottomland hardwoods and forested swamps of tupelo gum and bald cypress trees, capable—thanks to their organ-pipe trunks and rings of breathing "knees"—of withstanding the fiercest winds. The forests give way to freshwater marshes of wild rice, bull-tongue and myriad other grasses and reeds, then brackish marsh full of straw-colored wire grass, salt marsh with its endless oyster grasses and finally barrier islands rimmed by wax myrtle and black mangroves.

This diverse landscape has supported an equally diverse human culture, with regionally distinctive languages, traditions, music and food. That diversity is a product of history: the voluntary and involuntary movement of Europeans, Africans, Caribbean islanders and French Canadians (Acadians, truncated over the years to "Cajuns") through this crossroads of the world. And of natural history: with Houma Indian, Cajun and newer immigrant communities choosing to settle along the bayou where the salinity suits their favored creature: this one where oysters are happy, that one crawfish, another blue crab—their cultural identity and traditions emerging from the rhythms of living off this land. As the ground under their feet dissolves, those communities are already splintering. Without action, some forecast a permanent migration out of this region as large as any in the nation's history. New Orleans itself is in jeopardy: once fifty miles from the Gulf, it now has just twenty miles of land to buffer it from the next hurricane's surge. If the degradation of the remaining wetlands continues unchecked and it becomes a peninsula of leveed land—a fortress surrounded by open water, the regional economy and communities outside those levees gone—this extraordinary city will not survive.

Even in the early days of reengineering the river, some foresaw what it would bring. In 1897, a civil engineer by the name of E. L. Corthell described in *National Geographic* the evidence that the levees were causing the delta to subside: an old Spanish weapons storehouse in Belize Bayou had sunk as much in the previous nineteen years as it had in the two preceding centuries. Corthell did not, however, suggest a change of course; he had such great faith in future engineers' ability to protect people from the Gulf that he proposed carrying on even if it meant sacrificing the coast. "No doubt the great benefit to the present and two or three following generations accruing from a complete system of absolutely protective [river] levees . . . far outweighs the disadvantages to future generations from subsidence of the Gulf Delta lands below the level of the sea."

By the 1960s, Corthell had been proved half right: the delta was rapidly subsiding, but without the wetlands no advances in engineering had proven sufficient to protect those "future generations." Hurricane Betsy in particular, the first billion-dollar storm, made clear what a terrible trade the Mississippi basin had made, at least for Louisiana: though the levees had stopped the river from inundating communities in spring, that had come at the cost of far worse flooding from late summer Gulf storms. Concerted study of wetlands loss in the 1970s gave rise to coalitions to protect the coast; the passage of the 1990 Breaux Act provided $40 million a year for Louisiana wetland restoration; in 1998 the first long-term proposal was laid out in the Coast 2050 report.

But what had been a mostly regional concern (and even in Louisiana, known to very few) became something entirely different in 2005, when Hurricane Katrina made landfall east of New Orleans with the largest storm surge ever to hit the continental U.S., killing 1,577 Louisianans, displacing 250,000 and—with Hurricane Rita a few weeks later—wiping out 220 square miles of marsh. All of Merritt's neighbors now understood how vulnerable they'd been made by the stripping away of natural protections. That was especially true in the poorest neighborhoods, like the Lower Ninth Ward, where death-by-saltwater of a once protective cypress swamp had left residents at the full mercy of the flood when the Corps's levees failed. And the storm's impacts were felt across the nation: as Merritt and his peers struggled to keep goods moving in and out of Louisiana, fuel prices spiked across the country and grain exports slowed, badly hurting Midwest farmers. Louisiana's vulnerability, America realized, had become its own.

For Merritt, Katrina inspired a redoubling of his already deep commitment to his community, throwing into relief yet again the "extreme social stratification and corruption in government" that has long stained New Orleans. For years, Merritt has worked to address those inequities, serving in leadership roles in numerous civic organizations including Baptist Community Ministries, Tulane University and his daughters' school. Katrina reinforced the convictions that sustain him in that work: the belief that privilege like his carries with it an obligation to those who are more vulnerable, and the recognition that none of us, no matter how privileged, can get to safe ground on our own. All of this led him to shoulder his biggest civic commitment to date: helping to shape a comprehensive $50 billion plan to protect and restore the Louisiana coast and the people whose lives depend on it.

CBC had dealt with hurricanes many times, but Katrina was complex, says Merritt, beyond "our wildest imaginations." Between Friday, August 26, when the storm was still headed for Florida, and Sunday, August 28, when the mayor ordered everyone out of New Orleans, Canal Barge had to completely remake itself over a vast geography, getting 350 families and 700 marine assets to safety—and with most communications destroyed. It succeeded only because every employee rose to the challenge, demonstrating the deep sense of mutual responsibility and capacity for autonomous decision-making developed over decades by Merritt and this company.

Within two days, Merritt's staff had moved headquarters to Houston, operations to Memphis and customer service to Joliet, secured or moved 250 vessels out of Katrina's strike zone, and improvised a communications strategy based on Blackberry PIN-to-PIN messaging, handwritten logs and the cell phones of deckhands who lived far enough upriver to still have service. Within a week, they'd found and paid every employee and arranged housing and schools for the dozens of families now living as "CBC-in-exile." When, three weeks later, Hurricane Rita hit, landing directly on their Sulphur, Louisiana, facility, they righted the ship all over again. For 105 days, they functioned in this dispersed and improvisatory mode, fully returning to New Orleans only in December. Twenty-three barges wound up stranded inland, half in need of salvage. But they never missed a sailing, never lost communication with manned vessels and finished the year with the best safety and financial performance in company history.

Most importantly, every employee was safe and no one missed a paycheck or lost their job. "Most of our mariners live inland and their homes were spared, but some of our employees lived in hard-hit areas—eleven lost everything—and they needed more time and help to get their lives back together," Merritt says. (Though water got within four blocks of Merritt's uptown home, it was undamaged.) "The average post-Katrina turnover for New Orleans employers was 20 percent. We lost just five of 350 employees."

That loyalty derives from the way Merritt and his forebears have managed this company: with an expansive sense of family that is also at the root of his efforts on coastal restoration. Merritt's father, in particular—who worked at CBC until his divorce from Merritt's mother—taught his sons "that the equipment is nice but it's the people who make the difference," Merritt says. "Dad came from a modest background in Tennessee. He preached against snobbery, taught us your word is your bond. He made clear to us that the most important thing is respect and support for the mariners. If you take care of your people, they will take care of you."

That core belief—that most people will rise to the challenge if given the opportunity—made all the difference for Captain Belcher.

When he began his life on the river at age nineteen, Captain Belcher could barely read. "I could get through *See Spot Run* but I couldn't take a newspaper and understand what the hell it was about." He'd been orphaned at age three while sitting in his mother's lap, when his father shot her in the head; the orphanage he was sent to still haunts his memories: "I'll get this picture, like a mile long, two rows of single beds. The only way you knew who got adopted was by who wasn't in their bed at night." Finally his turn came: a family from a small town in northeast Arkansas took him home. The years that followed left him scarred and, like many on this river, not entirely at home in ordinary life ashore. Though big and brawny, with receding hair and deep creases around his eyes when he smiles, he has a boy's vulnerability, an urgent need to tell his stories. And, as Captain Williams says, "John Belcher's got more stories than Carter had liver pills."

Some of those stories—told while out for a smoke on the ship's deck, the engines and khaki-colored waters of the Mississippi roaring below—are as darkly comic as Twain's wildest yarns. "One time, when I was about nine, this neighbor asks us kids to find his sow with her new piglets; back then they just ranged 'em out in the woods. I knew where one was, so we all got us a tow sack—I was the youngest; it's the first time I ever done this—and sneaked up over the bank. Sure enough the sun was shining on a big pile of 'em, and the momma wasn't there. They said, 'Go on, Johnny. You found 'em. They're yours. It's only fair.' I'm thinking, 'They're never that nice; they're going to beat me up and take 'em. But I'll outsmart 'em. I'll grab them baby pigs and take off running where they can't catch me.' So I go acrost that creek and right up there to 'em, thinkin', 'I'll just throw four in. He said he'd give us 25 cents each so, hell, I'll have it made.' I reached in and got one by the back leg; that pig's kicking like a jackhammer; I'm trying to get it in the sack, and all of a sudden I hear, 'Hunh, hunh.' I look up and here come that sow. Well, I dropped that pig and took off runnin'. If I'd fell down she woulda eat me. I run and run and she chased and chased me. Finally, she quits. But meanwhile them older boys all ran down there an' loaded up them piglets. They used me for bait. They didn't know or care if that sow eat me."

Most of Captain Belcher's boyhood stories are as shadowed by loneliness as Huck's own beginnings. "Whenever we'd go to family functions, my grandmas would be like, 'This is Smokey's son and daughter. And this is Johnny, the one they adopted.' You know, put this separation." He lived every day at the mercy of a capricious cruelty. "I've been beat with a stick and had to cut sprouts—bushes—all day long. You beat a dog, it runs off; well, any chance I got, I hauled ass, but never made it far before they caught me. At night they'd chain me to my bed. We weren't allowed to go to the fridge. If you wasn't at dinner, well, you just missed it. My dad had been in the Navy so they'd go to the commissary and bring home fifty loaves of bread and these little peach pies, lock it all in a shed. I learned how to pick that lock to get me one of them pies, but I wasn't too smart. One day I went in and there was only one pie and I'm thinking, 'Yesterday there was a bunch,' when I heard a click. They locked me in there for three days. The building was corrugated tin; they'd come out at night with sticks and go round to scare the living daylights out of me. You'd think that when your parents died that'd done that to you, you'd say, 'Good.' But the day my dad died it just broke my heart; I lay down right beside him and cried."

The river, as it has for so many, provided John Belcher his escape. "Back then, if you had a strong back you had a job. Some of the best towboaters years ago were mean sons-of-bitches. Guys used to come on the river to avoid the law, just ride and ride and ride." It was John's good fortune to find in their midst protectors and teachers. "This old bosun, who'd been in the Navy, his wife was a grade school teacher and he brought books to the boat and helped me learn to read. I kept at it and got an education out here that I would never, ever, been able to get at home, doing any job. Most you can't even get from a book, not even at Harvard. We did training in job safety analysis, and I got third place in the whole company and I was like, 'Hell, yeah.' But I didn't do that on my own. Canal Barge taught me a long time ago, I could go to whoever I needed—from the deckhand to our CEO. Merritt Lane knows my name; he knows my wife's name; he knows my kids' names. That's what Canal Barge gave me.

"When I was a kid I always knew in my mind this ain't my fault, and when I get old enough I'm outa here, somehow, someway, and I'm going to have my own family, my own home. One reason I love this job: I never would have made anything out of myself at home. I would have been cutting firewood, feed somebody's cows, work in one of them little stores making minimum wage. Instead, I'm one of the top-paid captains in the fleet. Out here you can start at $30,000 and make $250,000 a year if you work at it hard. That's completely unheard of at home unless you're a lawyer or doctor or selling crack. That's pretty good wages for a man of my education. This is one job you can actually be nothing and make something out of your life."

To this day, the company remains committed, as Merritt says, to "growing most of our folks from the hawsepipe inside." The company spends several million dollars a year on mariner training, including the demanding steersman program that has brought Chris aboard. "I spent four years getting my tankerman* license, loading 400-degree asphalt onto barges heated up to 600 degrees—it melts the bottom of your boots; then a couple years as a mate and another half year ashore in Sulphur on night operations. Then I joined the steersman program: six weeks of school, a 120-degree summer in the engine room, time in the simulator. Finally, after five years, I got into the wheelhouse. I've now spent eighteen months piloting the boat with a captain supervising."

---

* A tankerman oversees transfer of liquid cargos on and off barges.

That long apprenticeship costs the company at least $250,000 for every pilot trained, says Merritt—a risky investment, since "that license is highly portable; there's nothing sticky about it." But few ever go. Attendees at CBC's seventy-fifth anniversary celebration included Captain Bill George, who'd served thirty-eight years, his wife Lavon, who'd joined him as a cook for the last fourteen of those years, and their son Shorty, who'd been captaining for thirty-seven years. "We try at every turn to demonstrate our commitment to our mariners and their families," says Merritt. "I try to personally make every training session for our vessel officers, and we invite spouses to join."

That human-centered approach makes all the difference on the water, where the strains of living far from home, trapped in close quarters with eight people for twenty-eight days, strictly regimented and sleep-deprived, can be more intense than any challenge posed by nature or engineering.

There are moments of profound tranquility. Twain described sunsets that turn the river to blood and unseen birds creating "an atmosphere of song which seems to sing itself"; the captains love spotting black bear galloping along the bank. There's camaraderie: sharing the grits and pork chops or chicken with cream of mushroom soup the cook gets up at 3 a.m. to prepare; telling tales of earlier days on the river—pirates attacking boats laden with Kentucky furs, steamship deckhands stuffing mattresses with Spanish moss, even their own olden days, before they had their Rose Point electronic charts showing the precise location and speed of every nearby vessel and cell phones to confer with those other captains. "We had to blow the whistle," recalls Captain Williams, "one time to say, 'I'll meet you on the one-whistle, port to port,' or twice, 'on the two-whistle, starboard to starboard.'" There's even the occasional practical joke: at 2 a.m., as Captain Belcher sits in the darkened wheelhouse steering through the dark night, telling stories of the ghosts he's seen on the water, Captain Williams (supposed to be sleeping) sneaks up outside, rattling windows and slamming the door shut.

But the long separations are hard on the crew and their families. Both captains saw their first marriages collapse under the strain of riverboat life. "It takes a special woman to live with a man who's gone half their lives," says Captain Williams. "They gotta take care of everything at home while you're out here—kids, home, car. And sometimes, you know, your buddy done come along to help her change a lightbulb and shit happens."

"You have these long hours, your personal thoughts every day, wishing I was here or there," says Captain Belcher. "I've seen grown men, strong men, just bawl out here. Especially back in the day, when guys would stay out on the boat for months, 'cause when they got off the boat, they didn't have a job. And the only way you could talk to home was by VHF radio. On Friday and Saturday, that was our entertainment: we'd listen to people make marine phone calls. You could hear the conversations on other boats too. 'You don't need to call me anymore,' she'd be sayin', 'cause now I'm with Joe, 'cause he's here and you're not.'" Sometimes those rivalries would come on board. "We had this one deckhand, he'd get out on the tow and tell this other guy, 'Hey, I just saw your wife; she wants you to call her.' And they'd wind up in a fight, though you get fired for fighting out here so you'd never see it. It was like Popeye and Brutus; did you know they worked on a towboat? They were brothers from Chester, Illinois.

"You not only got to be captain, you got to be mama, babysitter, marriage counselor, judge, jury, referee; sometimes you want to be the executioner." When a couple of deckhands scheduled to come aboard at Vicksburg call to say they're running late, the captains give them a paternal chewing out, coming down especially hard when the young men try to foist responsibility onto somebody else. "Half of these boys on here have some kind of college background," says Captain Williams. "That was completely unheard of back in the day, but that's what this economy's giving you. We've got deckhands with degrees. Sometimes that don't help a lot out here; if you're not gifted with common sense this can be a real hard job."

"But it's our job to grow 'em," says Captain Belcher. "It costs $7,000 to bring on a deckhand. If they wind up fired, then somebody somewhere didn't do their job. They didn't take that man and use him at his potential, whether it be mentally or physically, and it's a failure for everybody. A lot of them come from small towns where there's no jobs. I love to see people come up, especially if they worked for me, if I brung 'em up from deckhand to mate to steersman. I feel like I'm giving something back not only to the industry but to the community where he lives."

That sense of family commitment, both up and down the line and to the broader community, comes straight from Merritt, as Captain Williams had the tragic occasion to discover. "What they done for me," he says. "It's not all these companies out here that's like this. A couple years in a row I had a really tough time. First my dad, who also worked on the river, got killed in a van on his way to catch a boat. They helped me through that time, stayed with me. Then two years later my daughter got killed. She was seventeen years old, had a car accident. I was in the boat near Cairo. We never stop these boats but they tied up at the boat store and said, 'Don't worry about it; you go home.' I'd only been out here five years and I was the only one making a living in the family but I needed to be there with them and for them. So what do you do? You go broke to be there? There's a lot on a man's mind right there if he's the sole provider. And Merritt took all that pressure off me. He was in Washington D.C. at a meeting, but he called me. My mind was of course, shht, gone, and he said, 'Donnie, man, you've got a lot on your mind to worry about.' He said, 'There's one thing you don't have to worry about.' I said, 'Man,

I appreciate it, I really do, but I'll be fine.' He said, 'Donnie, shut up and listen to me.' He's the big guy, you know, but he'll talk to you very down to earth. He said, 'Look, take three months, take six months, take a year. I don't care how long you take, however much time you want or need, and you'll never miss a paycheck.' I coulda taken a year off and they'da paid me my captain's pay just like I was sitting right here."

Asked about that history, Merritt shrugs off the suggestion that he did anything extraordinary. "These are not inputs to production," he says. "These are human beings, and their lives have every bit of the complexity of any of ours and sometimes more, and if we can find ways to relieve that pressure . . . There's happy stories too that are part of our family culture. One of our captains had a son playing in his first collegiate football game. With no interaction with the office, he and his relief rearranged their schedules so the captain could be at his son's game. At some companies that would be forbidden; at ours it's encouraged. The more we can get our mariners to be self-managed and look out for each other, the more high-functioning our team becomes. We're a family-owned business but my brother and I are the only active management from the family. A lot of the key people weren't born into the CBC family but have become part of it. It's the connection our mariners have with one another that causes them to stay with us for decades. We're trying to have a more orderly and systematic and serious way of running a business while being as humane as possible and having all those important connections with your people. It's a harder but a better way to go.

"Most of the great businesses were family businesses. It starts as what you do; it's who you are. You can actually have noble motives if you're looking long-term. We're not allergic to innovation but the core value system is deep and wide about integrity, taking care of our people. We survived the horrible downturn of the eighties when a lot of the industry collapsed. We survived Katrina, the passing of my uncle, because we've built our company for continuity. We plan ahead not for years but for generations."

Merritt inherited his sense of civic duty from a long line of illustrious soldiers, doctors, ministers and lawyers, whose history he knows well. "Merritt has great respect for his ancestors," says historian Melissa Smith, who oversees the family archives, housed at Tulane. "Oftentimes people don't understand that they're in a position of privilege because of the work of their forefathers and foremothers. Merritt is fully cognizant and appreciative and wants future generations to understand that too."

Merritt can trace his maternal family history back to the Revolutionary War and Major John Jones, an aide-de-camp to Brigadier General Lachlan McIntosh, who fell before the British lines during the siege of Savannah in October 1779. Major Jones's son, Captain Joseph Jones, served in the War of 1812 (the final engagement of which, the Battle of New Orleans, gave the Army Corps its first experience in fortifying the Louisiana coast). The captain went on to become a successful Georgia planter, acquiring three "low country" plantations—Montevideo, Maybank and Arcadia—and more than one hundred slaves. Captain Jones became guardian there to his orphaned nephew, Charles Colcock ("C. C.") Jones, whose life in the Presbyterian Church has inspired several award-winning histories. The Reverend Jones wrestled deeply with the moral stain of slavery, devoting his career to the religious instruction of slaves ("Apostle to the Blacks," he was called) and helping several families secure their freedom and return to West Africa. The collection of his papers contains an 1851 letter from a former slave named Henry Stewart, requesting materials to help him in his new work as a minister spreading the gospel in Liberia.

When C. C. Jones's children came of age, "the first-born got the plantation, in the way of the South in that day," says Merritt. "The rest got some education and went into one of the honorable professions: soldiering, medicine, lawyering." Charles Colcock, Jr., a graduate of Harvard Law School, was elected mayor of Savannah in 1860 and became an ardent secessionist, serving as General Beauregard's chief of artillery during the Civil War. His sister Mary kept diaries describing the heartlessness of the Union forces. While giving birth she had to fend off searches and thefts by Sherman's army on its march through Georgia to the sea.

C. C. Jones's younger son, Dr. Joseph Jones, grandfather of CBC's founder, hewed more closely to his father's ideas of public service. Commissioned as a surgeon-major in the Confederate Army, he published a famous three-folio volume on conditions in the notorious prison camp at Andersonville, Georgia. He was also an extraordinary scientist, making the first microscopic identification of typhoid bacillus, discovering the protozoa that cause malarial fever and, says Merritt, getting "to the cusp of discovery that yellow fever was borne by mosquitoes."

Settling in New Orleans after the war, Dr. Jones helped establish Tulane Medical School and in 1880, in the midst of a yellow fever epidemic, was made president of the Louisiana State Board of Health. He spent four years fortifying quarantine stations and guards along the coasts, overseeing inspection of 10,000 vessels. Though yellow fever ravaged the coast from Texas to Florida, Dr. Jones kept New Orleans and the lower Mississippi Valley free from the scourge (while leaving a budget surplus of nearly $10,000). He also published dozens of papers in medical journals, on such topics as the "Black Vomit of Yellow Fever" and the efficacy of quinine against "bilious fever in the unhealthy climate of the rich lowland swamps of the Confederacy." His account of his "Explorations of Aboriginal Remains in Tennessee" is at the Smithsonian.

Merritt picks up the story. After his first wife died, "Dr. Jones married the daughter of Leonidas Polk, who was President Polk's cousin and a Civil War general who served under Robert E. Lee. He was also Episcopal bishop of Louisiana; he'd gone to West Point and theology school." Their daughter Caroline, Merritt's great-grandmother, was a suffragette who published one of the first Creole cookbooks, which Merritt keeps on his shelf; her elegant Garden District home now belongs to actor John Goodman.

It was Caroline's son Joe, Merritt's maternal grandfather, who founded Canal Barge at the age of thirty. "Joe Jones found himself at an interesting time in this country's history—the bottom of the Depression and run-up to the Second World War—with energy and connectivity to a lot of people and resources," says Merritt. Though the Joneses "were professional but not moneyed people," Joe had married Eugenie Penick, an heiress to a cane syrup fortune; he'd also built a successful law practice (Jones Walker) that remains one of the largest in the South. "So when the Jordans—they were cotton merchants, with a foot in the early stages of the oil industry— had the idea to start this company, they came to Joe for capital, and he invested $10,000." National unemployment was at 25 percent and just six years had passed since the Great Flood had caused several banks to fail, ending the reign of New Orleans as the wealthiest city in the South. But Joe was an "entrepreneur and optimist," says Merritt, "with an expansionist view of America. He knew it would lead in industry and foresaw the industrialization of the river."

The company began with a single asset—the world's first all-welded steel tank barge. Six years later, CBC secured its position as an industry leader by commissioning the M/V *Bull Calf,* a towboat with a revolutionary clutch that enabled the vessel to go from full speed ahead to full speed astern in eight seconds. That boat became the model for the landing craft used on D-Day. "Our culture of service and integrity," said Merritt, at a celebration of the company's eightieth anniversary, "comes directly from our founders."

Joe Jones's service to the nation continued after the war. In 1946 he became an assistant secretary of state, writing speeches championing the

Marshall Plan for Dean Acheson and Harry Truman. The following year he became a Tulane trustee, serving for many years as board president. Tulane had been founded by a segregationist and was led in its early years by a retired Confederate officer, so it was a thunderbolt in 1963 when Joe announced that it would begin admitting black students. When he died with his wife soon after in a fire, *Tulane Law Review* pronounced it "the end of a life and beginning of a legend," lamenting the loss of "an affirmative man, an imaginative, restless, bold human being, willing to do and dare . . . The ability to give of himself was perhaps his finest endowment."

"Joe had a worldview," says Merritt, describing his grandfather in language that could equally apply to himself. "He was not a man of small, narrow focus. I have a letter in my drawer he wrote to my father in 1960, saying that he understood how important race relations were going to be in furtherance of the South's future. We are all a product of where we came from. And then you learn, if you can. As he went through his life, he broadened his perspective. He was not a flame-thrower but a modern thinker with the strength of his convictions, even if that meant going against the perspective of lifelong friends. He was firm but reasoned with people, trying to make progress without kicking out the whole foundation. He was forthright and strong but also aware that Rome wasn't built in a day."

Merritt's own initiation into the company came primarily through his father. "Dad loved the business, loved that in this litigious society you could still make multimillion dollar deals on a handshake. He never pushed us into working here, but my brother David and I both felt there could be no more noble work than this. Our family has a tradition of trying to make wherever they are better, to be more givers than takers. Knowing their history has given me a definite sense of place, of legacy and continuity."

It was to honor that family legacy—and to protect the family he defines so expansively—that Merritt agreed in the wake of Katrina to help lead the effort to rescue Louisiana's vanishing coast. He well understood the underlying cause of the crisis: Canal Barge navigates from the living part of the coast in central Louisiana, where the Atchafalaya still feeds it, to the starved and dying regions around the leveed-off river's mouth. And after Katrina, he saw an opening for change unlike any in his lifetime. "Katrina created opportunities to accomplish things that were unattainable without the catalyst of a major crisis," he says. "We are the only American city where every last citizen had to leave, and then every one of us had to consciously decide to come back. That's incredibly profound: having faced the possibility of losing our hometown, we came back energized to rebuild it into the city it could be. Instead of everyone trying to preserve their little piece of the pie, all the pie got pulled out of everyone's hands anyway. When you lose something as precious as this, almost lose it entirely, you figure out what you care about."

In December 2005, with Baton Rouge still filled with Katrina and Rita evacuees, the Louisiana legislature gathered together the half-dozen disparate agencies focused on hurricanes, fish, water quality and development to create the Coastal Protection and Restoration Authority (CPRA). Their mandate was to produce a comprehensive master plan that would be updated every five years; the first was released in 2007. In 2008, the new governor, Bobby Jindal, appointed Garrett Graves his "coastal czar," heading CPRA. Graves had served on the staff of three Republican congressmen, including Louisiana senator David Vitter during Katrina. In 2014, he himself won a seat in the U.S. Congress representing the Louisiana coast.

Graves laid out a highly inclusive process for developing the 2012 plan, inviting onto the core team representatives from every coastal parish and every major industry. Merritt represented navigation. That team in turn sought as much as possible to tap local expertise: drawing heavily on scientists, economists and engineers from Louisiana institutions, and holding hundreds of public meetings along the coast to learn from the people living at the center of the collapse what was happening to their communities and to the shrimp and oysters and alligators.

Merritt was initially skeptical. He'd been to more than his share of "stakeholder meetings," often finding them to be little more than occasions for him to get beaten up. "People aren't always participating in good faith. Without that intellectual honesty it can be hard to see sometimes how you can really engage with people." He also, like Dusty Crary, struggled with the extremes. "I find it stunning how polarized these kinds of conversations can become. The radical fringes at both ends are really hard to connect to."

At one end (as in Montana), Merritt saw environmentalists who seemed to live in an imaginary world. "The issue I often have with the environmentalist conversation," he says, recalling long battles over how to address the ecosystem harms done by locks and dams without disrupting navigation, "is that it doesn't complete the sentence. It's just 'Stop doing this.' OK . . . and then what? It's not a solution if it kills this company. We have one of the world's most glorious waterway systems; it is one of our nation's greatest natural endowments. And we've invested billions and billions in these arteries of commerce, providing a fundamental source of our global economic advantage. If you shut us down, either the stuff doesn't move and you cripple the economy, or you shift it onto trucks, which are far worse for the environment.* The higher ground doesn't exist just because you call yourself an environmentalist. You have to articulate a stronger value proposition, a way that says, 'We don't want you dead; we want you better.'"

At the other end, Merritt saw peers in industry who "manage their lives and careers and businesses in little short intervals"—to some-

---

* Trucking generates a thousand times more emissions per ton-mile than barges.

times devastating effect. He understands the temptations. After graduating from UVA in 1983, he spent three years on Wall Street. "It
was exciting, lots of instant gratification; everyone was impressed with
you. When I left it—because I missed my family and had romanti-
cized working for Canal Barge—I spent a few years wondering if I'd
made the right decision. My buddies were making lots of money while
I was back here breaking rocks. But I realized Wall Street wasn't a
life for me. What earned you regard was being shrewd; people looked
down their nose at their clients. And the money piece becomes the
whole thing. My uncle and dad had taught us to think long-term:
'Don't be a quick-buck artist.' They'd taught us how lucky we were to
do work that adds real value every day; to have partnerships that span
three, four, even five decades; to have what we do feel aligned: doing
the right thing, doing things well, building teams, delivering a service
that is fundamentally important for this country. Some problems have
to get solved over life spans, certainly more than career spans, or to the
next trade or my next promotion. That focus on short-term profitability is what pushes people to take shortcuts."

Just how costly those shortcuts could be had hit home with the
failure of the levees and emergency response after Katrina, and again
with unprecedented force in April 2010, when BP's *Deepwater Horizon* drilling rig exploded off the Louisiana coast, killing eleven men,
causing the worst accidental oil spill in history and doing profound
and enduring damage to marine life and fishing communities (all
addressed at length in the following chapters). "How could you let
folks poke a hole in the ocean floor half a mile down and fifty miles
out and not have a strong contingency plan? What they did after the
fact was extraordinary, the equivalent of landing a man on the moon
in the middle of a crisis. But they had to make everything up after it
happened. I've got more of a plan for a tank barge than they had for
this giant thing.

"With our cargos, we have to be supervigilant 24–7, 365 days a

year. If we lose containment of even a minor amount of product—
onto the deck, with zero danger of it going into the water—we have a
remarkable number of protocols: calling the Coast Guard, the state,
the customer. The world stops spinning, at an immediate cost to us
of fifteen to twenty thousand dollars. So the consequences of a cut
corner are significant, but not only because of all the ways you can
be punished. Worse is the broken windows effect: your people are
watching you closely. Being careful about the environment is just a
better business practice. You attract a higher-caliber employee; your
customers have more faith; your underwriters don't have to charge you
so much; your shareholders get a better return. And living in Louisiana, and through Katrina and BP, we see the nexus between this
fragile ecology, and what we do for a living, and flood protection for
our families. This is all one ecosystem. We need solutions that work
across all those intersections."

Like Dusty Crary, Merritt did finally find what he calls "the middle 80 percent" that can forge such solutions. "The first-order challenge was getting the right people in the room and having candid
conversation. If you can get responsible people to sit down together
and figure out how to do responsible things, you can get somewhere.
That includes the leading thinkers in the environmental community,
who *are* trying to complete the sentence, to come forward with a
stronger value proposition that doesn't just make industry the enemy
and try to cut us out of the equation." Merritt had steeled himself for
the worst when at his first leadership meeting he found himself seated
next to Jim Tripp, a long-time lawyer at the Environmental Defense
Fund. "But I came to appreciate his substance, his genuine desire to
do the right thing."

For four decades, since what Tulane University law professor Oliver Houck has called the "predawn of awareness that all might not be
well along the Gulf," Tripp had played a central role in efforts to slow
the damage. It began in 1973, just a year after passage of the Clean

Water Act, when a New Orleans lawyer named Stanley Halpin called Tripp, one of the few people in the country with any experience in the new world of CWA litigation. Halpin told Tripp that the Army Corps had Congressional approval to spend $33 million of taxpayers' money to expand a channel across 900 square miles of the lower Atchafalaya so that two Morgan City–based oil rig manufacturers could move larger platforms to the Gulf. The Corps' plan included covering 8,000 acres of wetlands with dredged spoil. Would Tripp help represent Terrebonne Parish in a suit to stop them? Their principal witness, Halpin said, would be Dr. Sherwood ("Woody") Gagliano, a coastal geologist from Baton Rouge who had just published the first paper quantifying Louisiana's land loss.

Houck, a legend in his own right,* wrote a wonderful account of the events that followed for the *Rutgers Law Review*. "Tripp called Gagliano, whom he had never met, out of the blue. Could he stay the night and then go down together? It was a transformative evening. . . . Tripp showed up at his door with a briefcase in his hand and a toothbrush in his pocket, [his] standard traveling fare. . . . 'After several hours of talking to him,' Tripp recalls, 'I was stunned by what was happening in the delta, fascinated by delta geology, and hooked on the problem.'"

Though the judge hadn't yet ruled on his request to be admitted as counsel in the case, Tripp stood up and questioned witnesses anyway, quietly laying out the basics about sediments, salinities and water flows, most of which "he had only begun to learn himself the night before." Judge Sear was unmoved, dismissing what he called the plaintiff's "unchoked shotgun blast across the marsh . . . not unlike the hunter who fires without aim from the duck blind in the hope of hitting something that may, by chance, be flying past." But Tripp was undeterred, finally finding his opening four

---

* Houck was one of the most memorable characters in McPhee's *The Control of Nature*.

years later at a public hearing. Held in the rig manufacturers' home-town, it was a rowdy affair. A canal enthusiast, banging her shoe Khrushchev-style on the podium, shouted, "When God made environmentalists he should have had an abortion!" A botanist from the Audubon Society got booed off the stage. Then Tripp stood up, and in his deliberate, noncombative fashion, unearthed a solution that satisfied all: the Corps would be allowed to dredge but would have to abide by the provision in the law requiring them to choose the least harmful site for the spoils. "He had reduced the hearing from shout-fest to legal challenge," wrote Houck, and from then on, was "consumed by the Louisiana coastal zone." In the years that followed, Tripp helped establish a permitting process for coastal uses, create the Coalition to Restore Coastal Louisiana and develop and pass the Coastal Wetlands Planning, Protection and Restoration Act. His dedication and pragmatism eased Merritt's alarm. "Jim's also seen that my walls aren't up, though there's still a level of caution. What gets you past that wariness is taking the time to build a relationship, developing trust that you all can take information and not misuse it, seeing that what we all have to gain is worth any loss we may have to take.

"I find I'm becoming a person who can speak multiple tongues, understand not just my industry's issues but our region's issues. It's intuitive to me that you take one dollar and leverage it as many ways as you can." Merritt has carried that message to the Business Council of New Orleans. He also, as chair of the national Waterways Council, created a task force to focus on the Army Corps's two missions *beyond* navigation: flood control and ecosystem protection. "To be effective we've got to work with people focused on those other missions. We're not sitting here trying to figure out how to do it wrong, but how to do it better."

Merritt has brought to the coastal restoration planning process everything he's learned from the river and his family and CBC: how to act in fast-changing and dangerous circumstances; the *obligation* to act, even in the face of uncertainty; the conviction that openness and mutual respect can pull even the most diverse group together to face crisis. "I'm an optimistic person, I believe in abundance, gratitude: that you can bring out the better good in someone if you give them opportunity. I know when you don't give that benefit of the doubt nothing good is going to happen, so I try not to be overly guarded. We've got to legitimately hear other people's point of view."

Beyond the sheer force of good will, the Master Plan team also invented a tool to help them bridge divides. Rather than look at projects and their impacts in isolation, it allows them to model portfolios of projects and their effects across communities and industries. "We think it's a model that will be useful across the nation," says Spencer Murphy, Merritt's VP of risk management and sometime stand-in at the meetings. "We could run it to find the projects that maximize navigation, then run it to maximize land-building, or oysters, or flood protection, or to minimize community impacts. At the end of the day projects that made it into the Master Plan scored high across all those criteria."

The plan arrived at through this arduous process is remarkable for its reliance not mostly on walls but on nature, and not on mastering nature but on restoring its life-sustaining functions. The overarching strategy is to reconnect the river to its delta so that once again its sediments can build and repair and nourish the marsh (and barrier islands and oyster reefs), restoring their protective and biological functions. Though dredges and pumps will move some of that sediment, most will be carried by the river itself, released into its estuaries through a half dozen or more new controlled outlets, called "diversions," at times of high, muddy spring floods. Like Merritt's captains, who kick the towboat out of gear in tight turns to let the current pivot the barges, the idea is to cease fighting the river and instead ride its incredible force to safety. Within twenty years, the 2012 Master Plan forecasts, the river will bring the coast back to a condition of no net land loss; within thirty, it will be growing Louisiana again. (The 2017 Master Plan, taking account of new science from NASA showing still-faster rates of sea level rise, will set those turn-around dates further into the future.)

Using nature to protect people transforms what had always been a zero-sum game. Building a levee to safeguard one town invariably sent the flood bearing down on its neighbor. But a wetland shields a community not by deflecting the water onto someone else but by absorbing it. The plan, in other words, isn't just a parceling out of pain but a way to reconcile seemingly irreconcilable conflicts. Graves provided an example. "More than 80 percent of coastal Louisiana is in private hands, most of that owned by the oil and gas industry. We had a wetlands project we wanted to do with one of those owners, but they were opposed. The environmentalists were saying, 'Go take the property.' Instead, we sat down and explained: 'In Louisiana you lose your subsurface rights when your surface erodes; it becomes a state water bottom. But if we build these wetlands it's going to protect your property; you're no longer going to be the edge of the eroding coast; you're going to have land in front of you.' And all of a sudden they're on board. If you plan it right you can accomplish so many things at once with the river: use the sediment to build new delta, the nutrients to enhance the growth of wetlands vegetation, the water to manage salinity for your oysters and the whole estuarine ecology. You're cleaning water, keeping navigation open, attenuating storms, providing habitat for fish, protecting people and property— all with one investment." An April 2015 NOAA study quantified that range of benefits. It also noted that while seawalls provide benefits only during storms (and do damage the rest of the time), natural systems provide benefits continuously, can repair themselves after storms and, in some cases, grow quickly enough to keep pace with sea level rise.

Merritt has worked for years to advance regional economic development, serving on the boards of Idea Village, an incubator for entrepreneurs, and the New Orleans Board of Trade. Now he and fellow members of the Greater New Orleans Business Coalition—executives in real estate, sports, oil, banking—laid out the business case for restoration. Not only was the Master Plan critical to protecting existing multibillion-dollar assets, it would also give rise to a new restoration economy: creating both new local jobs (to build out the projects) and exportable equipment and expertise. More than 120 million Americans living in coastal counties in thirty states face similar challenges, or soon will. Worldwide, more than three billion people live as close to the sea as New Orleans. One trillion dollars' worth of vital water, power and transport infrastructure sits within three feet of sea level. And virtually all of the world's great deltas—the Ganges in India and Bangladesh, Pakistan's Indus, China's Yangtze with Shanghai square in its mouth—are on course to drown for the same reasons Louisiana is going under: too little river, too much sea.

The interests that lined up in fervent support of the Master Plan have rarely (if ever) all been on the same side of anything: its champions include the world's biggest shipping, oil and petrochemical companies, major environmental groups, and the entire Louisiana legislature, which approved it in a unanimous vote. The plan won a significant source of funding when the U.S. Congress passed the RESTORE Act, committing 80 percent of BP Clean Water Act (and other) fines—finally settled at $20.8 billion—to environmental and economic recovery of the Gulf Coast states. An additional $140 million a year will come to Louisiana from 2017 on, when Gulf states begin sharing in offshore royalties.

With construction started on first-priority projects, Merritt and the Master Plan team immediately began work on the 2017 revision, incorporating new research and modeling and evaluating projects suggested by villages and parishes. Merritt also signed on to help lead a second effort focused more tightly on the river but also across a longer (one-hundred-year) time frame. Set up as a competition to elicit more cutting-edge innovation than a government process allows, "Changing Course" invited multidisciplinary teams of engineers, coastal scientists and designers to answer an essential question. The Master Plan had found a way to capture up to half the river's sediment. How could Louisiana make use of it all?

After a global response from twenty-one teams and months of intensive work, in August 2015 Merritt and partners chose three winners: Baird and Associates (with offices spanning the Americas, Australia and Oman), Moffatt and Nichol (California-based but in thirty-one cities from Rio de Janeiro to Abu Dhabi), and Studio Misi-Ziibi (a mix of talent from Missouri, Louisiana and the Netherlands). Though each team had assembled a unique combination of skills and approaches, all had converged on several fundamental principles.

First, all amplified the Master Plan's optimistic finding: that with the river's power, the Mississippi River delta can be saved for the long term. But these global teams faced more squarely what had been soft-pedaled in the Master Plan. Not everything can be saved. The bird's foot and lowest reaches of the river will survive a few generations at best, no matter how heroic the interventions: given the high rates of subsidence and sea level rise and waves attacking the skeletal remnants of land, pouring more valuable sediments into those fast-sinking basins would be a futile exercise.

Instead, the best hope, the winning teams agreed, is to get ahead of the inevitable change. Not just wait, as one participant put it, "for the next big storm to wipe out communities, erase the shipping channel in the lower River and throw commerce into chaos," but over the next decades begin building the delta that *can* endure—smaller but thick and dense enough to be defensible—and then gradually transition communities and industry into that new landscape.

Their proposals to restore the river's natural processes, therefore, not only went well beyond the Master Plan but also focused those efforts much closer to New Orleans: shifting sediment diversions upriver and also (somewhat mind-bogglingly) shortening the river itself.

The Moffatt and Studio Misi-Ziibi plans shift the Mississippi's mouth—the place where it begins to release its sediments—some fifty miles north of where it is today. Both also open, near that new mouth, a navigation shortcut into the Gulf, to get ships quickly in and out of deep waters. Below that, they allow the river to branch and meander however it pleases across the shallow shelf, merrily silting up everything.

Baird, with the most radical plan, would move the river's mouth eighty miles north, nearly to New Orleans. It would also create something far grander than diversions, which it calls "managed distributaries" or "faucets." More like the Atchafalaya than the Master Plan's diversions, each would carry 15–75 percent of the river. Turned on and off at multi-decade intervals, the on-faucets would build new subdeltas ringing the southern flank of New Orleans. The off-faucets would leave saline marshes undisturbed, to sustain oysters, crab, shrimp and small fishing villages. With all the river water carried off into these different courses, the current channel below the city would be left to the Gulf to fill in, creating a deep, sediment-free, slack-water entry to the port.

Beyond their visions for managing profound changes in the land, the three winning teams also ventured proposals for easing the dislocations of people and industry. The communities know what's coming; many are living it already today. So the task is to face that forthrightly, and soon: to create an "equitable and transparent" community engagement process (easier said than done, as the next chapter makes clear), and then to let each family decide which generation will make the move (or when they've had enough of hurricanes) and where in this future delta they want their children or grandchildren to land.

It is a measure of how intense the jeopardy has become that Louisiana is taking the leap into another giant government project. Many Louisiana citizens and politicians oppose big government on principle. But they've also had more than their share of bitter experience. Merritt and others call Katrina the "federal flood" because the storm by itself would never have drowned New Orleans; the city wound up underwater only because the Corps's badly engineered levees failed. Katrina's aftermath, too, was made far more tragic by the Federal Emergency Management Authority's utterly inadequate evacuation plan and virtual abandonment of the 100,000 residents left behind, many of them trapped in barbaric conditions in the Superdome and Convention Center. As Windell Curole says, referring to then director Michael Brown: "Brownie ran horse shows. I'm sure if we'd had a horseshit emergency he would have been fantastic." Louisianans have also weathered more slow-going disasters: from the Corps's destruction of their wetlands to what many in industry here experience, as Merritt puts it, as federal efforts to "regulate us to death."

And yet, says Merritt, "you simply can't do something of this magnitude without the government driving it. There are just too many moving parts," and too many interdependencies that cross state boundaries. Merritt also sees reason for optimism in the emerging government–shipping industry collaboration around volatile weather. "The Army Corps and Coast Guard have learned over time that we have no stake in being unsafe but don't want to be dictated to with limited information. After the '88 drought, when the river was shut for long periods of time, we created a public–private River Industry

Task Force, so we could have open dialogue about how to orchestrate low water, high water, hurricane closures. Rather than government just telling us what they're going to do, *they* can make better decisions and *we* can operate in a more self-regulated mode, without a rulemaking—we don't have time for that—but with the regulators in the room with the power to hold us accountable for the commitments we make. It's one of the better examples of government and industry working together."

After years of fighting for reinvestment in river infrastructure, Merritt also sees proactive public investment as the more conservative, fiscally prudent path. The Katrina recovery effort cost $150 billion. But a $10 billion investment in wetlands restoration ahead of the hurricanes would have saved thousands of lives and many of those billions. Looking forward, a failure to invest now will bring still worse economic disaster to Louisiana, according to the Master Plan: a ten-fold increase in annual flood and storm damages by 2061, to $23.4 billion each year. Nationally, more than 6.6 million homes are threatened by storm surge; the National Flood Insurance Program, covering more than $1 trillion in assets, is one of the nation's largest fiscal liabilities.

Beyond scrambling the usual partisan stances toward big government, the far-ranging support garnered for the Master Plan is also the plainest example yet of a community stepping up to address climate risks, whatever its politicians might be saying on the national stage. Louisiana senator David Vitter has called the evidence for climate change "ridiculous pseudo-science garbage." In May 2015, he wrote a letter to FEMA denouncing as "ideological-based red tape" a new Obama administration requirement that states must account for climate risks in order to receive federal disaster funding. When Obama visited New Orleans to commemorate the tenth anniversary of Katrina, Governor Jindal asked him to drop any reference to climate change, saying "there is a time and place for politics, but this is not it." Louisiana has also historically been one of the friendliest states to the fossil fuel industry. When Jindal blocked a lawsuit brought by the Southeast Flood Protection Authority against ninety-seven oil and gas companies, seeking recompense for the portion of the damage done to the wetlands by their canals, critics saw proof that the "flag of Texaco still flies over the Louisiana capitol."

And yet the Jindal administration Master Plan incorporates the leading science on rising sea levels, including projections from the often excoriated U.N. Intergovernmental Panel on Climate Change; the 2017 Master Plan will incorporate new findings by NASA showing still more rapid rates of rise. Already, Louisiana is grappling with the fastest pace in the world, because its land is subsiding even as the oceans rise: a combination called "relative sea level rise." While in most of the world seas will be two feet higher by century's end, in some parts of Louisiana the Gulf will rise as much as five feet. Most of southeast Louisiana has an average elevation of three feet. "People talk about climate change and say a hundred years from now we'll be dealing with this, but we have Russian roulette every hurricane season," says Graves. "I believe that if somebody's holding a gun to your head, you can't just pretend it's not there. I know North Carolina passed a law saying they can't acknowledge sea level rise, but it's absolutely fundamental to the work we do. We're the canary in the coal mine, experiencing now what thirty coastal states and many countries will experience soon enough. Rising sea levels, potential changes in hurricane intensity and frequency: those are the parameters of our future."

Beyond designing in adaptations to those climate impacts, the Master Plan also laid out ways in which wetlands restoration could help mitigate the underlying causes. Mangroves, swamp forests and seagrasses take up carbon two to four times faster than mature tropical forests, and store up to five times as much per acre. Among the entrepreneurs supported by Idea Village (where Merritt serves on the board) is Sarah Mack. In summer 2015, Mack's Tierra Resources completed a three-year pilot program, with Conoco Phillips and Entergy, seeding mangroves from crop dusters; they are also working with the state to develop standards for measuring the "blue carbon" stored in such ecosystems, for eventual sale into carbon markets like California's. For the fifty landowners in Louisiana who collectively own 2.3 million acres of wetlands, says Robin Barnes, head of Greater New Orleans, Inc. (the regional economic development alliance serving the ten-parish region of southeast Louisiana), that could represent a huge source of new wealth.

Seeing those changes unfold at close range, Merritt has begun adapting his business accordingly. "It's inescapable that in my three decades in the business, weather patterns have become more extreme. I've seen far too many hundred-year floods and hundred-year droughts, more extreme back-to-back highs to lows to highs again, hurricanes, sea level rise: that's observable, impossible to deny. So we've thought about, how do you adjust to the possibility of more extremes? There are no ready answers. Our contracts certainly need to apportion the risks of uncontrollable delays from high or low water. And we might build a towboat that could shallow a little more."

He sees no purpose, however, in joining the fights over what's causing the change. "It's hard to argue that if you put something into the atmosphere it has no effect. So we keep searching for more efficient ways to operate. If we can repower a 6,000-horsepower towboat with a new $2 million engine with lower emissions, that's a good thing. Though I do it because it also has better fuel efficiency. It's too many dollars; it's got to have economic payback.

"But *global* climate change, well that's a big platform. We're talking about geologic time. The planet has cooled and heated before, for lots of reasons: volcanoes, seepage out of the ocean floor. That doesn't deal us out of the equation. It just tells me there's more to it than the simple view that if we just throw a little less into the air all will be fine. I see nature as being a lot stronger than mankind in that regard.

"That said, in my lifetime, I've seen a river in Cleveland catch fire and the Illinois River near Chicago so polluted it didn't freeze. And I've seen both restored; nature responds quickly when we take better care of it. So rather than getting caught up in the debate, I think we can simply say, 'There's some stuff going on and we ought to be better than we used to be, make it better for the next generation.'"

4

SHRIMPER

WHEN SANDY NGUYEN wants to understand the Louisiana coast, she heads down to the docks when the shrimp boats are in: to Buras, sixty miles downriver from New Orleans, or to Venice, ten miles further, nearly to the Mississippi's mouth. Her step quickens as she approaches the noisy, bustling scene. On the boats just in, vacuums suck a steady flood of silvery shrimp onto chain conveyor belts and from there into scales to be measured and weighed, while deckhands scramble about repairing the damage done to gear and nets from a week or more out in the debris-strewn Gulf. On the boats preparing to leave, men shovel ice into the holds and load crate after crate of groceries. But everyone stops for a moment when Sandy arrives, to call out a warm greeting, flirt and banter. "I can't tell them I'm coming," she says, "because they'll do all kind of crazy cooking for me." The fishermen love Sandy because she is sexy and effervescent and bawdy, her arched eyebrows, jewelry, cleavage and heels hard to miss amidst the shrimpers shuffling about in their flannel pajama bottoms and flip-flops or rubber boots—though the heels don't stop her from clambering onto a boat or navigating its slippery deck. They love Sandy even more because she has seen them through the worst, time and time again, her outsized personality matched by her brains, savvy and seriousness, a devotion and work ethic as relentless as their own. As head of the not-for-profit Coastal Communities Consulting, Sandy is everything to these fishermen: respectful daughter, unflagging cheerleader, "approved second wife"—their primary guide through the complex world of U.S. small business regulations, taxes, loans, immigration and citizenship. Most critically, she has been their tireless and canny advocate through a decade of disasters, each one compounding the damage to the long-suffering coastal ecosystems they depend on.

Restoring America's working landscapes requires seeing the big picture, the long view. But that larger vision is incomplete without the small, immediate, local, human picture: the families who have to get through this week, this year. Sandy's shrimpers live and work on the furthest edge of vulnerability: most hurt by the ongoing degradation of the coast but also most likely to be dislocated by the ambitious plans to staunch that loss. Having been rolled repeatedly, in their lifetimes, by world-scale events beyond their control, they know what it is to feel overlooked, devalued, helpless to shape their own fate. But they also know the depth of their own reservoirs of strength and remain intent, with Sandy's help, on securing the self-determination so central to the American identity.

Sandy teases the deckhands about the "six-pack abs they've built pulling net all day." But it is the elder fishermen, the captains with their big bellies, she has come to see. Like her father, these men have fished all their lives in two of the world's great deltas: the Mekong and the Mississippi. Like her father's, their fishing boats once saved and now sustain their family and community's lives. Perched on a boat deck or the concrete dock (or in one of the many meetings she persuades the shrimpers to have with environmentalists, or the state, or the Army Corps of Engineers), Sandy asks the captains what they've learned and seen in the decades they've been on these waters. They tell her that for thirty years they've watched the land disappear. "The piece of little island where I could hide out in the marsh with my boat in thirty, forty-mile-an-hour winds blowing at us, that's not there anymore," one says. "It's now just eight, nine feet of water." Without the barrier islands, another says, "the sand now comes in and covers and kills the oysters." And a third: "When I first came here seventeen years ago there were big bushy reeds that braved the wind, but now they've disappeared." They show her driftwood that nearly made it over the levee and next time will. "We're in a bowl; they built the levee so high above our heads. But when that east wind blows you can't stop that water; you could build a wall a hundred feet high."

Sandy listens intently to the men she calls "my fishermen . . . the smartest people when it comes to southeast Louisiana water." Under their influence, she has become an avid champion of coastal restoration, for the sake of both the national and local economy. "All tell me that we have to save our coast. Instead of having to send billions and billions of dollars down for recovery after each hurricane, we need to restore the wetlands. And the marsh areas are where the babies come from. All the shrimp and crabs and seafood that we catch, their life cycle begins in this estuary. So if we lose the estuaries we most likely will lose our industry with it."

But the fishermen also share with Sandy their deep fears about the state's Master Plan to restore the coast. A captain at Venice shows her a flyer left by the Louisiana Shrimp Association, warning that letting the Mississippi River flow again into areas sealed off from it for generations will deliver the "ultimate kill" to the shrimp—exterminating the babies, driving the big ones out beyond their boats' reach—ruining these families as war, flight across an ocean, Katrina and the worst oil spill in history haven't been able to. Another captain says that the freshwater has already left "the oystermen crying, because they can only go out once a week and that's not putting food on the table."

Both needing and fearing coastal restoration, the fishermen look to Sandy to reconcile that dilemma. They trust her to build the needed bridges across language and other barriers to reach the people who can explain to them what's coming, and to educate those people in turn, so that as the restoration projects are given final shape and built, the fishermen's deep knowledge and skills are brought to bear. They count on Sandy to answer the question pressing on all of their minds, as voiced by one Cambodian captain: "What if there are no shrimp? What if we can no longer earn, what should we do?" Like Merritt Lane, Sandy's engagement in restoring the Louisiana coast is driven not by any great feeling for nature but entirely by concern for people. For her bayou communities, nature is safety, sustenance, a future for their children. Asked what she sees, standing on the deck of her family's boat looking out over the water turning colors in the setting sun, birds sailing on the salt breeze, Sandy doesn't hesitate. "I see money."

Learning from and helping this community of fishermen, which includes her parents and husband, has been Sandy's calling from the time she was a girl.

Sandy was born in 1973 in the village of Phuoc Tinh, seventy miles southeast of Saigon, in the final years of a war that had darkened her parents' entire lives; among the two million civilians who died were members of her own family. "None of my grandparents came out of Vietnam. My dad's dad died from a U.S. bomb." The fall of Saigon, when she was two, ended the war but not the dangers. Because her family had been "with the Americans," she says, they faced new threats from the victorious Communist regime: the possible expropriation of family property, forcible relocation to a reeducation camp, even extrajudicial execution. So in 1978, when Sandy was five, her parents packed up their five children and joined the million people fleeing Vietnam in small boats never meant for the open ocean. "Daddy owned a fishing boat and was a great captain; he raised a lot of deckhands into captains, there and here. He gathered his close friends and said he's leaving, but won't know when till the day comes. When you're trying to escape a Communist country you have to pick the worst, stormiest day of the year, so you won't get shot. You can't plan it: when the captain calls you, you just got to go, leave people behind if needed. It's your only chance to get out of the country. He told them bring sixty dollars a family, in gold, to the boat."

Sandy wouldn't understand what happened that night until a few years later, when she began having recurrent nightmares and asked her mother what they could mean. "In my dream, there was always thunder and lightning; I'd be sitting in black dirt, in a little brown sweater. Finally, one day, my mom took that sweater out of her drawer and told me the story. We'd hidden behind dumpsters in the dark and pouring rain, then all rushed down to the boat on a muddy levee, with six feet of water on each side. The plan was that my dad would lead the crowd, followed by my two oldest brothers, then my mom carrying my baby brother, who was just six months old, then two guys carrying my sister and me. Well the guy that was holding me, his fiancée—who left him the moment she got to America—she slipped and fell into the water. So he dropped me to pull her out and carry her to the boat; just left me sitting there on the levee, all by myself in the dark, crying for my mom, everybody rushing past me. Everyone's on the boat; the engine's running; the guys with the guns are going to catch up with us and Mom said, 'Wait a minute. One of my kids is missing.' My dad looked around and said, 'Then we're not moving. If you gentlemen don't get off and find my little girl we're all going to be dead.' As soon as we left, gunfire started. I'd have died there.

"The boat was tiny and crowded, with about a hundred people; the guy that slept next to me got peed on every night. The lucky boats bumped into the U.S. Navy, got canned food or transport into camps. Others just drifted and hoped they landed anywhere but back in Vietnam. There were so many deaths out there; that's why Daddy won't give interviews. But my dad knew what he was doing; he timed it right, brought enough food to keep everyone going. We were starving but not to where people died."

Crossing hundreds of miles of the South China Sea, "he got us to Malaysia. There was some kind of diplomatic policy: any country that was a friend of the U.S. had to take in the Vietnamese. But fifty feet from land, Malaysian guards turned us away. They said, 'Our camps are overcrowded. We can't take you in.' Dad knew we were going to die if we went back to sea; we were out of diesel and food. He had to think quickly, do something to make them take us in. So he took us back out a mile offshore, turned off the engine and sank the boat. Everyone was in the water: husbands grabbing wives, wives grabbing kids. Now, naturally, Malaysia wouldn't want the U.S. to hear that it let a boat full of people drown; Dad had made it a humanitarian thing. So they brought us in, though they beat my dad so badly he thought he wasn't going to make it. We spent six months there, and lost everything. They stole our jewelry, money, gold: gone."

Like most of the boat people from their village—the majority of them Catholics who had opposed Ho Chi Minh and fled from North to South Vietnam after the 1954 partition—Sandy's family wound up in southeastern Louisiana. New Orleans archbishop Philip Hannan had visited refugee camps ringing the South China Sea and invited Vietnamese priests to resettle their communities in his diocese, offering the help of the Associated Catholic Charities of New Orleans in securing federally subsidized housing. The families adapted readily to the landscape and climate of the Mississippi delta, which reminded them of the Mekong. Village d'Est in east New Orleans soon became one of the largest settlements of Vietnamese outside of Vietnam.

"You had to have an American sponsor. One of my dad's friends had been here since 1975 and saved a little money, so brought us to his house, stuffing fifteen people into three bedrooms. He had a small skimmer boat, so dad immediately had a deckhand job. My mom got a job shucking oysters. She'd wait by the trash can at 2 a.m. for somebody to pick her up, come home at eleven to cook lunch for her kids, then go out again at three for her second shift and be back at midnight. One sack is five gallons of meat, hundreds of oysters, but she only got $1.25 each. With dad offshore and mom working, my ten-year-old brother pretty much raised me. Mom would give him one of her oyster checks and on the weekend he'd catch a ride with the neighbors to the local Winn-Dixie to get us bread and jelly to go with the commodity peanut butter we got from the church. I don't know how my parents did it with no language, no education, but they raised five very good kids, and we all have our own small business today." When, in 2007, Sandy was named a Woman of the Year by *New Orleans CityBusiness* magazine, she recalled how "in those first years, I'd see my mom have to pay a translator fifty dollars to go to the food stamp office with her," often waiting four hours. "I said, one day I'm going to be my mom's translator and I'm going to be the one to help our people out whenever they need it."

After a few months crammed in with her father's friend, Sandy's family moved into a Section 8 (low-income) apartment of their own, where they spent the next eight years. "Five of us kids shared two bedrooms but sometimes dad would put us all in one room, so he could bring more friends over. I have so many memories coming out of that little house: riding tricycles and skating around the neighborhood; my sister crossing the levee to that jungle to catch fish, or climbing trees to steal peaches. We played Vietnamese games: spinning tops, my brothers' cricket fights. They'd catch them in jars—they said the bigger the head the smarter they were—and when friends came over they'd pull a hair from my head, tie it around the cricket's leg and spin it dizzy. When they untied the hair, the cricket was crazy and would fight. They don't kill each other but the first one that leaves the circle is the loser. My mom hated it because we'd have tons of crickets in the house chirping all night long, and then some would escape and be all over the house. We had a garden in the little space under the stairs, where we'd grow our mints. But then mom got robbed, lost the eight thousand she'd saved for three years, so we moved out."

By high school, Sandy was balancing her new outside-of-the house identity—"all-American-girl Sandy, with my Budweiser and skimpy spaghetti straps and cutoff jeans"—with her inside-of-the-house "traditional chick—which, you know, is hard. A lot of kids my age took the wrong route in life because they couldn't balance that outside and inside person." Sandy saw her share of trouble. "I was a wild child, a bad child. I dated Americans and hung out with Hispanics, both no-nos." She often felt the wrath of Father Dominic at Mary Queen of Vietnam. (Though her parents are observant Buddhists, the Catholic Church was and remains the hub of community opportunities.) "Father D was strict and his word was law. And the nuns were so mean. If you didn't do your homework or your handwriting's bad, they'd turn a ruler on edge and hit you like that on the hand."

But Sandy was also president of the student body, salutatorian and a star athlete: the first Vietnamese basketball player and cheerleader at her high school and the first Vietnamese girl ever recruited for the New Orleans Saints' cheerleading squad. (She got sidelined by a broken leg but remains an avid fan, flying a Saints flag alongside the American flag on her boat and "loving our Drew Brees.") Offered softball scholarships to two colleges, she opted for an academic scholarship to study business at Tulane.

As the oldest girl, Sandy had been expected from age fourteen to handle the paperwork for the family fishing business. "I grew up having stuff thrown at me. My brother would say, 'Look, Sandy, you got to get this done for Dad. This is his license, renew it.' And I would call up to NOAA [the National Oceanic and Atmospheric Administration, which regulates fishing in federal waters] and figure it out, then apply what I'd learned to help my dad's friends. And then it became not only fishing, but other issues in their life. You know, 'Can you take me to court?' 'Can you help me enroll my son in this school?' And I would. I started helping my dad's deckhand, and then all our neighbors started coming to me, and I was so happy, because every time I translated, I learned something new, another trick of the trade. People would ask, 'How can you know what to do for incorporation, or immigration? I'd tell them, 'I learned early: don't be scared to ask; the person behind that window is going to help you.' But it became a flood. Tulane was twenty minutes from my house but Mom let me stay in the dorm because the requests for help never stopped."

At twenty-two, fresh out of college, Sandy became captain of her father's shrimp boat for a month; lacking the requisite U.S. citizenship, he'd been shut down and fined $25,000. "I never would respect and love my dad so much had I not gone fishing with him that month. I finally appreciated what he and all these fishermen had been doing for thirty frigging years to put us through school: the hard, stinky work, the danger and sleepless nights. You pull up the nets every five hours and before you can sleep you have to sort, clean and ice the shrimp; if you're good you can finish in two hours and get three hours' sleep before you have to pull 'em up again."

At age twenty-four, she met and married her husband, Phuoc Nguyen ("Mike"). Mike had left Vietnam in 1983 with his grandparents, waiting nine years before he could speak to the parents and siblings he'd left behind. He spent more than a decade in California before moving to New Orleans. (When, over a family dinner, Sandy begins to tell the story of how they met, her daughter Hana pipes up: "I want to hear this; I wasn't even *born* when this happened.")

"I was working at a po'boy restaurant my mom had opened, and after work most days, I'd go with my friend Amy to get a beer. We'd follow the fun, sometimes winding up at B-girl bars—where single men pay twenty dollars for a girl's companionship, just to have her sit with them, then blow the rest of their money buying her overpriced drinks. Amy and I were sweaty and greasy from the restaurant but also young and hot; all the fishermen had crushes on us, which the girls hated because the boys were too busy having fun with us. One night, I caught sight of Michael and was like, 'Damn, who's this new kid?' He spoke English, was quiet but hot as hell, dressed sharp, not the typical shabby fisherman in slippers. And he never bought for these girls. I'd planned my life: this is when I graduate; this is when I get married and have kids. He showed up at the right time, and five months later asked me to marry him. We had a big traditional Buddhist wedding, in front of an altar to our ancestors, with four hundred guests. In Vietnamese culture, our weddings are investments; the gifts are all money. You put out thirty thousand dollars and get back seventy thousand. Especially with my status, after helping people for ten years: the typical envelope is a hundred dollars, but mine were three to four hundred dollars. And my parents gave each of us kids twenty thousand dollars as a wedding gift, to start a business. So I opened my own consulting firm, Sandy Nguyen Enterprises."

Over the next several years, Sandy kept inventing new ways to aid her community. She helped found the first Vietnamese nonprofit in the state, the Vietnamese Initiatives in Economic Training (VIET), and began working for the Louisiana Small Business Development Center, often putting in hours far beyond what the LSBDC could pay her for. "People were telling me how dumb I was, undercharging."

But her calling became undeniable and all-consuming in 2005, when Katrina wrecked the entire shrimping fleet and destroyed many homes, including her own. "When the storm went east to Pass Christian, Mississippi, they said we dodged a bullet. But I lived way out east there; I knew my boat and house were gone. When I got home I couldn't even find my street." With her parents, husband, four-year-old Hana and three-month-old son Dylan, Sandy evacuated to her sister's house in Tennessee, but only because the mayor ordered everyone to leave. "My phone was off the hook, with fishermen calling me from

Houston, Atlanta, just breaking down on the phone. I would hang up and lay down flat on the floor of the garage, crying. I couldn't bear not being able to answer their questions. My dad said, 'I don't think any doctor's going to help. I think the best thing is for her to go home and be with her people.' So I left everyone in Tennessee, drove home alone, stayed at a little room in my aunt and uncle's, set up in my brother's insurance office and started working fourteen hours a day. That got me back to normal.

"The whole fleet was damaged or sunk. Everybody was in disaster mode. And because everyone was knocked out the same way, we didn't have the resources we typically do; you couldn't just go to family or friends for a loan. I didn't worry about my own house; I couldn't recover myself until I had everyone else recovered. It was terrible to see those men break down and cry. But my mom put it in perspective for me: that as horrible as it was, it did not begin to equal what we went through there and getting here. The Vietnamese and Cambodians had gone through so much worse, been to hell and back. That to us is normal. So we picked up right away and helped each other out. All of us became overnight carpenters; people were on rooftops who had never done roofs before. If you're a welder you're going to help weld my boat, so I can get back to work; the key was to get back on the water immediately to make money to rebuild your family and home, which you can't do on a $2,000 check from FEMA. Then when this one got his boat running, he shared his earnings with you.

"My own focus was to get the docks going. If the docks don't work, then boats don't work." The owners of Ditcharo Seafoods in Buras, who buy five million pounds of shrimp a year from Sandy's fishermen, were better prepared than most. Following an earlier storm that had knocked out roads and power and cost them $100,000 worth of shrimp, Sandy had helped them develop a business continuity plan. But D & C Seafood in Venice, started in the 1980s by Chan and Duong Tran ("who everyone calls Sugar 'cause he's so sweet," says

Sandy), suffered $1 million in damage to their dock, seafood cooler and trucks, some of it uninsured. Sandy helped them get a small business loan to rebuild.

"But nobody else was helping us. No one. None of our fishermen had insurance because it would have cost $40,000 a year, half their income. And without insurance, the banks wouldn't go near them." Asked by Senator Mary Landrieu to speak at a town meeting, Sandy talked about the desperation of fishermen who were being denied SBA loans because they didn't have documented incomes: the punched walls, the splintering families. "Senator Landrieu teared up, which made me tear up and get embarrassed and run straight out the door." She was followed out by Robin Barnes—then head of a New York–based nonprofit lender called Seedco Financial—who slipped her a card and whispered, "I can help."

"I called the next day, and took Robin down to Sugar's dock, a drive that should take an hour but took six because you had trailers on the highway, horses on top of light poles, graves everywhere, boats everywhere. By the time we got to Empire Bridge and she saw fifty boats all crumpled up, each one a family's business and life, she said, 'OK, I have enough. Turn back. Sandy, I want to help you and all those little boats. Help me help them.' She shut her whole office down for seven days to come learn about us, and assigned Jeremy Stone, a senior loan officer, to our case. Every day for months Jeremy and I made that drive to Venice, climbing boats to assess damage. The fishermen didn't have financial statements so we just went back and forth to the docks gathering up trip tickets* to see what kind of money they made, building that documentation from scratch. Our first round landed $100,000 each for five boats. The state then followed with zero-interest loans, and we created a revolving fund so as

---

* The records that dealers are required to keep, reporting which shrimpers they bought from and the species, size and quantity.

one loan got paid back that money could go to the next family. My boss at the SBDC allowed that collaboration: when you bump into a girl like me who goes above and beyond and beyond, and a group like Seedco that will take that kind of risk, it's a perfect match. We wound up helping 129 boats finance repairs, checking back a few months later to make sure they got it done." Of the thousands she has by now helped to secure loans, just two have fallen behind on payments; they can't even look at her, she says. "If you do it right, the whole economy picks up right away." By Christmas, just four months after Katrina, two hundred Vietnamese families were back and at work; within a year, more than 90 percent of the community had returned to New Orleans.

Jeremy became an honorary member of Sandy's family. "He had a hard life growing up, and when we became friends asked my parents to be his mom and dad." When he sent them a picture of his first child, they hung it alongside their other ten grandchildren "in the only white frame," laughs Sandy. "My kids call him their white uncle."

As he saw how much help her community still required, Jeremy began encouraging Sandy to "duplicate yourself" by creating a nonprofit. Byron Encalade, a third-generation oysterman from Pointe à la Hache and head of the Louisiana Oystermen Association, joined in the urging. "With Katrina we saw so much money coming to people who didn't understand the fishermen and what they needed," he recalls. "They took that money out of here or wasted it and the fishermen were left holding the bag, the empty bag. These guys are proud. They never did ask the government for welfare; all they wanted was a hand up, help to get back on their feet. We needed people that understood the culture and were rooted in fishing and are trusted and can reach and educate and help the grassroots fishermen, and Sandy stood out. She has such passion for the fishing people." With Robin, Jeremy and Byron making up her board, in 2010 Sandy founded Coastal Communities Consulting. "It was a way to organize the work that she was already doing without funding," says Byron, "just out of the will of her heart."

CCC committed itself to pulling down "the largest barrier for these rural entrepreneurs . . . a lack of capacity to navigate formal systems to access resources or communicate their needs." All the rules and regulations, says Sandy, "are so complicated; my husband graduated from high school and still can't maneuver through them. But I know the guys up at NOAA, so when they come out with new ones I can call up there and say, 'Simplify it, so I can call a meeting and explain it to them.' She helps her clients pay taxes, get insurance, diversify their business, even secure scholarships and tuition for their kids. "My mom thinks I'm crazy. She says you could charge them a percentage and get rich. But after all those years of hell they went through, I want to give them whatever I can."

The demand for her services has proved never-ending. "The more I go out, the more need I see. We have 6,000 fishing families across the coast, 75 percent of them in Louisiana. We'll see 1,400 fishermen in three months; they all have my number in their phone. I don't begin to have enough workers to fully serve them; every dime I have goes back to capacity building." First-generation Asian Americans account for about three-fourths of Louisiana's shrimping fleet, so most of Sandy's work is in those communities. She calls the Cambodian fishermen in Buras "my Khmer family," and in 2015 sponsored their New Year's festival. But she feels responsible for every coastal fishing family, who have been "underserved for the longest time, regardless of their ethnicity or language," as CCC's mission statement says. "I can't leave my white folks behind. The Asians and Croatians have the language barrier but the Americans have literacy issues; they can't fill out their own grant applications after the storm, or put together a written contingency plan for their business. I could never deny a fisherman that walks in. We all thrive the same; we all hurt the same. And a lot don't budge without my nod. That puts a heavy weight on me."

With Byron's help, Sandy has also been working to understand and support the oystermen, most of whom are of mixed African, Spanish and Native American lineage. "We got her plugged into GoFish," Byron says, "a consortium of groups representing fishing families of African, Cajun, Creole, European, Native and Asian descent. All these people have lived together along the river and bayous for generations. There was never segregation on these waters: we ate out of the same plates, slept on the same boats, have the utmost respect for one another." Byron himself feels a special tie to Sandy's people. "I served in the Wolfhounds regiment in the Vietnam War, was part of Operation New Life in the refugee camps. I saw people more courageous than any I've ever met, families separated, people struggling for their lives. That's what's in my heart."

These small fishing enterprises are completely entwined with family and community—dependent on relatives and neighbors for both financing and labor. So Sandy considers it an essential part of her job to "build these families," investing six years or more in each one. "When I look at a boat, I see a family. If you have the business but not the family, it doesn't work. Men need a wife. A single male, with a lot of cash money on hand, they just blow through it. But a deckhand, no schooling, can't *get* a wife over here. He goes back to the home country to get a wife. I help him get her over here and through immigration; that can take three years. I put them in a trailer first because all you got to do is pay like $275 for the land and you don't have a mortgage. If they make $4,000 in a month, and have $1,000 in bills, the other three thousand goes in the bank; we start building their credit. After three years, they can get a little boat, though if the wife's good at running a nail salon, I help them decide whether they should buy that instead. Then I help them through a title search permit, and whether to be an LLC or sole partnership, and now they have a business. Two more years, they've saved enough for a house and can start having kids. This is how I build up a family." The captains on the premier boats support her efforts, training the young deckhands into the skills they need to run their own boat.

For the Vietnamese families, the goal is to succeed enough in the first generation on the water to give their kids an entirely different life. "They bust their tails fishing so their kids don't have to," says Sandy. "You're in America, right? Land of opportunity. So they grasp onto that and work and work and work so their kids don't get on the boat. They want us to live the American dream, go to college, have our own business or work in corporate America. Mike won't let our son Dylan fish; my dad never let my brothers near the boat." The Americans, by contrast, "like to keep it going for generations. They hop on boats as early as sixth grade, can make $500 a day. Most hate the government, say, 'Don't bring 'em down here; leave me alone.' It's a pride thing, a way of life."

Because the women often handle the business side of these family enterprises, Sandy has created a fishermen's wives association that meets in three languages ("which can take five hours or more," she says). Her Gretna office hosts classes for them: in English as a second language, basic computer skills, how to get permits and licenses online, make a business plan, do profit and loss statements, work with banks.

All of that groundwork proved valuable in 2008, when hurricanes Gustav and Ike hit, knocking out power for weeks at peak shrimp season. "You lose a month of shrimping for a big boat like ours," says Sandy, "you potentially lose gross income of $60,000." But it proved absolutely vital when her community was hit dead-on by the worst environmental disaster in American history (or at least the worst to begin with a sudden bang): the explosion on April 20, 2010, of BP's *Deepwater Horizon* drilling rig. The spill closed more than 88,000 square miles, 37 percent of Gulf waters, and contaminated more than 1,300 miles of coastline from Texas to Florida. In a conversation in

2014, Congressman Garrett Graves (then head of the state's coastal restoration authority) traced the accident to the 1996 royalty holiday granted to oil companies by Congress, which sparked a huge rush in the Gulf "at a rate that outpaced oil spill prevention and cleanup technology." That failure of leadership was compounded by others, including a total disregard of requirements for industry to have plans and resources in place to respond to big deep-water spills, and EPA's ultimate acquiescence in BP's decision to apply two million gallons of the dispersant Corexit, sprayed on the surface of the Gulf and injected into the oil plume. The dispersant made it harder to recapture oil and greatly expanded the zone of exposure to toxic chemicals as they rose through the ocean's zones of life; when mixed with the oil, the Corexit increased total toxicity fifty-fold. "In the evening time," recalled one of Sandy's fishermen, "you could see a fog from the dispersant. We worked in it, with raw headaches and sinuses. They dropped it on us from airplanes. We was guinea pigs. They said, 'We don't know what it's going to do, but let's hide this oil.'"

The destruction of marine and bird life was unprecedented, and five years later still unfinished. The oil killed more than a hundred thousand critically endangered and threatened juvenile sea turtles. It killed a million birds, which when oiled lose the weatherproofing that keeps them warm. Few who saw them will ever forget the grievous sight of pelicans soaked head to toe in the black viscous oil, beaks wide open in desperation, waiting to die. Hitting in spring, the oil bathed millions of eggs and baby sea creatures, including shrimp, as they floated on or near the surface from their offshore spawning areas into coastal nurseries; more than two trillion larvae died, including the larvae of Atlantic bluefin tuna (which breed only in the Gulf and Mediterranean). Snapper, grouper and spiny lobster also drifted through the "kill zone." Five years later, bottlenose dolphins in Barataria Bay had developed lung disease and adrenal lesions and were dying or becoming stranded at four times normal rates. Stillbirths became commonplace; of nine observed pregnant females, only two produced live calves. Heartbreaking pictures circulated of mothers pushing their dead newborns through the Gulf surf. Mahi-mahi hatched in oily waters were unable to swim properly, and exposed bluefin tuna had irregular heartbeats and were dying of heart attacks.

As the oil came ashore, Louisiana took the hardest hit of the five Gulf states, accounting for more than half of the total oiled coastline. Its filigreed estuaries and marshes and seagrasses were also far harder to protect than the smooth beaches of its neighbors: once tides, waves and winds moved the oil into the wetlands, it was impossible to get out. Byron Encalade watched as the spill moved into the east bank, killing fish and oysters; though he knew it would disrupt salinity in the oysteries, he supported the decision to open up freshwater outlets full blast, to flush out as much oil and dispersant as possible. Still, rates of erosion doubled in oiled marshes and mangroves. On Cat Island—an important nesting site in Barataria Bay for brown pelicans, roseate spoonbills and least terns—all life was abruptly erased. Five years later, 25,000-pound tar mats were still washing up on barrier islands and concentrations of hydrocarbons harmful to birds, fish and wildlife remained high in Barataria and Terrebonne marshes, where scientists predict they will persist for decades.

At least as terrible were the unseen impacts, with damage at every layer in the water column and level of the food chain. University of Georgia marine scientist Samantha Joye has taken the deep-diving submersible *Alvin* down to the spill site several times. In the immediate aftermath, she found the ecosystem obliterated, as if by a nuclear bomb. The damage included ancient deep-water corals that can take centuries to recover and multiple species of mud-dwelling organisms. Those that survived ate contaminated sediments; eaten in turn, they moved the oil, dispersant and DNA-damaging heavy metals up the food chain, where to this day they're being found in sperm whales and the eggs of migrating American white pelicans in Minnesota, 1,400 miles away. Returning in March 2014, Dr. Joye found 2,900 square miles around the wellhead still befouled by oil and dispersant, in places two inches thick. In these freezer-like conditions, oil and dispersants scarcely break down, leaving behind toxic polycyclic aromatic hydrocarbons. Beyond a few eels and crabs, a rare vampire squid and several foot-long relatives of roly-polies, almost no life had returned.

The spill shut Sandy's shrimpers down for the rest of the year. "If a boat is down and out for months like that, divorce rates go up; it will always change the dynamic of the family," says Sandy. When allowed back to work, their harvest was dismal and in some cases deformed: the shrimp eyeless, their gills stained black. Even where the shrimp were fine, consumer fears killed the market. "It's not like a hurricane, where you can get back out and make money right away," says Sandy. "It had an environmental impact, killing the babies, creating terrible uncertainty. We still don't know how long we're going to be impacted. In the third year, the crab was still not there; in the fourth year, the oysters were still dead. We're thinking eight years, maybe ten before we're whole again." Gathered for a meeting around the time of the fifth anniversary, Sandy's fishermen reported what they'd seen. "The oil killed a lot of the land, and the seagrass the oysters survive on," one said. "The dispersant sunk it to the bottom so it's still there and seeping out. We might have never felt the impact yet," said another. A third noted that the thousands of miles of wells and pipelines they work among chronically leak, compounding the damage. "There's an oil spill every month down here. You don't hear about it, but I'm fishing crabs, they full of oil when I pick 'em up."

Getting help after Katrina had been nearly impossible, but navigating the BP claims process was far worse. "If you worked with me after Katrina, you were ahead of the game," says Sandy. "If you've got your trip tickets, your taxes filed correctly, you could prove your lost income." But with billions of dollars of corporate money at stake, Sandy had a new gauntlet to navigate. "Guys were getting cheated left and right," says Byron. "Going to claim places and getting scared to death. These guys don't speak English, or didn't go past sixth grade. You got a lawyer writing stuff up and expecting them to understand, or wanting them *not* to understand." (The *New York Times* devoted a whole story to lawyers scamming the Vietnamese.) Many would wind up, desperate, in Sandy's office. "Lawyers would charge folks 25 percent just to file the claim, then tell the fisherman BP hadn't paid yet, so they could hold the money and earn the interest," she says. "I ended up doing more than 150 claims, $23 million worth. Every time I saved a fisherman $40,000 on attorneys' fees, that was a year's tuition for his kid." (True to form, Sandy also found new clients in unlikely places, including one American fisherman who couldn't read and therefore didn't know about the claims. "If I hadn't met him at a bar, [he] would have missed out on $400,000.")

As it often does in Louisiana, politics played an outsized role. "One of the saddest days I've ever seen was right after the spill," recalls Sandy, "a meeting in Lafitte of the five coastal parishes hit hardest by the oil. Ken Feinberg [administrator of the BP Oil Spill Victim Compensation Fund] was sitting up there and right behind him, like they're backing him up, were the parish presidents, all looking really hot in their nice suits. Their constituents were hurting so bad; they should have been sitting with them. But they didn't care about folks living in trailers."

"We immediately saw a big improvement, as soon as Sandy got involved," recalls Byron. "Here in Louisiana, when there's money or work to go around, it's impossible to take politics out of it. But without Sandy it would still all be based on politics." When in 2012,

Sandy won the state Star Award from the Small Business Development Center, they tallied her accomplishments over seven calamitous years: Sandy had secured more than $25 million in loans for her fishermen, saved 1,068 jobs, created twenty-two new businesses and 193 new jobs, and helped boost seafood sales by $333 million. Her clients find it hard to express their gratitude. They donate to her nonprofit, or slip envelopes of New Year's money to her kids. "I'll take Hana and Dylan out to the festival to pop their little firecrackers and they'll end up with a couple of thousand dollars."

Walking the docks, Sandy swells with pride showing off "my success stories." At Buras, she greets the families with the smaller boats—the little "butterflies" that go in and out in a day, or skimmers, out for a few days at a time. These boats, she says, are entirely family-run: the husband the captain, the deckhand his wife. Living year-round in these small rural villages near the river's mouth, they switch to crabbing and oysters when the shrimp season ends, for subsistence and to sustain their modest incomes. At Venice, she shows off the bigger boats, including her own family's 65-foot trawler, the sky blue *Lady Hana*. These boats go out for a week or two at a stretch, often working as teams, with one scouting the shrimp for the rest. Earning enough in the eight-month season to support their family all year, most of these captains head home (between trips and off-season) to New Orleans. "Mike goes offshore for twelve days, comes home and throws me a chunk of cash." Sandy takes particular pride in the 90-foot *Miss Kandy Tran*, making one of its rare visits to dock: "It cost $1.2 million and produces more captains than anybody." Equipped to freeze shrimp right on board, it can stay out thirty days, "bring in $45,000 worth of shrimp in a single trip, and earn $300,000 a year."

As she passes each boat, Sandy provides that income tally. For these families who have repeatedly lost everything, money is a critical measure of success and a topic openly addressed. With money, they can protect their family, help relatives in need, share hospitality, give the best of everything to their children. Sandy and Mike, like many here, send their kids to a private Catholic school. And they have created for them a home that is elegant and ultramodern but clearly designed for big gatherings and fun. A karaoke machine sits alongside the big-screen TV; in the backyard is a brand new swimming pool.

Most importantly, money permits these families (many of whom blend Catholic and Buddhist practice) to properly honor their ancestors. Visiting Venice on a day when an approaching storm has brought in the whole fleet, Sandy is thrilled to find one of her families hosting a boat blessing. Having caught $30,000 worth of shrimp in two weeks, they are now thanking their ancestors: decorating their boat with flowers and fruit and generously welcoming every deckhand and captain and even strangers to a feast. Everyone helps to drag picnic tables into a long banquet table, pass out paper plates and beers and set out the meal: an immense, golden-crisped whole roast pig; platters of lettuce, mint, basil, scallion noodles, bean cake and ginger; bowls of a spicy, vinegary sauce. Using lettuce leaves to roll it all together into delicious wraps, the fishermen eat and laugh and chatter, warmly affectionate to one another and as always, especially, to Sandy.

This is clearly what sustains Sandy: seeing these stories of safety and bounty unfold before her eyes. In the office she loves seeing a client master an unfamiliar concept or process. "They love to learn, man. When they've learned something new they feel really good, leave the office very happy." But she is never more content than out here on the docks, savoring the deep, joyful fellowship she feels with "her boys." It's especially sweet when, as on this visit, Mike is in. Clambering aboard the *Lady Hana*, she finds him sixty feet up the mast, without a harness, struggling to fix a jammed pulley. He climbs down to greet her, exchanging the traditional sniff kiss (a quick inhalation against the cheek), then joins his deckhand, who is sitting cross-legged and barefoot on the deck, to mend bright green shrimp nets with a swift, rhythmic knotting like crocheting.

Sandy is equally in her element when she and visitors catch a ride in a seaplane to visit a bright red shrimp boat out working in the Gulf. The deckhands pull in the heavy nets, dump a mountain of shrimp on deck and squat to sort out the croakers and occasional eel and squid, which are all swept back into the water where gulls and dolphins wait. (Reducing this "bycatch"—a major problem in shrimping—is the focus of efforts by several of Sandy's clients and partners, as discussed below.) Sandy merrily joins in, sitting on a low stool in her leopard-print rubber boots and bright orange gloves, a jade Buddha dangling around her neck, to deftly pull heads off shrimp and toss them into waiting baskets. Soon brimming, the baskets are handed down into the ice hold, where another member of the crew rakes them into thin layers, sandwiched between layers of ice. A clothesline hung with pajama bottoms leads the way into the cabin, where a football game is on the big-screen TV and the crew is preparing lunch: boiling a huge pile of the just-caught shrimp to set out with rice crepes, romaine lettuce, chunks of pork belly, sticky rice noodles, the ever-present assorted mints and basils and red chili sauce. Sandy shows how to soften the rice crepes in a bowl of hot water, wrap the peeled shrimp and other ingredients inside, and devour them.

Back ashore and driving home along the lowest reaches of the river—past orchards of banana and grapefruit trees, a client's soft-shell crab operation, gardens filled with water spinach, oregano, bitter melon, guava and bamboo ("we eat the roots and use the stalks for fishing poles")—Sandy jokes that "white folks have just discovered slow food, but we've been doing it forever." She wants to retire down here, where "you can go clamming or just grab a fish out of the water and grill it right there. And oh my God, the satsumas are ridiculously juicy and good, and like four dollars for a whole bag."

Even within the city, in the "little Saigon" of New Orleans East, her community harvests the bounty of the land. "The elders consider the levee their farm," says Sandy, walking through the town's block-long center. "They walk or ride a bike out there, clear away enough jungle to grow some greens, package them into little bundles, sit here in the parking lot—right on the ground with the bundles laid out on an open newspaper—and sell them for 75 cents each." What they don't sell on the street they bring inside to the grocery store, to be sold alongside the lemongrass, lotus and fennel seeds, dried chestnuts, tapioca strips, palm sugar, eel sauce, soy wrappers, dried pork bits and bird's nest drink. "They can make $50 in a day; $3,000 in a year, which they keep under their bed." Across the street, at her fishermen's favorite bar, Sandy settles in with family and friends to enjoy the cook's masterful preparation of the day's catch: squid, crab claws, crawfish. Soon, the karaoke begins. "The boys like hard-rock hair bands like Scorpion and Def Leppard," says Sandy. "I like Billy Idol, Blondie, Madonna. I kill all her songs."

As devoted as she is to her community, nothing compares to the devotion Sandy shows her parents. Her children are with them every day after school, and every evening Sandy joins them for a meal prepared by her mom from the bau squash, tomatoes, onions, basil, hot peppers, dragon fruit and peaches she grows in her little suburban backyard. "We eat the peaches when they're still small, green and crunchy," says Sandy, "dipped in red pepper and salt." The seafood comes from the *Lady Hana*: "Every trip, Mike brings home 200 pounds of shrimp and 100 pounds of fish for the family." For drinks, they squeeze the juice of aloe vera (good for the liver, Sandy says) or the lemons they've preserved in an enormous jar. When she has time, Sandy does the cooking herself, posting to Facebook pictures of her swordfish-head hotpot, bamboo duck soup over rice noodles, or oxtail *bun bo hue*, a spicy noodle soup she made for her parents when both were ailing.

As Sandy's parents age—her dad's hair and narrow beard gone long and white, his sea-weathered arms growing ever more sinewy—Sandy's devotion only deepens. For more than a decade, she and Mike have sent them for a month each year back to Vietnam. Whatever she's doing, Sandy interrupts it when her mom calls, often for help communicating with a doctor or nurse. "My dad would die before any of his sons could take him to the hospital. He only wants his girls. When you take care of your elders, if they smell frustration in you, then they're just going to die. So you have to be really sweet." Her children are just as respectful and adoring: heading out the door with their mom, they rush to each grandparent in turn to say a charming goodbye in Vietnamese. "Culturally, you have to say goodbye to everyone," says Sandy. "That's the hardest thing in America, to keep the kids culturally connected."

The cross-cultural mix of experiences Sandy and Mike provide their children is as varied as the inside-of-the-house, outside-of-the-house dichotomy she grew up with, though without the secrecy or any apparent dissonance. On Saturdays, Sandy sends the kids to Catholic catechism and Vietnamese language school. She holds Hana responsible for her brother's homework; "That's culturally a must. So when he doesn't do well, she gets punished too." But she also gives Hana full rein to emulate her mother's wild energy and worldliness. Hana is in a dance troupe, hoping to wind up a cheerleader for the New Orleans Pelicans basketball team, and did her eighth grade science project on wetlands loss and the coastal Master Plan. She has also inherited her mother's precocious business savvy. In elementary school, she drew up contracts specifying when her mom was allowed to go out at night (on Fridays and whenever Mike's boat was in). "And that fresh new money she gets for New Year's," says Sandy. "If one of us borrows $100, she'll say, in Vietnamese, 'Remember, when you pay me back it's $110.'"

With their grandparents, the children are docile and dutiful, which includes attending to the Buddhist rituals, most of which center on honoring their ancestors. Sandy shows a picture of Dylan with other little boys all in red, performing for the lunar New Year. On a table set before a wall of pictures of Sandy's grandparents, they have set out incense, "which you burn to bring home the ancestors that take care of you" and a small Tet tree, covered with yellow blossoms and red envelopes filled with lucky money.

For all the solidity Sandy has helped her community build, these shrimping families remain deeply afraid of what lies ahead. As Louisiana's wetlands disappear, they face growing dangers at sea and at home: their boats exposed in open water, their urban neighborhoods flooded by hurricanes and storm surge, their rural hamlets sinking into the sea. And as the estuaries go, so go the shrimp, which depend on those sheltered wetlands to reproduce. The status quo, in other words, will doom Sandy's community.

But the enormous changes needed to move forward also frighten them. As so often before in their storm-tossed lives, these families feel caught in the path of forces beyond their control, at the mercy of plans made by powerful people seemingly indifferent to their fate. The Master Plan decision-making criteria include "support of cultural heritage and fair distribution of risk across socio-economic groups." But to Sandy it seems to "overlook the folks on the coast. It has one little paragraph about transition, which people think means they want us to leave our industry. It's like, 'Let it be the lower end that pays the price, because who cares?' I'm going with history; usually we get the bad end of things. The older ones feel like this will be it for them; they're done."

Sandy is not one to stand by as the world steamrolls her shrimpers, however, so she is doing what she's always done: diving in to learn all she can and forging relationships with everyone who might have useful insight and influence. "My goal is to learn more from the fishermen about the water and the sea life and the land and the coast and hopefully bring it to the state, bridge that gap and have the state engage our fishermen and move forward together. Historically, fishermen and government, there's just not a mix there, but this is where I want to be, connecting everybody. Get everybody to learn from each other and then move forward quickly to fix our coast."

Sandy began by engaging a group of people her constituents mistrust at least as much as government: environmentalists. "For the longest time, environmental groups and fishermen haven't worked together well," says Sandy. "The fishermen think they're out to shut them down, and often the regulations they set forth do impede their ability to work." She offers the example of turtle excluder devices, designed to allow endangered turtles to slip free of shrimp nets and required in federal waters since the 1980s. "TEDs had been our biggest hurdle. Fishermen were against them," complaining they allowed much of their catch to escape too. "If you develop and implement a new technology and throw it at 'em, they're going to rebel. And some environmental groups are just about the land and animals; they don't come out and understand the people."

But the fight over TEDs had also shown Sandy a different possibility. "We want to save the turtles as much as the government does," she says, "so we worked with NOAA to design a device that allows us to work, and now our fishermen are used to it."

Among those leading the work with NOAA (and other partners, including the Audubon Nature Institute) was Lance Nacio, one of Sandy's "American" fishermen. Raised in a bayou trapping camp built by his great-grandfather, Lance was taught by his father not to waste: anything you kill, you eat or sell. By age four, he was running his own traplines for muskrat and mink, working alongside a great-aunt who trapped well into her seventies. There was solid enough land then, Lance says, to keep a milk cow, grow a garden and go by horse and buggy to the coast. But by the time he was fifteen, all was gone: "there was places you could walk in tennis shoes that within a year was open water." So he turned to his shrimping roots on his mother's side: her grandfather, from the Philippines, had been smuggled into Grand Isle in a wooden barrel.

Lance has now gone beyond turtles and is also working on ways to reduce the accidental bycatch of other species, which has historically made up as much as 80 percent of a shrimper's catch and included high numbers of juvenile red snapper. Though he fishes primarily in state waters, Lance voluntarily uses the "bycatch reduction devices" (BRDs) shrimpers are required to use in federal waters, including a "fish-eye" that props open an arched passage out of the net, giving top-swimming finfish a possible escape route. He's also continually innovating improvements, which he shares with the National Marine Fisheries Service and fellow shrimpers. He's modified the TED, for instance: narrowing the space between bars to try to keep fish and animals smaller than turtles but bigger than shrimp out of his nets, and adding a length of pipe at the front of his net to create what he calls a "swirling action" to suck shrimp down while spinning some other fish up and free. He dumps his catch first into a salt tank, keeping much of the accidental catch alive till he throws it back. And, honoring his father's admonitions, he freezes any bycatch he can legally keep (including flounder, squid and whiting) to sell to farmers' markets and restaurants. Altogether, he says, he's cut his bycatch to below 10 percent.

Lance has also worked with NOAA to show skeptical shrimpers how those conservation efforts have actually boosted his bottom line. He has run boats side by side with and without excluders, to show how few shrimp are lost. He has also demonstrated, through his own branded business, that "clean" shrimp can bring a premium: because they're less bashed up in the bag by crabs and other non-target species, and because they're fresher—moved straight into his on-board flash freezer without time lost to sorting.

With Lance's and Sandy's efforts succeeding in easing fishermen's resistance, in July 2015, Louisiana shrimpers and environmentalists joined in support of a new law, signed by Governor Jindal, enforcing federal TED rules for the first time in state waters. In response, the Monterey Bay Aquarium's influential consumer guide, Seafood Watch, moved most Louisiana wild-caught shrimp off its red "avoid" list to a "good alternative." Sandy also sees a deeper change: "As our economy, ecosystem, community changes, these environmental groups are changing too, developing new ways of thinking. I have to be careful about which environmentalists are really for us here in the community. But the only way to restore the coast is to have everyone sit down and work it out together."

It's still not an easy conversation. An alliance of national and local environmental groups—the Restore the Mississippi Delta Campaign*— is among the strongest advocates of what frightens Sandy's fishermen most: the "diversions" that will periodically release massive flows of muddy Mississippi river water into the wetlands nearest their homes and docks. The shrimpers are most alarmed by the planned Mid-Barataria Diversion, which in October 2015 was advanced to the next stage of engineering and design. In high spring floods, the diversion will release as much as 75,000 cubic feet per second of the river into the marsh west of Myrtle Grove. That's as much water as flows on average down the Missouri River, the sixth largest in the United States.

More times than she can count, Sandy has driven visitors from government or NGOs down the river forty-five minutes south of New Orleans to show where that river water will land: in her families' back-

---

* Comprised of the Environmental Defense Fund, John Lopez's Lake Pontchartrain Foundation, the National Audubon Society, the National Wildlife Federation and the Coalition to Restore Coastal Louisiana, this local–national alliance is yet another example of people overcoming historically strained relations to marry local strengths (intimate knowledge of the ecosystem, multigenerational relationships with neighbors) with national reach (including the capacity to help shape federal law).

yards and critical shrimp nurseries. "That freshwater is our number one concern," she tells them. "Shrimp, fish, crab, oysters, all depend on saltwater. And spring: that's the incubation period, everything's being born. For the babies, that freshwater's going to kill them; for the ones that are already big, it's going to push them way offshore, where our smaller boats can't get to 'em. Do it once, you don't have a season. Keep doing it, they're going to die out. That's why everyone is so freaking out about the diversions." The nature of the Mississippi River adds to their fears. "The freshwater they're planning to pour into our spawning grounds is very different from back in Vietnam," says Sandy. "You're talking about cold water from Canada with who knows what's in it right into our backyard." Struggling already with the dead zone in the Gulf, where their shrimp nets come up empty, they worry that reintroducing nutrient-laden river water into the estuaries will create a dead zone there too.

All those apprehensions erupted with full force at the first ever meeting, in Buras in May 2015, of Sandy's fishermen with a representative from the Army Corps of Engineers. Or rather, the first three meetings, since Sandy always does everything three times, in three languages (English, Khmer and Vietnamese), helped by her trilingual staff member, Christina Duong.

Just getting the fishermen to come, and then to speak, required all of Sandy's persistence and charm. "They're scared to talk in public. This is all the fishermen, not just the Asians. They think, 'Who am I? No one's going to listen to me.' For years and years and years, they've been looked down upon. And so when you do stuff that hurts them, they're not going to say much. They're just going to deal with it." For the Vietnamese and Cambodian fishermen, survivors of their own governments' efforts to kill them, speaking up is harder still, says another CCC staff member. "They've been quiet for so long. And what the land has to offer us today, that's what we're thinking about.

If the land is here we get up and work; put food on our table; support our family. Tomorrow's not promised."

But Sandy never lets them hide. As the fishermen trickled in to the Lucky Ryan restaurant, she reminded them why this conversation had to happen. "When my daddy took me down here when I was eight years old, we used to play football next to these docks. Now that water is just right across the highway from you; you guys are protected by just that one levee. Couple years, that water'll be right on your house." She cajoled them: "Listen guys, I've worked very hard all my life to get you guys to be open and to be able to talk to government. If outsiders come from Washington to meet with you, do not be intimidated or fearful." Throughout the day, she was their cheerleader, their confidence builder: "You guys know more than anyone. I need you to talk from your stomach, your heart and your mind. OK?" The Corps representative, she promised, "will bring your voice back to the people who are making these decisions."

So the fishermen spoke, though not without first expressing how futile it seemed:

"We don't know what to say to make it really matter, you know?"

"A lot of us feel we don't have the power to do anything about the Master Plan; it's a take-it-or-leave-it deal. I want to speak to somebody who's able to overturn the decision and believes the fishermen, from what they've seen, life experience."

"We know what it's about; they want to protect New Orleans. You got to have something to slow the water down . . . but what about us?"

The conversation soon began to circle their prime preoccupation. "Mr. Sen thinks diversions will impact us greater than what the spill did, and it's scary," said Sandy, translating. "What good does it do if you're saving the coast and killing the community?" asked another fisherman. "You going to make thirty or forty acres of land over so many years and ruin thousands of families? You get a hurricane, it's going to be eating up them little acres, but where we going to live?"

A few talked about how they'd seen the Army Corps try releasing the river before, and it only made matters worse. "Right there by Pilottown was one of my honey holes," says one, referring to a place near Head of Passes, another ten miles below Venice, where the river divides into the three claws of the bird's foot and finally dumps into the Gulf. In 2002, "the Corps cut a hole in the riverbank and let freshwater go where it's not supposed to be going. Now, when river water rushing out pushes against the [incoming] tide, it creates a hole, a dropoff. They made the bottom so dirty we can't fish. They said the land going to grow up and I haven't seen any land or marsh come out of the water yet. I think they lost the land more." They tell her no one's listening, no one cares. "We want to be heard because this is the last line for us."

Even as they went round this circle of fear, however, the fishermen kept underlining how little they actually know about the Master Plan, and how badly they want to understand. "They don't know anything about the restoration process," Sandy told the social scientist dispatched here by the Army Corps. A year ago, "many didn't even know what a coastal Master Plan is. Two years ago, I didn't know. All the different parts; it's overwhelming. How much water? How long? We don't understand the logistics. It's the lack of communication by government with the people they may impact. And if I meet with the state and scientists, with their terminology—if I don't understand, how will they? If there's a public comment period, they wouldn't know what to comment, because they don't know where the state is heading with these projects. Right now they're having a hard time even speaking to you. If the government would do a better job at communicating and educating them about it, there wouldn't be a lot of, you know, whooping and hollering like there is now."

The Army Corps representative's response was, at least initially, stunningly unhelpful. Asked again and again to explain the diversions and how they fit in the bigger picture, she finally sighed, "I wish I could draw." It was a baffling comment given that this Master Plan has been drawn and modeled countless times, and all of those renderings are readily available, especially to someone from the Army Corps. Pictures would have also gone a long way toward bridging the language barrier: a transcript of the meeting revealed that a great deal had been lost in translation. And despite all she'd heard from the fishermen about how they didn't know enough to have informed opinions, she kept pressing them to voice a view, using jargon almost no one would understand. "How do you feel about marsh creation, saltwater intrusion, shoreline stabilization?" Eager to be cooperative but provided no new information, they retraced the same circles, just reinforcing one another's misunderstandings and fears.

Sandy gently tried to redirect the conversation. She made the case, somewhat heartbreakingly, that her community deserves consideration because of their economic value to the state. "When you guys couldn't shrimp for eight months, the welders went down; Anna's grocery went down; the private schools went down. We provide 25 percent of the country's shrimp. If something happens to us, it will impact the region. If we don't have seafood, what happens to tourism? Their businesses and harvested seafood products are vital to the state's economy."

She turned to an American fisherman: "Earl, why is it that what scientists are saying is not matching up with what you guys are saying?" He answered, "Well they smart people or they wouldn't be where they's at. But we pretty smart too. And we been out there every day of our lives for the last forty years. We begging to take people out; ain't a problem getting a ride out there and seeing firsthand what's happening. 'Cause if you ain't stood there and watched, you don't really know."

Slowly, the Army Corps representative started to gain her footing.

She shared that she too lives in southern Louisiana and worries what the collapse of the land will mean for her culture. "Culture doesn't go somewhere else; you lose it." She reassured them that the Master Plan is not set in stone but is still being modified in response to community input; that she will help them understand how they can affect the Corps's process, and bring in someone who *can* explain the plan; that the kind of knowledge they've accumulated in their years on the water is increasingly valued by the Corps. (Still not fully able to resist jargon, she called it TEK, "traditional ecological knowledge.") And with benefit of their firsthand knowledge and the Corps's own advances in modeling, she said, they can trust that future projects will be far better than the often disastrous projects of the past.

By day's end, Sandy seemed satisfied that her visitor genuinely wanted to bring what "our level of people are saying" to decision-makers in the Army Corps. "I'm going to be right here with my community folks. But we're at this bottom level. There's no way our voices could be heard from up there."

While the Corps will provide much of the muscle, it is ultimately the state of Louisiana shaping the Master Plan, so Sandy has invested far more time in building those relationships, and with greater success.

In July 2015, she got a close look at one of the state's priority projects when Chuck Perrodin, spokesman for the Coastal Protection and Restoration Authority, took her out by boat to see dredges rebuilding the Grand Liard Ridge. The ridge completes a chain of barrier islands resurrected to provide first-line defense for the Buras shrimp docks and Barataria Bay. Six miles out in the Gulf, Sandy and Chuck boarded a huge red and white dredge boat, the *CR McCaskill*, to watch its giant, pinecone-shaped cutter head burrow 35-pound teeth into the compacted sediments on the sea floor, suck sand and mud into a pipe and pump it back to Grand Liard. "When we started out here," the captain told them, occasionally slipping into French, "we hit huge cypress logs, forty foot long." Chuck nodded: "This used to be the shoreline of Plaquemines Parish. You could walk all the way here, through cypress forest that in the early 1900s made Louisiana the country's largest timber producer."

Traveling to the other end of the 50,000-foot-long pipe, Sandy watched the dredged sediments pour onto the ridges rising out of the bay and asked why the state is working from the outside in. "There's a sequence that makes sense," Chuck answered. "If we restore the interior basins without this perimeter protection, the next storm would do more damage, and we'd have thrown that money away. But building the perimeter first, we get surge attenuation so when we build marsh it won't get torn up." She asked if this new land will survive another Katrina. "The more we build, the less will wash away," said Chuck. "And the marsh is rooted now in dead vegetation, so it's easy to rip up. With hard mineral sediment and sand, it'll have something firm to root in."

Sandy's fishermen have nothing but enthusiasm for these dredging projects, which they see as a surer and safer way than river diversions to build land, because they bring in sediment without the freshwater. A favorite example is the Lake Hermitage Project, ten miles below Myrtle Grove. "Basically, the pump starts from the Mississippi, runs underneath this highway and then into the marshes. It's one of the most amazing projects I've seen," says Sandy. "It builds crazy, crazy beautiful land. They talk about you have to plant, right, to keep everything together? Well, the sediment coming from the Mississippi River for this project is so rich that it just kind of automatically grew. Now we have flamingos flying out of there. You could pretty much play soccer on it."

Watching the Grand Liard Ridge take shape, she put to Chuck the question her community always returns to: "When my fishermen pass one of these projects, they ask, 'Why doesn't the state just buy a couple dredges to shoot sediment wherever we need it?'"

Chuck's answer laid out the tradeoffs these stories invariably crash into.

The Master Plan, he began, does in fact allocate $20 billion to dredging, five times the amount budgeted for sediment diversions. And it frontloads that money, to capitalize on dredging's biggest advantage: it builds land fast, so is highly effective for triage, holding together key barrier islands and marshes until the natural system can be restarted and the river itself can begin to recover its delta.

What dredging can't do is build anything near all the land Louisiana needs. "Dredging can build hundreds, maybe thousands of acres, but we need thousands of *square miles*," Chuck explained. "There's not enough pipe anywhere to fill in Barataria Basin or Breton Sound. We have to use dredging for short-term protection and diversions for long-term restoration. We have to do it all." Dredging is also prohibitively expensive, costing more than seven times as much per acre built as diversions. And it doesn't last: to keep up with natural subsidence it would have to be repeated every few decades, at the cost of many more billions. Dredging also does its own share of habitat damage: that churning pinecone is not ripping through a vacant wasteland but destroying the home of bottom-dwelling creatures that play a critical role in the marine food web. Those bottom-dwellers include young adult shrimp.

The river, by contrast, is a self-sustaining engine, able to keep replenishing the land as it subsides and to repair ongoing damage from canals, saltwater and hurricanes, without investment of another dime. It's also powerful beyond any earthly machine: one dredge would take 300 to 450 days to pump as much sediment as the Bonnet Carré Spillway can deliver in less than ten days of flood. The river has other strengths, as well. Pumping sediments through the mouth of a pipe is like drawing with a fine pen, well suited for restoring a line of barrier islands or strategic dots of land. But the river is like a crayon on its side, spreading to fill broad areas, and building not dry land but the far more valuable wetlands that provide habitat for fish and wildlife. Its suspended clays make up 80 percent of the total sediment load, and are better for sustaining land than the larger sand particles dredges find offshore or at the river's bottom. Letting the river carry sediments out of the navigation channel will also reduce the costs incurred to dredge those shipping lanes open, which currently amount to hundreds of millions of dollars each year.

Ever the quick study, Sandy took it all in. "I get the argument. Dredging won't stay; it won't sustain, so you need diversions." She also, from her conversations with Chuck and others, was beginning to feel reassured that there will be latitude in how and when the diversions are opened: to maximize their value for habitats and fish and minimize their disruptive impacts on people. "I've heard, 'Sandy, we can adapt, but does it have to be full on at first? Can they start out smaller to allow us time to adjust?'" Engaging Sandy's community directly in such decisions is in fact a priority for those steering the Master Plan into its next stages. A newly formed Community Focus Group—made up of tribal leaders, faith groups and NGOs, including Sandy's Coastal Communities Consulting and the Mary Queen of Vietnam Community Development Corporation—will have ongoing input into the 2017 Master Plan. An advisory working group, including biologists, sociologists and natural resource economists, has also begun tackling operational questions head-on, aiming to maximize economic and social as well as ecological benefits. The group is also looking at possible compensation for those most affected in the near term, akin to the payments made by the Army Corps of Engineers to farmers if their land is deliberately flooded. Its initial focus is the Mid-Barataria Diversion, ground zero for Sandy's concerns.

Sandy's second goal, beyond getting her fishermen heard, is to broaden their "skill sets" so they can compete for the new coastal restoration jobs—in the four winter months when shrimping is closed, or as a full-time alternative should shrimping really go bust. The fishermen themselves, at every opportunity, volunteer for any work required, for the income but also because, as one fisherman put it, "I'm all about conserving also. That's why I follow the rules, because I'm pretty sure there is a good reason to try to conserve for the future."

"We have the small boats, the large boats," says another fisherman. "Just let us be a part of the government's plan and explain to us if we don't understand about the diversions." The docks, too, Sandy says, could play a crucial role, as supply centers for the restoration work. "I don't like to see folks from up north come down and just take it as a job. I don't think anybody will do this work with more heart than my fishermen."

Sandy is getting help preparing her fishermen for the new restoration economy from her first Katrina lender and board member, Robin Barnes. Now executive vice president and COO at Greater New Orleans, Inc., the region's economic development authority, Robin created GNO's Emerging Environmental Industry and Workforce sector, "to turn environmental threats into economic opportunity and job creation." With Robin's support, Sandy launched a training program at Delgado Community College, to certify her captains for both restoration work and as first responders in future emergencies. "Our pilot group graduated with nine certifications, in safety, fire, hazard, oil spill." Sandy would like to see them put on retainer by BP and other major oil companies, to be immediately available in the event of another disaster. She also pounced during her tour of the Grand Liard dredging project when Chuck Perrodin described state plans to plant thousands of mulberry and live oak trees. "I hear thousands of jobs potential here, Chuck. You think our locals can be part of this?" Chuck's answer was yes, especially as local restoration expertise deepens. The dredge boat they visited, he pointed out, was built right nearby, in Houma, Louisiana.

Like the Master Plan itself, Sandy's work on the ground with communities on the front line of environmental disasters is gaining recognition as a model with value worldwide. In March 2015, she was invited to Italy to present at a conference on urban resilience sponsored by the Rockefeller Foundation. The organizers' premise was that Hurricane Katrina, and the many crises cities around the world had faced in the decade since, "brought on by severe weather [and] slower, more protracted environmental threats," had shown that resilience requires "the direct engagement of people . . . living and working in communities. Large-scale [and centrally managed] efforts . . . are only useful when informed and supported by local approaches."

Achieving that marriage of top-down and bottom-up insight and innovation ("doing this with people, not to people," as EDF's Steve Cochran, a Louisiana native, puts it) requires, as in the Montana Rockies and Kansas prairies, acknowledgment that the world looks different depending on where you're standing, and that every choice entails tradeoffs that must be honestly faced.

The Caernarvon Freshwater Diversion, a concrete gate fifteen miles downriver from New Orleans that periodically lets Mississippi River water flow into the degraded marsh of Breton Sound, exemplifies how those complexities play out in this most complex of landscapes.

For coastal scientist John Lopez, the Caernarvon is a first step in restoring the natural order: returning freshwater and saltwater, and the fisheries associated with each, back to where they would have been had the wetlands not been lacerated with canals and the Army Corps not fundamentally deformed the river's normal functions. "Saltwater intrusion had shifted the oyster beds way upriver," Lopez explains. "So in the early nineties, the Army Corps built this outlet to moderate salinity and move the oysters closer to their traditional zone." It had the intended effect, he says, replacing highly dispersed but not terribly productive oyster beds with reefs occupying a smaller area but producing more oysters overall.

Caernarvon also, says Lopez, did something nobody expected. The diversion had been flowing on and off for more than ten years when, in the aftermath of Katrina, Lopez flew over to survey the storm's damage. "I could just see the emergence of a little skeletal framework," he recalls. "And I said, 'Oh my God, there's a delta there.'" The river, it seemed, had been quietly silting in the Big Mar, one of many square lakes in southern Louisiana, relics of settlers' ill-fated attempts to levee and drain marsh into farmland. It had taken a decade for the river to build the mud up to the water's surface, emerging (coincidentally) just as Katrina hit. Now Lopez watched in wonder as the bay turned back into marsh, with first a meadow of submerged grasses, then cutgrass and arrowhead and finally black willows and now (with his team's help) a young cypress forest. Since 2005, the river has given birth to a hundred new acres here each year, a thousand acres in total, material-ized out of empty water in the classic triangle shape that gives deltas their name. "So this is the birth of a new delta," says Lopez. "It's amaz-ing to see an ecosystem emerge right before your eyes."

This landscape is anything but pristine. Giant power lines run overhead; airboats roar deafeningly through the channels; hulking concrete walls and bedraggled metal buildings loom all around. But life, down here, doesn't seem to care; it explodes wherever you give it half a chance. Alligator populations, for instance, have more than doubled in the Caernarvon. "If you shine a light out here in August," says the airboat pilot who's brought Lopez out on his latest tour, "it looks like Christmas, there's so many red eyes." But the prime bene-ficiaries of this emergent ecosystem, Lopez says, are the people who live nearby. Just a quarter mile from the new delta is the $15 billion perimeter flood wall built after Katrina around New Orleans, Jeffer-son and St. Bernard parishes. Because vegetation slows down storm surge, having this new land "immediately in front of that protection system will buffer it from the onrush of water and wind, increase its resilience and help protect the people and assets inside." These new wetlands also protect a local levee that was overtopped in 2012 by Hurricane Isaac, "reducing the impact on it from wave attack and surge. So communities like Braithwaite that were completely flooded by Isaac will have a greater chance of surviving."

For John Lopez, the Caernarvon has begun setting Louisiana human and natural history back on their rightful course. For Byron Encalade's oystering community in Pointe à la Hache, on the east bank of the river some fifty miles south of New Orleans, it has done the opposite: putting an end to traditions and livelihoods that sustained their families for a century. It's not that Byron is opposed to bringing the river into the degrading wetlands. "It was the oyster industry that first asked for these diversions," he says. "We can't grow oysters in strictly saltwater, and there's no doubt the salinity line was moving further and further in and needed freshwater to push it back." But these communities have lived for generations in the world the Corps made; they may want the bayous and villages their grandfathers knew, but no one alive would recognize the world as it was before the leveeing of the river set the downward spiral in motion. And the Caernarvon, says Byron, overcorrected. "In those early years they operated it contrary to plan, running it wide open" and decimating the oysters which, unlike shrimp, can't swim away. "It was a disaster. My family had 2,000 acres of oyster beds. I was shipping two trailer-truck loads a day. Now I have 200 acres left. And they didn't give us time. We were starting to move our oyster beds further downriver. But someone decided to open it full on and we got destroyed."

Both views of the Caernarvon are valid and true. As one advocate puts it, "It's a Middle Eastern kind of dilemma: what year do you roll the clock back to?" That's increasingly being acknowledged by the most ardent champions of the diversions, who are recognizing that the time scales they think in don't always align with families and communities worried about tomorrow's harvest and tomorrow's bills. In a May 2015

"Open Letter to the Citizens of Louisiana," twenty-seven coastal scientists laid out both a defense of the diversions and a peace offering.

They began by reminding their neighbors that the current sweet spots for salt-loving shrimp, oysters and trout used to be cattle ranches and freshwater hunting grounds for ducks and alligators; when the Army Corps chained the river and so changed the course of Louisiana history, it had ruined *those* multigenerational livelihoods. And even with the Master Plan, they wrote, "saltwater will continue to replace freshwater throughout most of coastal Louisiana . . . [and] increase the fisheries associated with saltwater wetlands."

However, they conceded, "wetland-building diversions will immediately push [the saltwater fisheries] farther from some parts of our coast [and] economically harm some commercial fishermen." Recognizing that "the environmental cost of diversions will be borne by a few Louisianans in this generation whereas the environmental and economic benefits will be spread throughout coastal Louisiana for generations to come," they ended with a call for policy-makers and others to help commercial fishermen adapt to the coming changes.

Some of that easing into this new world will come from within the ever-evolving Master Plan: finding ways, as Merritt Lane says, "to engineer out most of the loss," avoiding extreme tradeoffs "so that rather than polarize you can find ways to optimize." Louisiana State University ecologist Robert Twilley has ideas for how to achieve that optimization, which he likens to advances in no-till agriculture like those Justin Knopf is helping lead on the Great Plains.

Like Justin's forebears, Louisiana's coastal fishermen have been living, Twilley says, on borrowed time: enjoying abundance but at the cost of thousands of years of accumulated fertility. Just as breaking the sod released an immense flush of productivity but in three generations burned through organic material accumulated over millennia, so the breaking up of the wetlands, Twilley says, has overstimulated productivity within the remnant wetlands: the opened edges of the marsh become accessible to the tiny creatures at the base of the food chain, who devour it, a fever of life swarming over a corpse. And just as prairie farmers sought greater predictability by clean-tilling and then chemically managing their fields, "the fishing industry sought to turn estuaries into managed ecosystems, with fixed isohalines" (bands of salinity). But in an environment that is defined and sustained by fluidity and dynamism, that effort to hold things still proved disastrous.

If the problems have essential commonalities, Twilley says, so do the solutions. As in Kansas, the key to sustaining Louisiana lies in emulating the native ecosystem: in this case the natural stages of the river, including muddy springtime floods. That's more difficult here, not only because communities have rooted themselves in the Army Corps's unnatural world but also because in Kansas the ecosystem itself is stable. Here, it is disturbance that maintains long-term equilibrium, a seemingly hard quality to square with the human desire for stability.

But not impossible, says Twilley. Because, in fact, all of the creatures in the estuary depend on the dynamism of this environment to thrive, not only tolerating seasonal salinity fluctuations but requiring them. Newborn shrimp larvae actually prefer freshwater; the introduction of river water into estuaries elsewhere in the country has not driven shrimp out. And oysters need a bath in the river's freshwater to rid themselves of a fungus and other pathogens that live only in high salinities; as long as that freshening lasts less than sixty days, they're fine. "We need to let the river control the system," says Twilley. "When it's at 800,000 cubic feet per second and muddy: step back, let it rip, deliver all the sediment it can—but for no more than sixty days; then let the estuaries return to higher salinities. As soon as the river goes to low flow, the salt comes right back because of tidal exchange."

Twilley's faith that big river flows are a boon to shellfish derive from his years of research in Four League and Atchafalaya bays, where a third of the river's volume has flowed for decades and oysters and shrimp both grow in abundance—tolerating not only the freshening, but also the nutrients delivered by the Mississippi. Far from being turned to dead zones, these estuaries use those nutrients like Miracle-Gro, becoming lush and beautiful even as they clean the water flowing into the Gulf. A study in the *Journal of Coastal Research* estimates that new wetlands growth will remove up to a quarter of total annual nutrient pollution in the Mississippi and Atchafalaya rivers. You can actually see that filtering: down near the Mississippi's mouth, south of the levees where the river flows free, the water comes into the marsh as mud soup and emerges just a few feet away so crystalline you can see the bottom.

Even with diversions operated to mimic Mother Nature, the Mississippi will remain a highly controlled system. As long as humans can stop it, this monster of a river will not be allowed to abandon its current delta, or to flood whenever or wherever it gets the urge. But as Lance Nacio says, "the healthier the ecosystem, the healthier the fisheries and the more secure our livelihoods."

Sandy's community takes immense pride in its extraordinary resilience. John Lopez recalls a talk he was invited to give in which he warned that the fisheries will change with or without the Master Plan—from collapse or from rebuilding. "They asked me, 'How soon?' I said, 'Five to ten years.' They said, 'That's enough time to adapt.'" The shrimp too, with their short life cycle, mobility, and incredible fecundity (Tidwell describes the November waters off Louisiana's coast "as a veritable soup of trillions and trillions of microscopic shrimp larvae") rebound from almost anything. Some species may have already benefited from freshened water: Lance says that the Davis Pond Diversion, built in 2002 above New Orleans, reduced brown shrimp numbers but increased populations of more valuable white shrimp.

But even if the diversions are managed, as Twilley suggests, for biology as much as geology—operated not primarily to build acres but for the benefit of fish, alligators, ducks, and the people who hunt them—communities like Sandy's will face significant challenges. So she and her board are advocating for various kinds of transitional help, though as always with the intention of paying everything back and preserving their self-sufficiency.

If the shrimp do move out into water deep enough that "our skimmer gear is not going to be compatible," for instance, Sandy would like to see low-interest loans or the kind of gear exchange successfully used in other fisheries. The added diesel costs of longer trips will shrink margins already severely squeezed by subsidized Asian imports, which have gutted prices—small domestic shrimp fell from $3.75 a pound in 2014 to 85 cents in 2015—and reduced U.S. Gulf shrimping by about 35 percent. Those imports are also ruining the reputation of America's favorite seafood species, says Sandy, including by false labeling: "Those small pink things in fried rice: that's not real shrimp. It has a rubber-band taste, from the chemicals they use to keep it from spoiling." And recent investigations have shown the great human and ecological damage done by some Asian shrimp farms, including destruction of highly productive and protective mangroves, water pollution from wastes and antibiotics, and labor conditions sometimes bordering on slavery. The greatest help they could get to weather the Master Plan transition, say many shrimpers, is simply better enforcement of trade agreements.

Finally, on behalf of both shrimpers and oystermen, Byron has been working to secure policy guarantees around the Master Plan. "I do believe that if they build and operate the Master Plan diversions exactly as they say, it will work. But do you think anyone on the bayous is going to believe that?" They've seen too often, he says, how "a few Louisiana senators dole out favors to special interests that contribute to their campaigns. Don't just give me some committee we don't even have a vote on that can control them, decide to run them to their benefit. If the Master Plan's so good, put it in law. Then, fine, I can live with you." The operations working group looking at the Mid-Barataria Diversion is also tackling these larger governance issues, acknowledging the "many examples from our existing freshwater diversions: that without clearly defined parameters and monitoring, their operation can be held hostage to political whims and specific user group concerns, whether or not those are founded in science. These issues have to be addressed head-on."

Byron and Sandy have also been working for legislation that would enable oystermen to move their leases as coastal restoration rewrites the salinity map, and create greater flexibility for fishermen to shift to the species moving into their backyards. As Jerome Zeringue, who succeeded Graves as head of the state Coastal Protection and Restoration Authority, noted: "speckled trout, reds, oysters and blue crabs might move further south . . . but the areas they might leave won't be without fish, it will just [be] different fish. Instead of specks and reds you might have bass, catfish and sac-a-lait."

Underlying all this effort is an abiding faith in America, an unalloyed patriotism. Sandy's community trusted America to be their ally in Vietnam and in flight from the Communists, to give them safe harbor when they arrived with utterly empty hands, to reward their unflagging hard work and entrepreneurial energy with the realization of the American dream. One of the most beautiful pictures on Sandy's Facebook page shows her standing with a woman holding yellow flowers, a small American flag and proof of her new citizenship, her face lit with an ecstatic smile. "To help somebody who's been here twenty years get U.S. citizenship," says Sandy, "and see them stand right there and cry in front of the office. That's the bomb." Even in the parade of meetings about the Master Plan, these fishermen grow frustrated but never cynical. With Sandy, they know that a refugee who can't speak English and works with his hands can still contribute meaningfully to fixing this vital, beautiful, bountiful piece of the nation. "If you seriously talk with people," says Sandy, "both to those who want to help and those who need the help, we're all humans; there's got to be ways to figure it out. Even when there's so much money involved, our country is all about the humanity."

5

FISHERMAN

ON APRIL 1, 2001, Deborah Werner was at home in Galliano, Louisiana, with her ten-year-old son Andy, getting ready to go watch his thirteen-year-old brother Nick play basketball. Her husband Wayne wouldn't make the game: a commercial fisherman, he was out in the Gulf of Mexico chasing red snapper. But as Deb and Andy headed for the door, the phone rang. It was Donnie Waters, a rival fisherman and one of Wayne's closest friends. "Hey, have you heard from Wayne?" he asked, momentarily confusing Deb, since Wayne could rarely communicate when out at sea. "No, why?" she asked. "Because I heard him on the radio send out an SOS," Donnie said. Deb laughed, used to Donnie's high-stakes jokes. "Yeah right, Donnie. I know what day it is." But then the phone rang again. This time it was the Coast Guard. They had found Wayne's boat sinking ninety miles out from shore, but no sign of the captain or crew.

Wayne had taken out three young deckhands, and it fell to Deborah to call their wives. "There were two kinds of deckhands' wives: the ones that wanted them to leave and the ones that didn't," she recalls. "We were all the second kind; every storm that came up you worried. So I didn't know what to say. I had to tell them they were missing. But you don't want to say they're lost at sea. And I really trusted Wayne on the water."

Wayne had been fishing the deep waters of the Gulf for twenty years. He'd been on boats since the age of ten, riding along sometimes when his dad, deployed with the Air Force to Guam, took military bigwigs on a break from Vietnam out fishing for blue marlin. By age thirteen, Wayne was helping his dad—by then retired, though still "mean," Wayne says—run a charter boat out of the Florida Keys, hitching a ride to the docks every day after school to bait hooks and clean private boats for a little cash. ("They were half-million-dollar boats," he recalls. "You always had your buffing chamois.") Every summer, he and his dad would take their 52-foot charter boat, the *Hawk*, to the Bahamas for a month or more, hiring out to fishermen hungering after big game: on one trip they caught a 682-pound marlin. Wayne had even, at age fourteen, gone missing on the water once before, surviving two nights out with next to nothing to eat or drink. "I'd bought a little boat and one day me and another kid went out for the afternoon. We broke down by Twin Keys, and had to spend the night out there in just T-shirt and shorts. The next day we push-poled partway back, but hit deepwater channels and had to stay out another night." Caught between a harsh wind hammering one side of the boat and swarms of mosquitoes on the other, "it was horrible," Wayne says. Finally, that third day, "here come a skiff and pulled us back." He rushed to find his dad. "I said, 'Look, I'm all right.'" Though more than forty years have passed, long enough for Wayne's shaggy black hair and thick moustache to become streaked with silver, the memory still evokes a pained sigh. "And he said, 'What? Were you missing?'" He repeats that last bit, in his low growly drawl. "I'm serious. Quote, unquote: 'What? Were you missing?'"

For all Deb's confidence in Wayne, however, she couldn't quell her terrible forebodings. She knew Wayne was distracted by grief, having lost his dad just six weeks earlier and his mom a year before that. "And you can't ever account for what the storms are like out there," says Nick, whose memories of that day remain vivid fourteen years later. His mom showed up at the game, he says, as if all was well, but at half-time told him they had to leave right away, "because something bad had happened." He remembers her calling the Coast Guard every five minutes, frantic for any word. "Mom was not calm. If you've got a loved one in trouble and there's a brick wall in the way and you don't know what's going on behind it, and you love them, you're going to be freaked out."

While Deborah was struggling not to let her young sons see the depth of her fear, Wayne was adrift in the middle of the stormy Gulf with his three young crewmen. They'd been anchored and asleep when at 2 a.m. they woke to 20-knot winds, nine-foot swells and the 42-foot boat swamped and sinking. Scrambling into a life raft, abandoning 1,400 pounds of fish, Wayne told one of the deckhands to grab the emergency positioning device, but in his panic the kid left it behind. So they drifted off into the vast emptiness of the Gulf with nothing but smoke canisters to signal for help, the seas churning and the crewmen wretchedly seasick ("which I made worse by smoking," Wayne says). Though they had emergency water, Wayne kept them from drinking as long as possible. "My dad taught sea survival for the Air Force and taught me basic stuff, like if you drink it just makes you want more."

The floundering boat was spotted after dawn by an oil company helicopter, which notified the Coast Guard. But it wasn't until late that afternoon that aircraft finally saw the little raft's smoke signal and sent an oil rig supply boat to rescue them. The $100,000 (uninsured) boat, aptly named *Wayne's Pain*, was lost.

The high seas and heavy winds had not come as a surprise. Wayne had ignored the nasty weather forecast, against all better judgment, because he had no other choice. By 2001, federal fishery managers had cut the commercial fishing season to so few days (creating a "derby"—a race for fish) that a man who needed to support his family had to go out no matter the dangers. "I wouldn't have been there given the forecast and all. I wouldn't have been out except the derby was on."

Those constraints didn't exist in 1982, when Wayne first began commercial fishing. Until 1990, Wayne could bring in as many red snapper as he could catch, 365 days a year. He and Deb, who met at Coral Shores High on Plantation Key, have blissful memories of the freedom of those years. Wayne figures he had spent 800 hours on a boat by their senior year. "You could miss forty-five days of school; I missed forty-five days." He caught everything under the sun: wahoo, tarpon, even a 45-pound spearfish—"the first fish I ever put in a boat by myself." Both have memories of swimming with sharks: Wayne and friends jumping off bridges into waters thick with them; Deb tying back her long curly auburn hair to dive headlong off the bows of boats while friends chummed off the back for tiger sharks and hammerheads. Sea turtles weren't yet a rare sight. Deb would visit the seafood cannery across the street from her house to watch green turtles in the corral (pronounced "crawl") come up for air. When they were fat enough they were canned as turtle soup. Wayne hunted them in the wild. "Cause we ate 'em all the time," he says, now wincing at the thought of the unwitting damage they did. "It was a way of life for us, every week turtle steak, black beans and rice. You pegged 'em, with like a harpoon." That tradition ended in the 1970s, when all five species of Gulf sea turtles were listed as endangered.

An all-American wrestler, Wayne turned down scholarships to the universities of Miami, Iowa and Nebraska because all he ever wanted to do was fish. "If I wasn't on a boat, I don't know what I'd do." Deb, who had been editor of the yearbook and active in the French club, art club, prom club and National Honor Society, as well as a contestant in the Miss Coral Shores pageant, abandoned her own college plans to be with Wayne. She made flashcards to prep him for his captain's test and went to Fort Lauderdale with him on the day of the exam. He passed, and at age eighteen, fresh out of high school, took the helm of a Florida-based boat belonging to Andy Anderson, owner of a Colorado coal mine. In 1979, Anderson persuaded the couple, now married, to drive Wayne's Gran Torino Sport out west to try life in the mountains. They were glad for a break from the Keys: "It was not the place to be in the late seventies," says Wayne. "Too many drugs were being smuggled through." In Colorado, they lived in the only occupied house in a town abandoned after a 1942 mine explosion killed thirty-four miners ("I kept expecting to see ghosts," says Wayne). Trains rattled the house all night carrying crushed coal from the tipple to the Coors brewery; it was lonely and cold and in 1981 they went home. Wayne spent the next two years building boats and in 1982 he and his dad bought *Wayne's Pain* and began fishing their way from North Carolina around the southeast coast in search of abundant snapper. They found the fish they were after off the coast of Louisiana—the rich bounty of its estuaries. For the next five years, while Deb waited at home in Florida, Wayne and his dad went off for weeks or months at a stretch to fish out of Griffin's Dock at Leeville, bringing in all the red snapper and other reef fish they could catch.

But even in Louisiana, it was getting harder each year to catch fish. In 1988, when the National Marine Fisheries Service made its first formal stock assessment, they discovered why: the Gulf's iconic fish—coveted for its lean, moist, firm texture, sweet nutty taste and the beauty of its rose and silver diamond-patterned scales, spiny dorsal fin and the short, sharp needle teeth from which it gets its name—had been so thoroughly overfished that its capacity to recover was nearly destroyed. Snapper are a long-lived species, able to grow to more than fifty pounds over fifty years. As in many such slow-growing fish, a female's peak reproductive years come very late. Though a red snapper may begin spawning as early as age two, a ten-year-old will produce thirty-three times as many eggs as a three-year-old. And only at about age fifteen—when she is called a sow or a "big old fat fertile female" (BOFFF)—does she achieve her full potential, which she then sustains for twenty-five years or more. Overfishing in the 1970s and 1980s wiped out most of those sows, reducing the "spawning potential ratio"—a measure of the stock's ability to reproduce itself—from 45 percent in 1950 down to 2.6 percent in 1990.

Federal management of the nation's fisheries began in earnest only in 1976, when Congress passed the Magnuson Fishery Conservation and Management Act, extending U.S. authority over the Atlantic, Pacific and Gulf of Mexico from its historic boundary, just twelve miles off-shore, out to two hundred nautical miles. The Magnuson Act created a management system that is both byzantine (layers of bureaucracy, mountains of acronyms) and uncommonly open to public input and influence. It divided the newly expanded federal fisheries into eight regions, each overseen by a Fishery Management Council account-able to the National Marine Fisheries Service (NMFS), which is part of the National Oceanic and Atmospheric Administration (NOAA), which is part of the U.S. Department of Commerce. While states have jurisdiction over near-shore waters, the Gulf of Mexico Fishery Management Council oversees federal waters off the coasts of Texas, Louisiana, Mississippi, Alabama and western Florida.

With passage of the Magnuson Act, the Gulf Council had to respond to the dire findings in the 1988 assessment: most of the red snapper gone, the stock sapped of its ability to reproduce. Though the Council stopped short of following scientists' recommendations that they cut red snapper death rates* by 75 percent, they did put an abrupt end to the fishing free-for-all Wayne had enjoyed all his life, replacing it with rules so crushing he feared he might lose the life he loves. In 1990, they set, for the first time, a "total allowable catch" (TAC): the

---

* Counting both landed fish and those thrown back that die.

pounds of fish they calculated could be harvested and still leave enough in the water to begin rebuilding depleted stocks. They then divided that TAC—51 percent to commercial fishermen, 49 percent to recreational anglers. When the commercial fishermen blew past their limit in August, the Council closed them down, the first time they had ever done so. The next year, fishermen were ready. When the season opened they raced so hard to catch the last fish before somebody else did that they hit their limit by February 22. Again, the fishery was shut down, this time staying closed until February 16, 1993. The government, in other words, shuttered the mainstay of Wayne's small business for a year.

Over the next fifteen years—as the Council kept trying to hit sustainable catch targets—Wayne and friends found themselves spinning in what one marine scientist calls a "blizzard of management," struggling to adapt to ever more stringent rules and still support their families. In the years after the shutdown, the fishery managers tried imposing first a "mini" and then a "micro" derby—eventually allowing commercial snapper fishing just the first ten days of each month. They also set a "trip limit" for each vessel of just 2,000 pounds. Wayne's boat could hold five times that, but every time he caught 2,000 pounds he had to burn precious hours and expensive diesel running back in to the dock to unload. Or he had to stay closer to shore, where the fish are scarcer and smaller. Any fish under the legal size limit—13, 14 or 15 inches, depending on the year—he had to throw back.

Though meant to rebuild the fishery, the derby approach to management did far more harm than good. Because regulators were trying to hit catch targets not by directly capping what was landed but by constraining "effort"—tying the fishermen's hands and hoping they'd bring in the right amount of fish—in more than half the seventeen derby years, commercial fishermen went over their allotted catch. Fishermen who could afford it got around short seasons and per-vessel harvest limits by buying multiple boats, increasing the pressures on the fish those limits were meant to relieve. Fishermen were also forced to throw back more than a million fish each year, because they were undersized or the boat had reached its daily limit. Some would "high-grade," keeping only ideal-size snapper and throwing back everything else. Fishery managers guessed that about four of every five fish caught were thrown back, but assumed most survived and set the harvest accordingly. The fishermen (and soon scientists) saw a grimmer, truer picture: two-thirds or more of those released fish died.

And the useless killing of snapper didn't end when the season closed. The commercial fleet went on fishing for other species still in season, sometimes moving into deeper water (300 feet or more) in search of tilefish and deepwater groupers. But they still caught lots of snapper, now *all* of which they had to throw back. And even more of those died, Wayne knew, given snapper physiology. To stay where they're happiest, near the sea floor, snapper have to control their buoyancy, which they do with an internal air bladder that works like a diver's vest. If pulled up fast from depths beyond fifty or sixty feet, the sudden reduction in pressure causes that internal balloon to overexpand and kill them, even if they've survived the trauma of being hooked. If those exploded snappers are big fat fertile females, the loss is more tragic still, says Wayne, because "those are egg values you're talking about to repopulate the fishery." All in all, "the waste of fish was incredible, two pounds killed for every one on the dock, just floating away behind our boats. When we finally came out with self-reporting," tallying discards to get a more realistic picture on which to base management, "we'd killed two million pounds a year while the season was closed. We were cutting our own throats."

The excessive landings, the vast waste: still that wasn't the full extent of the carnage. A black market in snapper was also flourishing, with as many fish being traded illegally, entirely under the Council's radar and off their books, as were legally moving through docks and fish houses. Red snapper, in short, were still in a death spiral; the Council was failing dismally in its legal duty to rebuild the fishery.

The vanishing fish and crazy rules combined to devastate fishing families and communities, not only forcing crews out into gales and heavy seas but also making it nearly impossible for a fisherman to reliably support his family. "The derby started when Nick and Andy were babies," says Deb. "That was your paycheck. If the weather got bad or you had any breakdown and couldn't fish those ten days, it was hand-to-mouth. There were times I didn't know if I was going to be able to buy groceries. I'd drag the kids in their little OshKosh B'Gosh overall matching sets to Council meetings in Mobile to say, 'You need to understand what this derby is doing to our poor family.' We were always scraping by and didn't see the light at the end of the tunnel and didn't have any reserve." From the kitchen, where he is boiling up a lunch of fat Louisiana shrimp in his own mix of spices, Wayne jumps in. "It was pretty simple; if you called my dad and said, 'I need to borrow some money,' he said, 'No.'"

Things were scarcely better when the season opened. With too many boats and all the fish coming in at once, glutted docks would pay rock-bottom prices, or take fish from Wayne only on consignment. If nobody bought the fish they were junked and Wayne got nothing. In a December 1999 field hearing before senators John Breaux and Olympia Snowe, Wayne described the market distortions created by regulators' decision to set sequential seasons—first for snapper, then grouper, then tilefish. "We target one fish at a time, one fish at a time, all the way throughout the year. We are not just driving down our prices on the one fish; we are driving down the prices on all of our fish all year long." The rules, he told them, trapped fishermen in awful dilemmas. "I may need a couple hundred pounds of snapper to finish my trip limit. Do I ride an hour or two farther from the dock" into waters deep enough to catch fish consistently larger than the minimum size limit? "Or do I kill 200 to 300 pounds of undersized snapper and stay competitive with the fleet?" The pressure to get in before the market price crashed, he says, grew so intense "that the season would open at 12, and at 12:45 guys were at the dock unloading their fish. [The Louisiana Department of] Wildlife and Fisheries would pull up and go, 'You're telling me you caught your snappers, gutted your snappers, iced 'em all down and brought 'em in here in forty-five minutes?' And they'd look at him and go, 'Yes sir.' Everybody knew they could only catch one of us."

Relations among fishermen grew cagy and tense. "It created an atmosphere of just a pure race," Wayne says. "'I want to catch more than the guy next to me.' So there was a lot of secrecy about where we were fishing, who you sold to, at what price. Any holiday, you'd have to watch because someone might sneak off and go fishing on you. In bad weather everyone'd be at the dock. If one guy left, it was almost like he had a rope tied to everyone else. We weren't going to let him go fishing without going fishing too." Wayne was as secretive as the rest, even with fellow Leeville fisherman and close friend Russell Underwood. "After [Hurricane] Rita, I found out the floodgates were opening so we could get back out. I didn't want Russell to know, so my boat could get there first. We were cleaning up the wreckage and rebuilding the dock and fish house. But I told Marshall [Wayne's nephew and fishing partner], 'Just slip out of here and go.' Two hours later, Russell's brother Ricky was squeegeeing down this floor, looked up and said, 'Where's Marshall?' I figured enough time had passed, I said, 'He's probably coming through the floodgates. You know they're open and all.' He dropped his squeegee, thwump, didn't say a word, just walked out the door."

Crews showed the strain. Every twenty-four hours they might run eight hours out into the Gulf, fish madly for an hour to catch their trip limit, run eight hours back to the dock, unload for an hour, then head back offshore again. "Crewmen have fallen overboard, asleep at the rail," Wayne told senators Breaux and Snowe. "I have had crewmen break down—they just couldn't go on—worn out."

For families, it was even worse. When Deb got pregnant in 1987, Wayne moved her down to Galliano, fifteen miles north of the Leeville dock. "I told her, 'You got to be closer to me. I don't want to hear that you're in labor and have to run twelve hours back to the dock and twelve hours driving to Florida.'" He made it to his sons' births but never to their birthdays, which were "on the first and fourth of the month," says Deb, "right at the beginning of the derby." And with most of the dads offshore all at once, on fishing boats or oil rigs, Nick remembers the bayou as a rough place for a kid, especially one like him, big as his dad but also bookish. Most of his friends had shotguns by the time they were eight or nine, and Nick would join them in the swamp to dodge poisonous snakes and blast at rabbits and ducks and occasionally one another: "I've eaten my share of birdshot-filled duck soup." It fell to Nick to defend himself and his little brother from bullies, like the school bus driver's nephew Odie, who every day as the kids got off the bus would swing his full pack at Nick's head. "I finally asked," recalls Wayne, "'Could you outrun him?' Nick goes, 'Yeah.' I said, 'Well then, hit him in the nose and run for the house.'" It worked, and grown-up Odie now works with Wayne. "I think between first and seventh grade I got in more than fifty fistfights," says Nick. Deb could never embrace the disciplinary strategies favored by some other families. She left Andy with a neighbor at age three and came home to find he'd been made to kneel on rice for forty-five minutes, finally falling asleep on his knees.

Most painful of all for Wayne was how much he missed of his parents' final days, and of time to grieve their deaths with his family. As his mother lay dying, Wayne was trapped out in a derby. He got home in time to say goodbye, but "as soon as she was buried, my dad said, 'You go fishing.' I was on my boat fourteen hours after her funeral, because I only had that weekend to catch my fish. Most seasons I could tell you where I went when and what I fished. By the end of that one I couldn't even tell you what I'd done. I just went through the motions, put myself on autopilot. Offshore's not a place to be dealing with that stuff; it's kind of like being in jail—confined, you know."

The death of Wayne's mom, Nick recalls, derailed his grandfather. "Without my Grandma Ruthie, Grampa Sarge kind of lost his sense of what to do with his life and time." He'd never softened. "He wasn't a guy to use a kind word," says Nick. "You wouldn't hear he was proud more than he was disappointed. If he was going to play a game with you, he would play ruthlessly. We spent hours playing checkers, and then chess, and he would beat you again and again until you learned how to beat him." But Wayne remained deeply devoted, driving to Florida and back between derbies to fill his dad's hours. "A year later, I was out in the derby and knew my dad wasn't doing well. I rushed here the day after it closed, and two days later he died." Others had it still worse. "A friend of mine's son died, he was done for six months. Almost lost his boat, his house, everything, 'cause you had to go. You had to choose between dealing with something like that or feeding your family."

Had life not been so harsh when his boys were growing up, says Wayne, "I might would have made them fishermen. But with that tough life and bleak outlook, you'd never want your kid out there. Russell's kids aren't fishermen. Donnie's kids aren't fishermen." Nick was barely allowed on the boat, though he remembers one trip spent at the rail after too many sodas and Little Debbie snack cakes. ("Guys up for twenty hours at a time," he says, "like to keep their energy high.") And another, at ten, watching in awe as a deckhand pulled in a tuna, fileted it on the spot and ate it raw, a scene that came back to him in high school when he read Hemingway's *The Old Man and the Sea*. "My dad broke his back every day and didn't want us to have the same struggle. And it was so risky. A bad run of weather, bad judgment call, bad crew: a guy's operation goes under. He wanted my brother and me to go to college."

It was in part to set his kids on a different path that, after sinking his boat, Wayne moved the family back to Florida, into the house north of Gainesville where his parents had lived, near good schools and far from hurricanes. Deb missed the coast terribly and decorated as if still in her beloved Keys, with Wayne's prized catch stuffed and hung on the walls and a blooming garden that invited scurrying little Cuban anoles, mud daubers, snakes and wolf spiders "with egg sacs the size of golf balls that would explode with babies." But for the next six years, "I was almost a single mom," always at this or that school event or working fundraisers to defray Nick and Andy's band and sports fees. With Wayne commuting again to a new boat in Louisiana, he seemed a figment: "Other moms would ask the kids if their dad was in prison."

As it became ever harder to survive the derby, Wayne decided he had to put up a fight. He began showing up at Council meetings, gradually summoning the confidence to stand and speak. At first, his mission was obstruction: to block rule changes he thought would compound the disaster. The most important of these came in 1993, when he got wind of a proposal to introduce an "individual fishing quota" (IFQ) for red snapper. Under an IFQ, sometimes called a "catch share" or more broadly "rights-based management," managers jettison seasons and trip limits. Instead, they figure the total allowable catch and then divvy that up among the fishermen, letting each one catch his share of fish whenever he likes.* Catch shares were by then in use in Holland, Iceland, and New Zealand, and in the works for Alaska halibut, but this was the first one proposed in the lower forty-eight for a fish highly coveted by both commercial and recreational fishermen.

Wayne was still making up his mind about this IFQ idea when

---

* Or nearly so; some catch shares include spawning season closures or closed areas.

one morning, at a Council meeting, he and Donnie were joined at breakfast by a marine scientist who'd served on the Gulf Council and also advised the Coastal Conservation Association (CCA), a powerful lobbying group representing recreational anglers.* As Wayne recalls it, he asked them a frightening question: 'Well, if we get this IFQ, what's to stop us from buying a million pounds of your quota?' We realized they had a bunch of money and were just going to buy up our fish to use in recreational fisheries, get rid of commercial snapper fishing like they'd done for redfish and speckled trout.† I spit my eggs back in my plate. I couldn't swallow 'em, I was so upset."

That breakfast impelled Wayne and Donnie, in May 1995, to show up at their first ever Senate subcommittee field hearing, in New Orleans. Congress was by then debating reauthorization of the Magnuson Act, now called Magnuson–Stevens and co-sponsored by Alaska Senator Ted Stevens. The debate around catch shares was also heating up in fisheries around the country, centered on exactly the fears then plaguing Wayne: that, as Donnie would testify in the hearing, the IFQs "were designed to put the commercial fishermen out of business . . . to take a fishing industry [with] 1,050 boats and put it down to . . . one corporation or one special-interest group."

"We didn't even really know what Magnuson–Stevens was about," Wayne recalls, "but at a Council meeting in Pensacola someone asked if we were going. I had to borrow a suit from Donnie's dad, which fortunately fit me perfectly. We were sitting in the back of the room, arms crossed, pissed off, when an aide asked, 'You guys fishermen? Ted Stevens wants you to testify.' So I stood up and said, 'Well, I'm one of the I's in the IFQ you're talking about it, and I don't like it." In

---

* The recreational sector is a broad category, encompassing everything from pier fishing to rowboats to yachts to charters.
† In fact, this could not have happened without undoing the 51–49 split between the two sectors, but the early days of IFQs spawned many confusions and fears.

the end "a bunch of us fishermen got language inserted in Magnuson–Stevens, putting in place a moratorium on *any* new IFQs." The law required the moratorium to remain in place until 2000. It also required that even when lifted, a majority of fishermen would have to vote twice, in a double referendum—the first to approve a catch share in principle and the second to approve its design.

The reauthorization of Magnuson–Stevens in 1996 was a landmark moment. Confronted with precipitous declines in many important American fisheries, the bill introduced serious marine conservation principles: requiring that councils end overfishing, rebuild all overfished stocks within ten years or as soon as possible for longer-lived species, minimize bycatch (the accidental catching and killing of species other than the target), and define and protect essential fish habitats. Soon after its passage, Wayne was asked to speak at Tulane Law School on how the law's new requirements would play out in the Gulf and impact snapper fishermen. He accepted, even though the invitation came from environmentalists: Pete Emerson and Pam Baker from the Environmental Defense Fund and Chris Dorsett from the Gulf Restoration Network. Wayne remembers the warning he got from a Panama City buddy, another commercial snapper fisherman. "He was telling me, 'You don't need to be dealing with those people.' I told him, 'Look, man, some have been at Council meetings for years.' He said, 'They're going to shut you down. You're going to be done. These people are dangerous.'" Wayne shrugged off the warnings. "Chris had been asking for everything to be cut to hell or even closed. I told him, 'Until you learn to play in the middle of the field, you're not going to accomplish anything.' He was asking for too much. It was unrealistic, never going to happen."

It was at that meeting that Emerson and Baker asked Wayne what proved to be a pivotal question: Would he reconsider an IFQ if the fishermen themselves designed it? And if, instead of more closures, it secured and even expanded the commercial fleet's ability to fish?

Yes, Wayne answered, "in that case, I'd be interested." It still took years, Baker recalls, "to gain their trust and confidence. But it never occurred to me that you wouldn't work with fishermen. I'd grown up around them, saw how much they know. And we were learning how IFQs bring people to the center." Little did she or Wayne foresee that federal fisheries policymaking would soon become as all-consuming for him as fishing, or that twenty years later he'd still be missing countless hours in the Gulf to sit in hotel conference rooms or congressional antechambers for days on end. But once set on a course, Wayne does not give up.

Like our heroes upriver, Wayne saw that the first step in protecting his livelihood was gaining deeper understanding of his ecosystem, including its human members. As in those upstream landscapes, this would require marrying deep local knowledge gained over lifetimes on the water with the expertise of marine scientists and managers. The weeks between derbies now became for Wayne as furious with activity as the sleepless hours on the boat. With Donnie, Russell and a handful of others gathered at his house, "we'd sit and look through stacks of documents back to the sixties and seventies," says Deb, trying to make sense of the ups and downs in the fishery. They began sharing their secrets. "We had to put our cards on the table," says Wayne, "say how many fish we'd been killing. The stock assessment panels said 30 percent of released fish died. We said, 'No, it's probably 80 percent.' We had to point the finger at ourselves, say, 'Look, this is what we're doing wrong and this is how we fix it.'"

From an occasional attendee Wayne became a fixture at every Council meeting—five week-long gatherings a year plus field hearings—reading every word of the presented studies, absorbing every minute of the debates. At first, he was skeptical. "You have to understand, in 1990, when I walked into my first scientific meeting, I was lost and said, 'These people are crazy. They don't know what the hell they're talking about.'" But the other fishermen began relying on him to dig into the history and science, and by decade's end he'd become part of the process. "I threw my hat in to be on the first SEDAR [Southeast Data Assessment and Review, the official process for assessing the stock, launched in 2002]. I was the first commercial fisherman to become part of that scientific process." His fisherman's perspective brought added rigor. He brought to SEDAR's attention, for instance, a crucial omission in the logbooks fishermen are required to keep. Though they were asked to record the depth of water they'd fished in—an essential piece of information for knowing both where the fish are and what fraction of their discards likely died—they were given room to enter just one number for an entire fishing trip. "I might have fished for vermillion snapper at 300 feet, yellow-edged grouper at 600 feet, red snapper at 130 feet. Within two months, I had the logbooks changed."

Though self-taught, Wayne eventually became versed enough in fisheries science to hold his own among the PhDs. Asked, for instance, how anyone knows how many fish are really out there, he frowns a moment in thought, then apologizes for not being able to offer anything simpler than a three-part answer. He lays it out chronologically, describing first the forty-year-old program of trawl surveys: "the entire Gulf is marked out in three-square-mile blocks; a computer picks random blocks to sample; then research vessels drag nets and count the snapper they catch to measure recruitment." (Recruitment is the number of young fish that have become big enough to enter the fishery.) Those findings, he continues, are combined with assessments of "catch per unit effort: how many snappers do you catch per hour of fishing?" assembled from logbooks. "Now, the more technical: Jim Cowan at LSU has studied the fish for fifteen years. He has a submersible with side scan sonar that spends 150 days at sea: they take one pass around a rig [where snapper like to gather] and count and estimate the size of every fish. You have to build up a database over many years to give it credence; everything has a confidence interval."

With that deepened knowledge in hand, the next step was to design a catch share as tailored to the local ecology and respectful of human uses and communities as the Heritage Act Dusty Crary and his allies devised. That required grappling, again, with the complexity of traditions: acknowledging that some—such as unlimited access for everyone to the ocean's bounty—might no longer be tenable, while others—such as local fish houses and restaurants and fishing villages built around small enterprises like Wayne's—might be worth preserving. Though on the water, as in Kansas, small does not necessarily mean better for the environment. Bigger boats often fish "cleaner" (with less bycatch), and are far easier to monitor than highly dispersed little boats near-shore. Designing the catch share also required tackling issues of equity: who would get access rights to the fish, and to what fraction of the total?*

One by one, Wayne and his allies arrived at their best answers to those questions, though with the derby still on "and all that rivalry," recalls Deb, "it was hard to get to agreement. The fishermen were suspicious of each other's motives: 'What are you getting?' 'Where are you going to screw me?' You had to convince them that what you were taking forward was in everybody's best interest, not just your own. But Wayne was the one the others wanted as their advocate; he'd take a consensus and carry that to the Council. These guys wanted the generation that followed to have the resource as well. They didn't want to catch baby fish. So we had to stand up and help ourselves."

Before they could go further, however, they had to clear the obstacle Wayne had helped erect: the moratorium. In 2000, two years before it was due to expire, Emerson asked Wayne and friends to help

---

* A catch share is an access right that can be revoked, not a property right.

EDF get a meeting with Mississippi Republican Trent Lott, then the Senate's leader on Gulf fishery policy. (Lott cared about fishermen but "didn't want to hear a word about environmental anything," says Wayne.) They wanted to ask for enough of an exemption from the moratorium that they could begin designing a catch share program, to be ready when the ban was lifted. The first time Wayne went to Lott's office, the senator's legislative director, Jim Sartucci, told him, "You have fifteen minutes and you won't tell me anything I don't know," Wayne recalls. Two hours later, the meeting finally drew to a close. "I guess he figured out I knew a bit more than he thought. A lot of people have asked me where I got my education." They also needed Ted Stevens's support. "All we'd heard from his office is no, no, no," recalls Wayne. "He was head of appropriations. People would say, 'Who's against you? Stevens? See you later.'" Finally, Wayne and Donnie got a meeting with Matt Paxton in Stevens's office. "Donnie just went off," says Wayne. "He started slamming the table. 'Snapper doesn't have a goddamn thing to do with Alaska; tell your boss to let go of our throats.' Even I was thrown back. Paxton got up. I thought he was going to get security to throw us out. But a few minutes later he came back. 'Mr. Stevens says you can work on your IFQ, and good luck.'"

Pam Baker, EDF's marine ecologist, was at that meeting and almost every other appearance the fishermen made, whether before the Gulf Council or Congress. She recalls often holding her breath. "Wayne has a good heart and ideas but he sometimes felt the best way to get what you need is by being forceful. That's the way he knew how to be, to work through conflict. So you weren't always sure what was going to come out of his mouth. But a surprising number of times, he came out with exactly the right thing, the right message for that moment. In the face of hostile regulators, hostile recreational fishermen, hostile NGOs, he'd stand up and say something from the heart that was just the truth. He'd say, 'This is important to all of us. We all see good and bad in it, but we can make it better.' All those years it took to get it going, he was one of a handful of fishermen out front and center. He gave up a lot of his own money and time and made a big contribution to bringing people together."

The solutions they all arrived at—wrestling through them on an "Ad Hoc Red Snapper Advisory Panel" made up of thirteen commercial fishermen, including several still dug in against the IFQ—are no more perfect than Justin Knopf's farming practices or the Louisiana Master Plan, and for the same reason: they're built on knowledge that will never be complete and on realism, the need to face and balance tradeoffs. But by the end, Wayne had helped design a catch share that won support from 80 percent of Gulf commercial fishermen, because it designed in protections for their small businesses and local fishing communities. To ensure that limited special interests couldn't monopolize the fishery, for instance, they capped at 6 percent the shares any one business could own and kept in place a requirement that at least half of every shareholder's income must come from fishing. They also grappled with the challenge of fair allocation. "When you go to divide the fish, thirteen guys will have thirteen ideas," Wayne says. "And the right one will be the fourteenth or fifteenth. And when you add up everyone's historic catch you'll come out with 120 percent of the fish. So you may end up giving up 20 percent of your historical catch. I'd been catching 2 percent but got allocated just 1.35 percent of the total." Like everyone else, Wayne had to compromise.

The last step was securing the necessary political support on Capitol Hill. "First it was just Pete and Pam and a couple of us fishermen, then four, then sixteen guys to DC. There we were in 2004, ten years after blocking the IFQ, after going through all this derby, mini derby, micro derby, back up there asking to use our referendum process to allow us to *have* an IFQ." Still, right up to the end, not everyone was convinced. A number of NGOs were highly skeptical of catch shares (some still are). That difference of opinion led to EDF's exit from the Marine Fish Conservation Network. And when the Council held the referendum to give commercial permit holders final say on whether or not to go forward with the IFQ, 20 percent voted no. Among those was Buddy Guindon, a red snapper fisherman famous throughout the Gulf for being a consistent "high-liner" (catching more fish than anybody) and owner of Katie's Seafood in Galveston; he "thought I was crazy," Wayne says. Some thought it crazier still that in the months just before implementation Wayne and Donnie became the most passionate advocates for a big cut in the catch—"we asked them to cut the total quota from nine to six million pounds"—because they now had security in the fishery and wanted it to last. "We'd been overcapitalized," recalled Coon Hadley, one of the fishermen who now regularly join Wayne at Council meetings. "Some people, including my dad, weren't awarded any fish. But he overcame it; he bought quota. We all paid the dues to rebuild this fishery. Anybody that thinks we haven't went through struggles is just wrong."

Since its debut in 2007, the catch share has succeeded beyond all expectations, making fervent believers even of early skeptics like Buddy, now its most ardent and public champion.

For the fishermen, improving stocks have allowed the Council to steadily increase the commercial red snapper quota, from 3.3 million pounds in 2007 to 7.3 million pounds in 2015. Wayne's share of those stocks, which allowed him to catch 28,000 pounds that first year, was in 2015 worth 60,000 pounds. With catch shares extended to additional Gulf species in 2010 (though not to mackerel, vermillion snapper and amberjack, all still being hammered under the old derby system), Wayne now also has quota for deep-water grouper, shallow-water grouper, gag grouper and tilefish, which gave him access to another 17,000 pounds in 2015. Altogether, he lands about 145,000 pounds of fish a year.

Their growing abundance has made those fish much easier to catch. Where Wayne used to struggle to land 500 pounds of fish a day, he now averages 1,500 pounds daily. In one remarkable two-day stretch in 2014, he and Marshall caught 11,000 pounds. "That's with hook and line," he says proudly, "one fish at a time." The price he gets at the dock is also far higher, and steady: $4.50 to $5 a pound for his snappers, a far cry from the $1.50 or consignment sales he faced under the derby; and $4.50 for groupers, up from $2.50 a pound. (Those fish still managed by derby remain as volatile as ever, says Wayne. "King mackerels, when we don't have market gluts, are up to $3 a pound, but when they open the season in state waters they fall to $1.20.")

Wayne has cut expenses, meanwhile, 75 percent, largely because he no longer has to burn diesel running back and forth to the dock every time he hits 2,000 pounds but can wait to fill his boat with 10,000 pounds or more, making one trip instead of five. Instead of traveling 150 miles in a day, he now might go that far in a week, if he's after deep-water groupers; less if he's after just snapper. "We can go thirty miles from the dock and catch all the red snappers we want; they're so prevalent with their comeback."

All of that adds up to a good and steady income. "Back in the derbies we struggled to catch $200,000 worth of fish a year; this year we'll catch $500,000 worth." That gain comes not just from increased quantity but also quality. Instead of fish that is hastily handled, sometimes beaten-up and landed all at once, requiring processors to freeze most of it—immediately slashing its value—Wayne and friends now supply American consumers with fresh, high-quality fish every day of the year. They no longer compete with black-market fish undercutting the price. Tracking systems built into the IFQ—matching up logbook data with dock and fish-house sales—shut down illegal trade.

And with their long-term stake secured, the fishermen are continually finding new ways to maximize the market value of their fish. Buddy Guindon, for instance, has led development of Gulf Wild: every red snapper sold at his seafood market is now tagged with an ID code that tells the buyer who caught the fish and where. As of 2014, more than forty fishermen collectively holding a third of the Gulf-wide red snapper quota had joined the program, signing conservation covenants that commit them to minimize bycatch and discards and to be monitored by video or observer. Beyond giving consumers a glimpse of their fisherman and assurance that the fish was cleanly caught, Gulf Wild protects them against the rampant mislabeling of cheaper or imported fish. (Recent tests in twenty-one states found that 93 percent of fish labeled red snapper were in fact something else.) It also provides enhanced safety testing, which helped ease consumer worries after the oil spill. Wayne, meanwhile, has focused on diversifying his offerings, bringing in a mix of fish to the New Orleans restaurants that buy his catch—the Redfish Grill, Court of Two Sisters—to please their customers and stabilize his income, now a reliable $100,000 a year.

As fishing incomes and markets have become stable year-round, fishing communities have also become more stable. "Instead of a lot of part-time type jobs with people running around the docks that are basically idiots," says Wayne, "we now have permanent good jobs, with people who are paying taxes." On a good trip, says Deb, "a captain's wage will be $6,000, a deckhand's $2–3,000. And we ration out the IFQs to make them last all year." Fishermen's kids have even resumed following in their dad's footsteps: Buddy's sons, who grew up under the catch share "when they could see a good life and future," have joined the family business.

That boon for people has not come at the expense of the fish, but the opposite: both overall fish numbers and bigger, older fish are roaring back. A 2013 stock assessment showed the red snapper fishery more than halfway to its rebuilding goal, "exactly where twenty years ago the scientists said it was supposed to be," says Wayne, to reach its target abundance by 2032. In recognition of that progress, Seafood Watch moved Gulf red snapper from its "avoid" list to its "good alternative" rating. Spawning potential has rebounded, nearly tripling to levels not seen since the 1970s. And where once hundreds of thousands of fish died for nothing, that waste has been vastly reduced. With the pressure of the derby lifted, Wayne and his peers can carefully target their efforts: fishing at just the right depths (thirty feet from the bottom) to find the right-sized fish, neither too small (below the current 13-inch minimum), nor too big (the old fat fertile females) but the ideal five-to-eight-pounder. They can also now keep almost everything they catch, because they can trade quota. If Wayne catches say, a lineful of grouper but has already used up his grouper quota, he can just call a buddy who still has grouper quota in the bank and lease* enough to cover the extra fish. "I usually get the quota I need before I get to the dock. It's not a problem." Wayne figures his personal discard rate at about 2 percent.

Even for those he does throw back, Wayne's careful fishing minimizes mortality. "If you bring 'em up fast you're decompressing 'em faster, but if you bring 'em up slow he fights hisself out of a lot of energy, so we try to bring everything up at a medium speed. We also fish with real little hooks and a single little leader, so a big fish will straighten out the hooks and pop off the leaders. A 25-pounder will slide right off and swim away. We never even bring 'em up; it leaves them right where they're at. They still have us listed for 30 percent mortality but we don't have anywhere near that anymore. It's probably closer to 12 percent."

Most importantly, with a secure stake in future fish harvests, commercial snapper fishermen have become champions of conservation practices that grow the stocks over the long term. "When you give a person a right to catch that fish, it's his; he's going to take better care of it," Wayne says. Catch shares realign economic incentives, replacing the "tragedy of the commons"—where it is in each individual's interest to maximize his harvest even to the detriment of the com-

---

* A lease is the use of that allotment for a single year.

mon resource—with a structure in which individual wealth grows on pace with the common wealth. So, after years of pushing back against constraints, these fishermen now support "marine protected areas" (off-limits to fishing) in spawning grounds such as Steamboat Lumps, and fight efforts to relax rebuilding timelines or raise quota too quickly. "They get a good year or two class," Wayne complains, "and managers say, 'The snappers are thick' and jack up the quota. Then in three or four years they realize they're hurting the population and have to cut back," a swing that hurts everybody. "More than once, in my public testimony, I've told them, 'I think you're raising the quota too quickly; you need to go a lot slower than this.' I'm never in a hurry to raise quota. I figure if they're out there swimming, they're swimming in the bank."

For the sake of the fish, Wayne and his commercial fishing buddies also now share information they once kept secret. "If you ask someone where they caught their fish," says Wayne, "they give you a straight answer, so you can go fish somewhere else. We don't want to keep pounding on one place; we want to let 'em build back up." As shareholders, they also welcome strict accountability measures: landing logbooks but also dock checks, on-board observers and rigorous enforcement. "If I brought in red snappers illegally and tried to sell them without reporting, they could take some of my shares," says Wayne, approvingly. Since 2007, the commercial sector hasn't once exceeded its quota. And Wayne says he'd welcome still more monitoring, including a camera on every boat, as in British Columbia. By giving fishery managers the ability to actually count every discard, cameras free them from the need to estimate a number and then subtract out those imagined discards before divvying up the allowable catch. "Every fish over the rail counts toward your quota, but if you can show you're not throwing fish back, they don't get taken out," says Wayne. "It's hard to say no to something that gives a guy 30 percent more fish to catch."

Perhaps most importantly, for fishermen once as buffeted by storms of rules as they were by storms at sea, Wayne and his fellow shareholders can now shape their own fate: helping to craft the rules they have to live by, developing business plans for the long term, executing contracts to deliver an agreed-upon poundage at a certain date. "People resisted the IFQ because they feared they'd lose their freedom," says Wayne. "I've never been as free as under an IFQ."

Like all these stories, this one doesn't really end but remains a perpetual work in progress. So Wayne has not returned, as he expected, from policy to fishing. Instead, he is increasingly consumed with the issue that came up at that long-ago breakfast and has only become more contentious, especially in the Southeast: how to balance commercial and sport fishing.

That question is particularly charged for red snapper, given its singular allure as both food and game. But even as the compliance gap between the commercial and recreational sectors has grown, management has been slow to respond. Between 1991 and 2013, anglers (or sport fishermen or game fishermen, the terms are used interchangeably) caught more than their share of fish two out of every three years, killing far more fish than the entire commercial sector, in some years double the commercial catch. But only in 2007, and under pressure from Congress, did the Gulf Council begin seriously trying to rein in those "overages." And even then, they turned to the very strategies that had failed so miserably in the commercial sector: ratcheting down the recreational season, until by 2015 it was just ten days long, and cutting the "bag limit"—the fish per trip each angler can keep—from seven fish to two.

The frustration that has erupted among recreational fishermen is all too familiar to Wayne. "They're where we were in 1992: fewer and fewer days to fish, smaller and smaller trip limits." He sees anglers being pressed into the kinds of risks he used to take. Testifying to

Congress in June 2013, Wayne noted that in that year's recreational derby, six anglers had died. He also sees them doing the kind of damage he and his friends used to do—catch limits blown, an ocean of fish tossed back to die.

That damage does not reflect irresponsibility or indifference. To the contrary, ocean sport fishermen have played as central a role as Dusty Crary and his fellow hunters (and trout fishermen) in conserving America's wildlife. Without the critical support of recreational fishers, Magnuson–Stevens would never have included its mandates to end overfishing and protect habitat. And the growing millions of anglers have become an increasingly powerful constituency for protecting living resources against the threats posed by coastal and seabed development; CCA (the Coastal Conservation Association, the main lobbying group for recreational fishers) partnered repeatedly with environmentalists to protect critical habitat in the Atlantic. Many anglers aren't even in it for the meat but love the thrill of the chase, the chance to test their wits and strength against powerful, canny fish; they fish with extreme care to leave the fish unharmed and let go nearly everything they catch. That includes Wayne, who every year goes out with his oldest friend fly-fishing for redfish (a.k.a. red drum), which they catch and release.

The problem is an antiquated management system that is failing recreational fishermen just as it used to fail Wayne. Managers fly largely blind: while the commercial red snapper catch is now counted in near-real time (Wayne must submit a full report within hours of landing), the recreational harvest is tracked only sporadically, through phone or dock surveys and crude estimation strategies. And regulators typically don't get even that spotty information for months, *after* they've had to make the call to close the season to avoid overruns. As Environmental Defense Fund chief marine scientist Doug Rader puts it, sounding like the March Hare, "They're trying to *anticipate* what actually *was* caught and killed."

Managers know even less about discards: neither how many fish anglers throw back, nor how many of those live or die. Wayne thinks they vastly underestimate death rates, by leaning too heavily on a study that assumes anglers stay in shallow water where most tossed-back snapper will survive. At one science panel meeting, he challenged the study's author to defend that assumption. "I asked, 'Where do you get your information? Do you have scientific evidence that they're fishing at sixty to eighty feet?' And in front of a hundred scientists he looked at me and said, 'That's where we fish.' I say, 'We know some charter boats go on twelve-hour, even two-day trips into deep water. You're grouping them in this category too?' and he says, 'Yes.' I'm stunned. And this is still part of the stock assessment today. Even though you see private boats with two, three, four engines, built to go sixty or seventy miles offshore, still they say they're fishing in less than one hundred feet." Rader agrees that "dead discard may be far worse than managers think. Because hitting their bag limit doesn't mean anglers have to stop fishing or killing them. They just have to stop keeping them."

At bottom is a problem of sheer numbers. The private recreational fishing sector is one of the last in the U.S. to be "open access," with no limits on who can participate. "It's a problem if you've got a three-million-pound snapper quota and three million anglers out there trying to catch them," says Wayne. And as quickly as the snapper are rebounding, they can't keep up with the anglers' growing might. Even as managers *doubled* the recreational quota in the five years after launch of the IFQ, anglers increased their daily catch *six-fold*. In a recovering fishery, big snapper are just much easier to find, especially with ever faster boats, better fish-finding electronics and bigger reels. Many of those bagged trophies are fertile females, magnifying the long-term impact on the fishery. "By 2032 you could have more head of anglers than head of fish," says Wayne. "It's like cow elk in Colorado. If everybody who wanted to was allowed to shoot one, there'd be no cow elk left. It's not the fault of the poor guy who goes fishing. It's the way they're running the fishery."

Still, for all that Wayne empathizes with the vise squeezing recreational fishermen, hostilities in both directions have only intensified.

Several anglers' groups have fought catch shares from the beginning, seeing them as the primary cause of their woes. They've called it a travesty that commercial fishermen can be on the water all year long while anglers can fish just a few weeks or days. They've derided Wayne and his friends as pawns of environmentalists, or as an "elite group of snapper barons." With their allies on Capitol Hill, they've tried every policy move: pushing to turn red snapper management over to the Gulf states, or to ease or rescind Magnuson–Stevens's conservation requirements, or to reallocate fish from the commercial to the recreational sector, or some combination of those three.

The increasingly personal attacks have roused an equally bristling response from the commercial side. Wayne and his friends dismiss the lobbying groups' claims to speak for the ordinary angler. "These aren't guys who just want their kid to catch a fish but the 1 percent," says Wayne, with outsize political influence and narrow concerns. "If you've spent a million dollars on a boat, you don't want it sitting in the dock." Most regular guys just out for a little fishing, he says, don't understand the fight and when they do, "when they see the big changes commercial fishermen made to have the fish they have today, and that they're talking about taking fish away from our family businesses and the docks and fish houses and restaurants and grocery stores all over the country, most of 'em go, 'We'd like to leave it the way it is.'" He finds especially laughable the idea that the commercial guys are the "haves." He holds out his wrestler's arms and thick, reddened hands to show the ugly scars and welts left by a thousand hooks and knife-sharp fins and gills, and offers a look into his cramped and cluttered boat: its captain's chair a stained, listing Barcalounger missing an arm, the toilet a bucket tossed over the side. He recalls an encounter with one

of the recreational captains who singled Wayne out for special deri-
sion. "I said, 'Look, man, I don't want to be called a snapper baron.'"
He pauses a beat, letting the full impact of his 250-pound frame
sink in. "I said, 'Please call me a snapper king.'" (His laugh, a flat
"heh-heh-heh," would suit a cartoon character.)

On his trips to DC, Wayne now calls not only on friendly Gulf-
state senators like Alabama Republican Richard Shelby but also inland
congressmen who represent "every American who enjoys eating fish,"
as Wayne says, "but doesn't have a boat or time or money to go catch
it themselves." State management, he warns the legislators, is a bad
idea: "The states doesn't have the capabilities to do the science. You're
talking about a fish that lives far offshore, not a trout or redfish in
the marsh. And the states'd just become cannibalistic on each other,"
setting policies to let their fishermen grab all the snapper first. Still
worse, he cautions, are proposals to ease the Magnuson–Stevens time-
table for snapper recovery. "The first meeting I went to about snapper
was in 1989, when they set a goal to rebuild by 2032. Forty-four years,
that's how long a rebuilding program we're on. And they're looking for
more relaxation on time?"

Wayne fully understands why people might think it nutty (or
worse) for managers to keep tightening the noose on anglers even as
red snapper become so easy to find that "you can catch your two
fish in fifteen minutes." But the seemingly limitless plenty, he says,
is misleading. Though the fishery is unquestionably rebounding, it
is still just halfway to its target abundance: the level at which it can
produce its "maximum sustainable yield" of food. For people used
to seeing a depleted fishery, that half-abundance might look like a
lot. But to go back to unlimited fishing now would undo the hard-
won recovery, especially given the still-overwhelming preponderance
of juveniles: born since the 2007 launch of the IFQ but not yet old
enough to replace the missing generation of sows that would have
been in their fertile thirties and forties today had they not been fished

out in the 1970s and 1980s. Stocks in Florida, Alabama and Mississippi remain particularly weak, Wayne says: "The eastern zone has just
25 percent of the fish, but anglers take 80 percent of their catch here,
and commercial fishermen 30 percent of theirs, so some are starting
to have a hard time getting their fish." David Walker, who fishes out
of Alabama and is the first commercial red snapper fishermen to serve
on the Gulf Council, told Wayne that in 2012 he "was catching 5,000
pounds a day, but in 2015 he's landing less than half that." Wayne
used to get calls every week from Florida fishermen who'd brought in
so many fish they needed to lease some of his quota. "This year I've
heard from one."

Frustrated that fishermen were turning against fishermen, in
2013, Buddy, Wayne and some two dozen others filed suit against the
party they hold responsible for the mess: federal regulators. The suit
charged U.S. Secretary of Commerce Penny Sue Pritzker—who in
her role as head of the National Marine Fisheries Service oversees all
eight Regional Fishery Management Councils—with failing to fulfill
her legal obligation to sustainably manage Gulf red snapper. In March
2014, a U.S. District Court ruled for the fishermen. The administration had violated Magnuson–Stevens's requirement to use the best
available science, wrote Judge Barbara Rothstein, "disregarding accurate and reliable data to avoid penalizing recreational fishermen." They
had also failed to keep the fishery on track to recovery. The Council's
decision to open the fishery as usual despite continuing recreational
overages roused the judge to a particularly blistering rebuke: "Administrative discretion is not a license to engage in Einstein's definition
of folly," she wrote, "doing the same thing over and over again and
expecting a different result."

Despite that ruling, in August 2015, before a packed room including Wayne and other commercial fishermen, the Gulf Council yielded
to political pressure and voted to allocate *more* fish to the failed recreational derby: shifting 2.5 percent of the snapper, about 380,000

pounds of fish, from the high-performance commercial IFQ into the poorly managed recreational fishery. The decision came despite impassioned testimony by renowned chefs and other members of Share the Gulf—a coalition of conservationists and businesses including seafood shops, restaurants, tourist associations and grocery stores from around the country who all depend on fish supplied by commercial fishermen. As Wayne says, "These aren't my fish; they're the American public's fish." Wayne was so upset when the vote came down, he could barely speak. Not only had the Council set a precedent that could imperil the fishery's recovery, it had also failed to solve the problem; the reallocation will do little to satisfy the recreational sector's desire for more days at sea. "They already had half the fish and caught those in nine days," says Wayne. "Even if you gave them *all* the fish, they'd get an eighteen-day season. We need to find solutions for anglers, but without hurting everyone else."

Such solutions are emerging, with important impetus from charter boat captains, historically managed as part of the recreational sector. Seeing how well the catch share was working for commercial fishermen, they asked to follow suit. In August 2013, NOAA approved the Gulf Headboat Collaborative Pilot Program, carving off for some Gulf charter operators a portion of the recreational red snapper and gag grouper quota (based on their historic landings) and letting the captains divide it among themselves. Like Wayne, they could now fish 365 days a year.

For individual anglers, solutions are moving more slowly, though smartphone-based apps that let fishermen report their catch in real time may help ease the problems of too little information coming in too late. The Angler Action Program was one of the first to enlist sport fishermen as citizen scientists, collecting harvest data (including releases) first on snook and now on one hundred inshore and offshore species. For the Gulf's most fought-over species, anglers can use Texas A & M's "iSnapper" to document (with photos) the red snapper they catch and throw away, as well as the depths and locations they fished in. Involving anglers in data collection can help build trust in the science shaping management decisions. It may also enable a decentralized system modeled on duck stamps for managing the catch. Already, the fishermen piloting these apps have been surprised, says Wayne, by their real discard rate. After one excursion, Galveston charter captain Scott Hickman asked a mate to guess how many fish they'd thrown back. The mate guessed 80. In fact, they'd thrown back 262.

Even as Gulf red snapper continues to be the proving ground for innovations in fisheries policy, other regions have been following its lead, and with similar success. Most began in equally dire straits: the fish depleted by overharvest and waste, the fishermen throttled by untenable regulations and bargain basement prices. Today, two-thirds of all fish landed in U.S. federal waters are under catch share management and one hundred species are on the path to recovery, including clams, crab, flounder, haddock, hake, halibut, rockfish, striped bass and whitefish. (Even McDonald's fish filet is Alaskan pollock caught under a catch share.) The design varies, tailored to regional ecologies and values. In Monterey Bay, for instance, the quota is community-owned and managed by a nonprofit Fisheries Trust on behalf of local fishermen, who can lease it on flexible terms. But the gains for both fish and people are consistent: NOAA's 2014 annual report to Congress on the status of the nation's fisheries found overfishing at a historic low, with thirty-seven formerly overfished species rebuilt since 2000; the agency's 2015 analysis of all twenty U.S. catch share fisheries found significant gains in economic productivity.

One of the most dramatic recoveries has been on the West Coast, where the "groundfish" fishery—some ninety species including lingcod, sablefish and many species of rockfish that live on or near the seafloor—was declared a federal disaster in 2000. Provided some protected waters and transitioned into a catch share in 2011, long-lived species like the brilliant orange-and-pink canary rockfish were expected by fishery managers to take more than four decades to recover to sustainable levels. Instead, just four *years* later, canary rockfish and petrale sole had recovered enough to be declared rebuilt. And while under the old system a Pacific fisherman might pull in 25,000 pounds of fish in his trawl but have to throw back four-fifths of them, carpeting the sea with belly-up casualties, those discards have been cut 80 percent. Needless deaths of people have also been dramatically reduced, even in the Bering Sea crab fishery in Alaska (made famous by *The Deadliest Catch*), where families used to mourn five fishermen a year.

The implications of catch shares extend beyond marine resources. Because they use economic incentives to achieve environmental goals, IFQs are held up as a model by many conservative economists for solving all kinds of conservation challenges. Some have been directly involved in their design: Donald Leal of the Montana-based Property and Environmental Research Center (PERC), who with Hoover Institution fellow Terry Anderson coauthored *Free Market Environmentalism*, served with Wayne and Pam Baker on the Gulf of Mexico IFQ advisory panel. Michael De Alessi at the libertarian Reason Foundation helped provide access on Capitol Hill and partnered on several congressional briefings.

The resilience and flexibility afforded by such rights-based systems will become ever more crucial, as marine life is stressed by graver and more long-lasting challenges.

As in Dusty Crary's Rocky Mountains, the appetite for mineral resources poses ongoing threats. The BP oil spill offered a preview of what may lie ahead in the gathering rush for seabed minerals and petroleum resources. Like Sandy Nguyen's shrimpers, Wayne felt the full brunt of the disaster. When the Gulf finally reopened for fishing, "we had to go sixty miles just to get out of the oil," Wayne recalls. "I could see it and smell it; it gave me headaches. A friend from Panama

City had to go forty hours to get anywhere he could kill a fish; those two days he saw not a bird or dolphin or any living thing." From an economic perspective, the IFQ was a life-saver. While Sandy's shrimpers lost the 2010 season, Wayne was able to delay his fishing until the waters reopened. Commercial fishermen in parts of the Gulf that never opened that year were able to lease their quota to fishermen in western Gulf waters that never closed, maintaining their income and supplies for retailers and restaurants.

But the harm has persisted, Wayne says: both 2010 and 2011 had the lowest numbers of juvenile red snapper seen in the eastern Gulf fishery since 1994. "Red snappers spend their larval stage, the first forty-five days, in top water, then settle into deeper waters," Wayne explains. "All the larvae got pushed into the spill by the loop current. So it killed the little ones year one. Then in year two, because they used dispersants and sunk the oil, the fish that survived the floating oil ate the sunken stuff, and we lost another bunch." The big snapper also suffered, because they eat mud creatures that were eating the poisonous sediments. Wayne was one of many fishermen reporting snapper with lesions or rotting fins. "Every time we talked about bad fish, everybody kind of went nuts on us. Just like, 'You're hearsaying,' you know? And we're saying, 'Well, they're there.'" Even five years later, as the long-lasting toxicants continued to be cycled through the food web, scientists were finding evidence of multigenerational reductions in female snappers' reproductive success.

Still more frightening to Wayne are the effects he's beginning to see from warming water and other global-scale changes. "I'm not going to say to another fisherman, 'Do you not believe in climate change?' because he has to; he's seeing it firsthand, things you've never seen before. Like the Sargasso weed that comes up every year, that for the last six years for about a month gets this green slime on it. And the box jellyfish, lots more of those, which show up when the water gets hot, because warm water can hold less oxygen." Not only are new things beginning to move in (including, as in Montana, pernicious invasive species), but Wayne's fish are beginning to move out. Every marine creature at every stage of life has a temperature band it can endure, as do their prey. So biologists are already seeing species like gray snapper moving north to find those proper temperatures; that asset flight will have profound implications for fishing communities. And when those species' bands of preferred water temperatures bump into a continent, Rader says, many could disappear. In general, tropical systems have greater diversity but lower productivity; as the Gulf becomes more tropical its total production will likely decline.

Many of Wayne's reef fish will also suffer, alongside Sandy Nguyen's shrimp, as rising seas and intensifying storms degrade their wetland nurseries. Like poisoned sediments, the effects of declines in key nursery habitats ripple through the food web: stressing even animals that live their entire lives offshore but prey on crabs or drums that depend on the estuaries. Offshore animals will also be affected as warming waters further reduce oxygen concentrations in the northern Gulf "dead zone," and as declines in wetland acreage reduce the natural removal of the nutrients that cause that dead zone.

Perhaps most damaging in the long term will be the growing acidity of the ocean as it absorbs excess atmospheric carbon dioxide. The last time the earth's oceans experienced this rapid a change in chemistry was 56 million years ago: that episode drove many marine species to extinction, and it was 100,000 years before the ocean recovered its balance. Acidification compromises the ability of "calcifiers"—from the tiny sea snails that form the base of many oceanic food webs to corals, oysters, lobsters, clams and crabs—to build shells, leaving them vulnerable to predators and disease. Fish fare somewhat better, though maintaining their internal pH costs energy that could otherwise go into growth or reproduction, and beyond a certain point they too may lose critical functions, including the ability to understand the chemical cues they need to navigate and dodge predators. Shrimp react in unexpected ways: making bigger and thicker shells, which costs energy but also makes the babies less transparent and therefore more vulnerable to predation.

All of these pressures—overfishing, habitat destruction, warming, acidification—are threatening fisheries around the world, especially in the developing world where they play a far more essential role in human survival. Three billion people rely on fish for their dietary protein and 38 million rely on fishing for their livelihood, 90 percent of them working at small (often subsistence) scale in tropical near-shore fisheries. Already, more than half of those fisheries are overexploited or collapsed, so depleted they can't sustain yields over time. Though even that may understate the crisis. A study published in January 2016 found catch underestimated by 30 to 50 percent in the developed world and as much as 200 percent in the developing world, masking still-steeper declines in global abundance. And population and demand for protein are growing fastest in exactly those regions, like Southeast Asia, where climate change will most suppress marine productivity.

Early experiments and new models suggest that long-term, well-designed access rights like those that rescued red snapper and Wayne also offer the best hope for these threatened fisheries, especially when combined with no-take reserves to protect nursery habitat. As in the Gulf, fishermen around the world readily support conservation measures when coupled with a secure access right because those protections build the value of *their* stocks. In low-governance regions, the most promising approach allocates not fish per se but territory: protecting for a community exclusive access to its own local waters. (Those waters are now often the focus of intense competition, if not poached or even leased out from under the locals, including sometimes to foreign fleets.) An early success is Belize, where illegal fishing was depleting spiny lobster and conch, deepening poverty and food insecurity and driving more desperate poaching. In

2011, the government agreed to experiment: granting communities dedicated rights to fish their local waters but also the responsibility for enforcement and for counting the catch, to build the scientific foundation needed to inform management and policy. The experiment succeeded—increasing catch and reducing illegal fishing by 60 percent—and fishermen petitioned the government to expand the program. In 2015, Belize began putting in place "territorial user rights for fishing" (TURFS) for all its coastal waters.

The oceans are an underperforming asset: half the world's fisheries produce less food, jobs, economic value and biodiversity than they could. Without action, those fisheries will continue to decline, leaving millions of people who rely on wild fish to starve. But a study of almost 5,000 fisheries representing three-quarters of global catch shows the potential for quick recovery if rights-based systems and sustainable fishing become the norm. In a decade, 75 percent of fisheries could be healthy, increasing the harvest of wild fish for food by 29 percent and fishing profits by 200 percent while doubling the fish left in the water for conservation.

In answering the question of how to feed a growing world, seafood will be a critical part of the answer. Fish and shellfish, both wild and farmed, could satisfy a much larger portion of human protein needs, and—if properly designed and managed—with a fraction of the impacts of "land food": on greenhouse gas emissions, water use, nutrients and displaced ecosystems. And that sustenance for humans needn't come at the expense of marine biodiversity. Even now, the essential building blocks of ocean life remain largely, inspiringly, intact: like Dusty Crary's world before the slaughter of the buffalo, or even earlier, when mastodons and wooly mammoths still roamed the American continent. A study led by UC Santa Barbara ecologist Doug McCauley has compared the fate of marine and terrestrial species, finding that in the last 500 years human activity has caused 500 animal extinctions on land but only 15 marine animal extinctions. An incipient "marine industrial revolution"—including mining claims staked on a million square kilometers of the deep sea floor—may have brought us to the brink of irreparable devastation of ocean life. But it hasn't happened yet. "While the tiger may not be salvageable in the wild," as the *New York Times*'s Carl Zimmer wrote, paraphrasing McCauley, "the tiger shark may well be." It is not, in other words, too late.

Even as the fishing years left to him dwindle, Wayne still devotes most of his hours to the seemingly never-ending struggle around red snapper management. He does so "because I'm a fisherman at heart; it's what I've known my whole life. I got to see this thing run through the right way. I want to see the fishery remain in the hands of fishermen, and to see the snapper thick, filling the top of the water like an aquarium. I went to my first meeting in 1989; that's thirty years. I'd like to see this fishery rebuilt before I die."

Still, he's never happier than on those increasingly rare occasions when he's in his Crocs and shorts, nosing out of the estuary at dawn, tucking his big belly into the curve of the wheel to steer so his hands remain free for texting. (On one early July morning, he is asking a friend in the Dominican Republic to get him some Larimar, a rare watery blue gemstone, as a birthday gift to match Deb's blue eyes.) As he winds out in the lingering dark past cypress snags, shrimp boats, tugs, cranes, bait shops and oil platform supply boats, crossing the Mississippi shimmering with the reflected lights of refinery flares to head toward the dozens of rigs glimmering like constellations on the horizon, he dials his cellphone: "I got to tell the government we're heading out." (Wayne and his peers have to tell the feds when they're leaving port and within twelve hours of their return.)

With the fishery and his life so dramatically turned around, Wayne nurses some regret that he succeeded in driving his sons away from fishing. As planned, they fell far—very far—from the paternal tree. Both went to college, the first on Wayne's side of the family to do so, both winning full "Bright Futures" scholarships—based on their grades, test scores and public service—to the University of Florida. Nick studied classics and ancient languages, pursuing a love born in the hours his mom spent reading science fiction, fantasy and mythology to them as kids. For his masters degree at Oregon State he wrote a thesis on Greek lyrical poetry. He now teaches Latin at a Catholic high school and may go for his PhD; to keep his skills honed, he devotes an hour a day to each language. Andy studied Japanese and computer science and is now the leading dealer of Japanese 1.0 versions of video games (which are coveted, Nick explains, by people who "speed run" and like to exploit glitches). "We had to fight with him that it's not the only thing in the world," says Wayne, "and Congress could change things on you. For a while our house was the delivery place, with fifty-pound packages from Japan of anime and crap like that; you go in his apartment, there's a chair to watch TV and the rest is boxes of inventory. They're apples and oranges. Nick, we have to stay on him about watching his money; Andy, we can't figure out how he counts it all. They both tell each other they're bums." Despite all these differences, it's hard to miss how tender Deb and Wayne remain toward each other after forty years of marriage, or the family's low-key, easy rapport.

Wayne *has* raised a young fisherman: his nephew Marshall Gross, who now frequently captains the boat when Wayne is involved with Council meetings or trips to D.C. The youngest of three children of Wayne's only sibling, Carla, Marshall grew up wild, in a family damaged by alcohol. Nick remembers as a kid thinking his teen cousins were "thugs, dressed in FUBU [a hip-hop clothing line], talking slang, listening to Eminem." Before he was eighteen, Marshall had a Florida warrant out for his arrest and had fled to Louisiana. "I never had a dad, was young and stupid and doing stuff that wasn't right," he says.

But Marshall summoned the courage, Wayne says admiringly, to turn that all around. At twenty, he began working for Wayne, married a local Cajun girl, Tricha, settled on the bayou twenty-five miles north of Leeville and had a baby. When Katrina and Rita hit, he spent four hundred hours volunteering, helping coastal towns rebuild docks and homes. Then Wayne told him he had to go back and face the warrant. "Uncle Wayne gave me a chance and I wasn't going to mess that up. Though I'm glad I ain't no fish; Uncle Wayne catches some serious fish . . . more than two million pounds in the years I been here."

Marshall's crimes were "juvenile stuff," says Wayne, but "when he turned hisself in, because he had ran, the DA wanted to give him forty-eight months. So me and Archie [the owner of Griffin's Seafood, who buys Wayne's fish] and another guy who runs a business on the dock all got on a conference call with the DA. We told him, 'This kid's cleaned up his act, really changed since he got here.' I said, 'He's been working for me for six years now; letting him run my boat is like putting $100,000 in someone's hands. And when he gets done you're not going to see him back here in Alachua County. He's staying in Louisiana.' He wound up with four months in county jail, including the two he'd already served. Boy, it was killing him to be there cause his life had already changed; he had that baby at home. But his past caught up with him."

His record cleared, Marshall returned to the bayou life he loves and had a second child. He and Tricha now hunt and fish for nearly everything they eat: duck, deer, crawfish, catfish, rabbit, gator, frogs. "You got to be quick to go frogging," says Wayne. "You gig 'em with a three-pronged thing, but they're fast." Tricha explains how to skin them: a knife across the back of the neck, a finger wedged in to pry it loose, then peel them like a banana. Marshall's accent has grown so deep, he sounds like a native and proudly calls himself a "coonass" (as bayou Cajuns call themselves). His multiple tattoos include

"Rest in Peace" down one calf for his mom, Carla, who died of liver illness, when Wayne and Marshall were out fishing two hundred miles from shore.

Marshall is still awestruck by what he sees out on the water: great white sharks "the size of Volkswagens," a black frigate bird soaring into view, spooky with its forked tail and motionless seven-foot wings, "like a frigging pterodactyl." Wayne, on the other hand, pays scant attention to the pod of dolphins that follow a few feet behind his boat in the silvery dawn water, or to the hundreds of egrets perched like huge white magnolia blossoms on the bushes along the shore. Though a conservation champion, Wayne is not what you'd call a nature lover. He does notice a flock of pelicans decorating a buoy, but only because they are a "classic fisherman's sign" that king mackerel are swarming below. But when he's not hauling in fish, he rarely steps outside. He's content in the dank cabin cluttered with heaps of food, tools, boots and rags that the two men share for a week or more, leaning back in his busted captain's chair while Marshall carefully breads and fries fresh-caught grouper on their beat-up little stove, or napping on a bare, tattered mattress half-sliding off a narrow bunk.

The fishing itself is less an encounter with nature than a step into a kind of abattoir of the sea. The first to meet the knife are the mackerel brought for bait: in two swift swipes the men filet each one, then skewer a chunk on each of the multiple hooks strung along weighted lines. Using his sonar as a guide, Wayne searches the depths around the reefs and rigs. When he finds snapper, he and Marshall drop their weighted lines down near the bottom where the snapper love to loiter, feeding on pipefish and pigfish, snake eels, sea robins, striped anchovies, sea lice, crabs, pinhead-size zooplankton and the bottom worms that gather on hillocks of sediment called mud lumps. It takes just seconds before the reels start madly bouncing as snapper hit. A few moments later the reels quiet, and they pull in the lines.

Occasionally, a canny fish has stripped the bait "like he had hands," says Marshall. "Trigger fish are the best at thievin'. Then we're not catchin' 'em, we're just *entertainin'* 'em." But most times the lines emerge with a fat snapper dangling from every hook, like shiny pink Christmas tree ornaments. Freed from the hooks and dropped on deck, the fish drum out their last gasps of life. Wayne and Marshall gut them, extracting livers, intestines, stomachs and kidneys, then toss them into baskets that soon brim with the gleaming, gorgeous creatures, ripe with a vibrant iridescence that will quickly fade. Gulls follow, diving to grab the entrails, then dogfight in midair to keep thuggish gulls from snatching the prize away. The pitching deck is soon flooded with crimson-stained saltwater.

The men work quickly and mostly in silence: baiting, dropping, reeling in, unhooking, gutting, icing—though occasionally they stop for a (usually funny) story. On a day when Tricha has joined them, there is sweet ribbing all around. Fishing with a pole off the back, Tricha complains about the gear Marshall has rigged for her: "I need a smaller hook." Wayne pipes up, "You need a bigger fish." Marshall prods Wayne to "tell about the time you were on the phone and walked right off the dock onto a piling and busted your ribs." Wayne in turn recalls the hazing he gave Marshall his first time out, persuading the credulous twenty-year-old he had to "grease up" the anchor so it would slip into the rocks. Worse was the seasick rookie they promised would be cured if he tied a string around a slice of raw bacon and swallowed it. The kid did and stood with string hanging from his mouth, asking miserably, 'Now what do I do?' They told him he had two choices for how the string could come out, and from then on called him "nightlight"—a dim bulb.

The Gulf of Mexico, where this journey ends, is at once as vast and wild as Montana and as industrialized as a space colony. Sixty miles out in a little fishing boat, it's impossible not to feel dwarfed by the inhuman scale of the human-made infrastructure visible in every direction. Wayne marvels as he passes a crane "that can pick up a whole oil rig. You could buy my house for that." And then a self-propelled jack-up, a drill rig that "can pick up its legs and run; no one needs to assist it," like the Imperial Walkers in *Return of the Jedi* (though a whole family of those futurist war machines could live on one of these oil behemoths). At night, the stars themselves are dimmed by the lights strung all over the rigs, scattering reflections like gold confetti across the chop.

It is just a few hundred miles from Wayne's favorite fishing spots here to Havana, Cuba, or the tip of Mexico's Yucatán peninsula. And just a bit farther still into the Caribbean and Atlantic oceans: distances readily crossed by sharks and manatees, or the larvae and eggs of grouper, snapper, parrotfish, damselfish, corals and shrimps. Here, the interdependencies traced by the Mississippi River flow past America's borders to touch the rest of the world: what the U.S. does in its oceans and atmosphere will foretell those nations' destiny, just as decisions they make will shape ours. These waters, in other words, that began in the mountains above Dusty Crary's ranch, sustaining his cattle and roaming grizzlies and shimmering trout; these waters that traveled through Justin Knopf's prairies, nourishing his parched soils and the wheat the world eats; these waters that filled the banks of the Mississippi, carrying on its writhing back America's history and prosperity, settling into its rich, bountiful protective bayous to produce most of the fish we eat; these waters that knit together the fortunes of everyone in this book, everyone in the Mississippi watershed, everyone in the nation, extend further still, linking our fortunes to every microbe, grizzly bear, crustacean, weed, finfish and human on earth.

NOTES

CHAPTER I: RANCHER

1 **"You'll be all right":** All quotes not sourced are from interviews with the author or with the makers of *Rancher, Farmer, Fisherman*, a feature documentary based on this book, by McGee Media for Discovery Communications, 2016.

3 **the Crown:** "Crown of the Continent," Northwest Connections, last modified 2014, http://www.northwestconnections.org/crown-of-the-continent/; and "Crown Managers Strategic Plan," Crown Managers Partnership, last modified 2011, http://crownmanagers.org/strategic-plan/.

3 **fifth largest wilderness:** With the contiguous Scapegoat and Big Bear wildernesses, it is the third largest; see "Bob Marshall Wilderness," Wilderness.net, a partnership of the University of Montana, the Arthur Carhart National Wilderness Training Center and the Aldo Leopold Wilderness Research Institute, http://www.wilderness.net/NWPS/wildView?WID=64, accessed October 27, 2014.

3 **biggest population of grizzlies:** Jim Robbins, "Grizzlies Return, With Strings Attached," *New York Times,* August 15, 2011, http://www.nytimes.com/2011/08/16/science/16grizzly.html?pagewanted=all.

4 **The air overhead fills:** State of Montana, Montana Field Guide, http://fieldguide.mt.gov/default.aspx.

5 **northern mixed grass prairie:** David W. Keller, *The Making of a Masterpiece: The Stewardship History of the Rocky Mountain Front and the Bob Marshall Wilderness Complex (1897–1999)* (Missoula, MT: Boone and Crockett Club, 2001), 20–21.

5 **Old North Trail:** Ibid.

6 **"that they might be informed":** Reuben Gold Thwaites, ed., *Original Journals of the Lewis & Clark Expedition,* vol. 2 (North Scituate, MA: Digital Scanning, Inc., 2001), 131; ProQuest ebrary, accessed June 15, 2015.

7 **"the deadly axe . . . civilized man"**: George Catlin, Letter No. 31, from *Letters and Notes on the North American Indians* (London, 1841), in *The Last Best Place: A Montana Anthology*, edited by William Kittredge and Annick Smith (Seattle: University of Washington Press, 1990), 260.

7 **Audubon's 1843 Missouri River journals:** John James Audubon, "Missouri River Journals," in *The Last Best Place*, 198.

7 **Generals William T. Sherman and Philip Sheridan:** David D. Smits, "The Frontier Army and the Destruction of the Buffalo: 1865–1883," *Western Historical Quarterly* 25, no. 3 (1994): 313–338, http://www.jstor.org/stable/971110.

7 **fewer than a hundred animals:** Philip W. Hedrick, "Conservation Genetics and North American Bison (*Bison bison*)," *Journal of Heredity* 100, no. 4 (2009): 411–420. http://jhered.oxfordjournals.org/content/100/4/411.full. Margaret Mary Meager, *The Bison of Yellowstone National Park*, US Government Printing Office, 1973, http://www.nps.gov/yell/learn/nature/upload/bison.pdf.

7 **Blackfeet died of starvation:** Keller, *The Making of a Masterpiece*, 21.

10 **"a little transition":** Kirk Johnson, "For Wildlife with Wanderlust, Their Own Highway," *New York Times*, December 2, 2004, http://www.nytimes.com/2004/12/02/national/02border.html?pagewanted=1&_r=0.

12 **islands . . . are simply not enough:** the understanding of the need for landscape-scale conservation evolved from the work of ecologists E. O. Wilson and Robert H. MacArthur, who in the 1960s studied the effects of island biogeography on wildlife populations. They found that the larger the island, the greater its diversity and resilience, but the smaller and more isolated the island, the faster its ecology declined, with predators the first to disappear. In the 1980s, ecologist William Newmark applied that insight to U.S. National Parks, realizing that they were islands isolated by human development and would doom wildlife to decline. That led to a new commitment to reconnect fragmented landscapes. As Jon Jarvis, National Parks Service director, put it in 2014: "the key to long-term conservation is connectivity . . . for migratory species . . . for waterways, for transition of species that are going to be driven by climate change, all of that." The Yellowstone to Yukon Initiative, a collaboration among more than 300 partners across state and national borders, is the boldest effort to realize this concept of habitat connectivity. See Dave Foreman's essay "Wilderness: From Scenery to Nature," reprinted in *The Great New Wilderness Debate*, edited by J. Baird Callicott and Michael P. Nelson (Athens, GA: University of Georgia Press, 1998); Jason Mark, "Conversation: National Park Service Director Jon Jarvis," *Earth Island Journal*, Autumn 2014, http://www.earthisland.org/journal/index.php/eij/article/national_park_service_director_jon_jarvis/; and Yellowstone to Yukon: y2y.net.

12 *higher* **rates of plant and animal biodiversity:** Jeremy D. Maestas, Richard L. Knight and Wendell C. Gilgert, "Biodiversity across a rural land-use gradient," *Conservation Biology* 17, no. 5 (2003): 1425–34.

12 **Dusty's family history:** Information from Emily's memoir, a cousin's website ("Harry & Bess Van De Riet: Outlaws, Indians, and In-laws," *Story Apples,* http://storyapples.blogspot.com/2011/10/harry-bess-van-de-riet-out laws-indians.html), and interviews with the family.

14 **Gannon retreated:** Accounts of Gannon's suicide vary. See Robert K. DeArment, *Deadly Dozen: Forgotten Gunfighters of the Old West, Vol. 3* (Norman: University of Oklahoma Press, 2010), https://books.google .com/books?id=bx7kCHlUg_AC&pg=PT404&dq=%22charles+ gannon%22+outlaw&hl=en&sa=X&ved=0ahUKEwjx7bWI2KfLAhVB VD4KHVOEDx0Q6AEIKzAB#v=onepage&q=%22charles%20 gannon%22%20outlaw&f=false.

15 **Coxey's Army:** The first of several "red," agrarian populist movements in the west, they were headed to Washington to protest the unemployment caused by the Panic of 1893 and to advocate for paper currency. L. Frank Baum's *The Wizard of Oz* is in part an allegory of that campaign.

19 **a center of mining:** The history of copper mining in Montana is from Larry Hoffman, "The Mining History of Butte and Anaconda," Mining History Association, http://www.mininghistoryassociation.org/ButteHistory.htm, accessed June 25, 2015; and "History," Butte Convention Visitors Bureau, http://www.buttecvb.com/history/, accessed June 25, 2015.

19 **40 percent of U.S. coal production:** Jim Luppens and Alex Dumas, "USGS Estimates 162 Billion Short Tons of Recoverable Coal in the Powder River Basin: New basin-wide assessment of recoverable resources and reserves," United States Geological Survey Newsroom, February 26, 2013, http://www.usgs.gov/newsroom/article.asp?ID=3518#.VDxl7mK9KSN.

20 **he wrote in *High Country News*:** Gene Sentz, "Montana's Rocky Mountain Front: Sell It or Save It?" *High Country News,* June 26, 1995, http:// www.hcn.org/servlets/hcn.Article?article_id=1144.

21 **the Friends won protection for the Bob:** Paul R. Portney, *Natural Resources and the Environment: The Reagan Approach* (Washington, DC: Urban Institute Press, 1984).

21 **putting more than 150,000 acres:** Coalition to Protect the Rocky Mountain Front, press release, January 2010, http://www.savethefront.org/assets/ docs/factsheet_rmflease.pdf.

23 **"most of our economic value":** In 2011, USFS lands generated $36 billion in economic activity nationwide, 72% from recreation, timber and grazing and just 22% from mining; see U.S. Forest Service, Budget Overview Fiscal Year 2014, http://www.fs.fed.us/aboutus/budget/2014/FY2014For estServiceBudgetOverview041613.pdf. Montana's outdoor economy sus-

tains 64,000 jobs and adds $6 billion to the state's economy, according to Senator Jon Tester; see Congressional Record, December 11, 2014, https://www.congress.gov/congressional-record/2014/12/11/senate-section/article/s6649-2. A 2012 study by Headwaters Economics found that rural counties in the West with "significant protected federal land added jobs more than four times faster than [counties] without protected federal land"; see "West Is Best: Protected Lands Promote Jobs and Higher Incomes," http://headwaterseconomics.org/land/west-is-best-value-of-public-lands. Hunters and anglers also play an important role in funding conservation, through excise taxes on gear and licenses; see Brittany Patterson, "Battlefield Conversions: Hunters See Climate Changing Their Traditions," *E & E News*, April 7, 2015, http://www.eenews.net/stories/1060016344.

24 **forty-three wells:** Rebecca R. Hanna, Consulting Paleontologist, "Summary of Oil and Gas Development Issues Along the Rocky Mountain Front of Montana," July 2002, prepared for The Energy Foundation.

24 **"the Lewis Overthrust":** Hal Herring, "Rocky Mountain Front Blues," *High Country News*, June, 24, 2013, http://www.hcn.org/issues/45.11/the-rocky-mountain-front-blues.

24 **U.S. demand for just three weeks:** Estimate based on USGS data analyzed by the Wilderness Society. See Pete Morton et al., *Drilling in the Rocky Mountains: How Much and At What Cost?* (Washington, DC: Wilderness Society, 2004), 7.

24 **less than a day's worth of natural gas:** Ron Selden, "Fighting for the Rocky Mountain Front: Montana Rancher Karl Rappold," *High Country News,* June 21, 2004, http://www.hcn.org/issues/277/14828.

25 **"crushed and sullen":** Myron Brinig, "Silver Bow," in *The Last Best Place*, 460.

25 **in the Bakken:** A newspaper series on western North Dakota traced its "rapid transformation from a tight-knit agricultural society to a semi-industrial oil powerhouse" and the toxic spills, rig and oil train explosions, waste dumps in the Missouri river floodplain, illegal discarding of radioactive materials, black tap water in farmers' homes and other prices being paid by residents and others downstream. See Deborah Sontag and Robert Gebeloff, "The Downside of the Boom," *New York Times*, November 22, 2014; and Deborah Sontag, "Where Oil and Politics Mix," *New York Times,* November 23, 2014.

26 **350,000 acres:** Todd Wilkinson, "Citizen Flora: The Rise and Fall and Resurrection of a Forest Service Whistleblower," *Orion*, September/October 2003.

26 **104 claims:** Keller, *The Making of a Masterpiece*, 53–54.

26 **Startech came back:** Gloria Flora, "Backing the Front, Fighting Oil and

Gas Development in Montana's Rocky Mountain Front," in *The Energy Reader: Overdevelopment and the Delusion of Endless Growth*, edited by Tom Butler, Daniel Lerch and George Wuerthner (Healdsburg, CA: Watershed Media, 2012).

27 **top ratings:** Interest group ratings come from https://votesmart.org/interest-group/1034/rating/1299#.VOySUvnF-So.

27 **a lease held by a Louisiana-based company:** A November 2015 decision by Interior Secretary Sally Jewell to withdraw the lease landed the case back in federal court. See Alex Sakariassen, "Badger–Two Medicine: Interior Moves to Cancel Lease," *Missoula Independent*, November 25, 2015, http://missoulanews.bigskypress.com/missoula/badger-two-medicine/Content?oid=2565828.

27 **Primary Petroleum:** The dramatic geology that makes the Front here so beautiful also makes it challenging to extract oil. Companies continue their search, investing in oil leases on millions of acres of private, state and reservation land, but so far have found only small volumes in an unyielding geological reality: "The [Bakken's] oil-producing layer is 20 to 80 feet thick, whereas the layer along the Front is around five feet thick." See Herring, "Rocky Mountain Front Blues."

27 **"the havoc . . . ever happened":** Much of the information about Yeager comes from George Black, "The Great Divide," *On Earth*, March 3, 2014, http://www.onearth.org/articles/2014/02/montana-energy-wars-pit-neighbor-against-neighbor-along-rocky-mountain-front.

30 **"gray, slimy, creeping cloud":** Pearl Price Robertson, "Homestead Days in Montana," in *The Last Best Place*, 532–43.

30 **"splendid farmers":** Joseph Kinsey Howard, "They Bought Satin Pajamas," in *The Last Best Place*, 524–31.

31 **almost no grizzlies left:** "Grizzly Bear Recovery," U.S. Fish and Wildlife Service, "Endangered Species," http://www.fws.gov/mountain-prairie/species/mammals/grizzly/, updated May 2, 2013.

31 **"We wouldn't move":** Darryl L. Flowers, "Ranch Family Fights For Montana Heritage," *Fairfield Sun Times*, May 7, 2013, http://www.fairfieldsuntimes.com/news/article_72d4e437-4361-53e8-ba44-05d264339144.html.

32 **"watched a bear":** Douglas H. Chadwick, "Grizzlies," *National Geographic Online Extra*, July 2001, http://ngm.nationalgeographic.com/ngm/data/2001/07/01/html/fulltext1.html.

34 **"easements on 295,000 acres":** U.S. Fish and Wildlife Service, "Rocky Mountain Front Conservation Area—Montana," U.S. Fish and Wildlife Service Refuge Planning, http://www.fws.gov/mountain-prairie/refuges/lpp_rmf.php, accessed January 11, 2016.

34 **"Once the wildness is taken away":** *Common Ground*, film directed by Alexandria Bombach (Red Reel, 2014), http://commongrounddoc.com.

35 **Rocky Mountain Front Heritage Act:** S. 364 113th Cong. (2013), https://www.congress.gov/bill/113th-congress/senate-bill/364.

36 **"This legislation was not generated":** A strain of conservatism has long celebrated this kind of community-based problem-solving and "place-based" legislation, often referencing Edmund Burke on how civic engagement starts with "little platoons." See https://conservefewell.wordpress.com/2013/01/22/societys-little-platoons-for-conservation/; and Bradley Anderson, "Green Conservatism," *American Conservative*, January 3, 2013; ". . . near the top of any list of things worth conserving: home."

37 **Citizens for Balanced Use:** Eve Byron, "Baucus Backs Heritage Act," *Independent Record*, October 29, 2011.

37 **Alliance for the Wild Rockies:** Rob Chaney, "Rocky Mountain Front Heritage Act Would Designate 6 New Wilderness Areas," *Missoulian*, May 29, 2010.

37 **Others excoriated "collaborationists" more broadly:** David Beebe, comment on Matthew Koehler, "Who'd Have Known? 47% of CFLR Projects Nationally Appealed," *A New Century of Forest Planning* (blog), May 14, 2013, http://forestpolicypub.com/2013/05/14/whod-have-known-47-of-cflr-projects-nationally-appealed.

38 **"negative impacts":** "About WWP," Western Watersheds Project, http://www.westernwatersheds.org/about/, accessed November 11, 2014.

38 **$1.69 in 2015:** Mark Mendiola, "BLM Public Grazing Fees Raised for 2015," *Western Livestock Journal*, February 9, 2015, https://wlj.net/article-permalink-11160.html.

38 **conflict in January 2016:** Phil Taylor, "Belying Militants' Claims, Ore. Ranchers and Feds Get Along," GreenWire, *E & E News*, January 13, 2016, http://www.eenews.net/greenwire/2016/01/13/stories/1060030565.

38 **16,000 ranchers:** Phil Taylor, "Bundy owes U.S. more than all other ranchers combined – BLM," *E & E News*, June 4, 2014, http://www.eenews.net/greenwire/stories/1060000713.

39 **"We had some core principles":** George Black, "The Great Divide," *On Earth*, March 3, 2014, http://www.onearth.org/articles/2014/02/montana-energy-wars-pit-neighbor-against-neighbor-along-rocky-mountain-front.

39 **"When we were working":** "Testimony in support of S.1774, the Rocky Mountain Heritage Act, Hearings on Public Lands Bills Before the Subcommittee on Public Lands and Forests of the Senate Committee on Energy and Natural Resources," 112th Cong. 51 (2012).

39 **environmentalists' indifference:** Author Rebecca Solnit made a similar criticism from within the ranks of that outsiders' movement: "I spent

some of the 1990s with and around activists in the public forests of the West, and a lot of the supposedly most radical had a remarkable knack for going into rural communities and insulting practically everyone with whom they came into contact. . . . Grubby, furry, childless pseudo-nomads who could screw up all they wanted and live hand to mouth until something went wrong and the long arm of middle-class parents reached out to rescue them scorned the tough economic choices of people with kids, mortgages, and no bail-out plan or white-collar options. Some of them did great things for trees, but their approach wasn't always, to say the least, coalition-building. It also wasn't ubiquitous . . . but the scorn was widespread enough to be a major problem. And it seemed to be part of the reason why a lot of rural people despise environmentalists." Rebecca Solnit, "One Nation Under Elvis: An Environmentalism for Us All," *Orion*, March 19, 2008, https://orionmagazine.org/article/one-nation-under-elvis/.

40 **Kerry White:** See statement of Kerry White, Executive Board Member Citizens for Balanced Use, Bozeman, Montana, Oversight Field Hearing before the Committee on Natural Resources, U.S. House of Representatives, 113th Cong., 1st sess., September 4, 2013, https://www.gpo.gov/fdsys/pkg/CHRG-113hhrg85208/pdf/CHRG-113hhrg85208.pdf. Rob Chaney, "Endangered Species Act's 40th Anniversary Draws Celebration, Critics," *Missoulian*, December 23, 2013, http://missoulian.com/news/local/endangered-species-act-s-th-anniversary-draws-celebration-critics/article_878c9174-6a96-11e3-894c-0019bb2.963f4.html. And Citizens for Balanced Use quarterly newsletter: http://balanceduse.org/wp-content/uploads/2013/11/May-CBU-Quarterly-2014-issue.final-proof.pdfhttp://balanceduse.org/wp-content/uploads/2013/11/May-CBU-Quarterly-2014-issue.final-proof.pdf.

41 **"Duck Factory":** Scott Yaich, "The Waters of the United States," Ducks Unlimited, http://www.ducks.org/conservation/habitat/the-waters-of-the-united-states, accessed March 11, 2015.

41 **demanding the transfer:** Laura Lundquist, "Montana Study of Federal-Lands Transfer Dies in Committee," *Bozeman Daily Chronicle*, March 13, 2015, http://www.bozemandailychronicle.com/news/environment/montana-study-of-federal-lands-transfer-dies-in-committee/article_3eff1966-d63c-525b-94c5-37b42d33064f.html.

43 **Set against that view of man:** A period of growing enmity between these two camps—conservationists who believe Nature should be defended for its immanent value versus conservationists who believe that the most secure protections will come from proving nature's worth to human beings, by quantifying the services ecosystems provide (water filtration and storage,

crop pollination, storm protection)—culminated in November 2014 with a kind of peace treaty, published in *Nature*, championing "a unified and diverse conservation ethic; one that recognizes and accepts all values of nature, from intrinsic to instrumental." See www.nature.com/news/working -together-a-call-for-inclusive-conservation-1.16260.

43 **"conferred nor revocable":** Michael Soule, "What Is Conservation Biology," *BioScience* 35, no. 11 (December 1985): 727–34.

43 **"most ominous . . . toward you":** David Quammen, "Jeremy Bentham, the Pieta, and a Precious Few Grayling," in *The Last Best Place*, 934–45.

43 **"a pushing inside":** A. B. Guthrie, Jr., *The Big Sky* (1947; reprinted New York: Houghton Mifflin, 2002), 150.

44 **"Who speaks for":** *Common Ground*, film.

47 **creation of the American character:** Ronald Reagan, whose administration protected a record 10 million acres as Wilderness, believed its preservation "aided liberty by keeping alive the nineteenth-century sense of adventure and awe with which our forefathers greeted the American West." Message to the Congress, October 3, 1988, http://www.presidency.ucsb. edu/ws/?pid=34951.

48 **"we're so implicated":** Michael Pollan, "An Environmentalist on a Different Path: A Fresh View of the Supposed 'Wilderness' and Even the Indians' Place in It," *New York Times,* April 3, 1999. Many made similar arguments around the Wilderness Act's 50th anniversary; see http://www.nytimes .com/2013/05/09/opinion/save-the-wolves-of-isle-royale-national-park .html?_r=2&.

48 **"ten year too late":** Guthrie, *The Big Sky*, 150.

49 **"watching Gene and Roy":** Leslie Fiedler, "The Montana Face," in *The Last Best Place*, 748.

49 **American Prairie Foundation's campaign:** American Prairie Foundation, "Frequently Asked Questions," http://www.americanprairie.org/aboutapf/ faqs/, accessed December 17, 2014.

49 **"three-million-acre bison ranch":** Emma Marris explores the paradoxical efforts to fabricate an authentic, untouched landscape in *Rambunctious Garden: Saving Nature in a Post-Wild World* (New York: Bloomsbury, 2013), 71.

49 **U.S. grasslands being lost to the plow:** Christopher Wright and Michael C. Wimberly, "Recent land use change in the Western Corn Belt threatens grasslands and wetlands," *Proceedings of the National Academy of Sciences* 110, no. 10 (2013), http://www.pnas.org/content/110/10/4134.full.

51 **"they kicked us out again":** Grazing leases that predated the Wilderness Act were left intact in wilderness areas. Leases were further protected by the Congressional Grazing Guidelines, which give the BLM and the USFS authority to suspend (but not terminate) overgrazed permits. To date, they

have not exercised this authority. Mark Squillace, "Grazing in Wilderness Areas," *Environmental Law* 44 (2014): 415–46, http://law.lclark.edu/live/files/17161-44-2squillace.pdf.

52 **"shovel rebellion":** Florence Williams, "The Shovel Rebellion," *Mother Jones,* January/February 2001, http://www.motherjones.com/politics/2001/01/shovel-rebellion. Protesters sometimes drive into a closed area to establish "historic" precedent and, by altering landscapes, make them ineligible for wilderness designation.

52 **Gloria Flora . . . resigned:** Jon Christensen, "Nevadans Drive Out Forest Supervisor," *High Country News,* November 22, 1999, https://www.hcn.org/issues/167/5393.

52 **all-terrain-vehicle protest ride:** Jonathan Thompson, "A Reluctant Rebellion in the Utah Desert: For ATVers at Recapture Canyon, Realpolitik Meets Out-of-Town Zeal," *High Country News,* May 13, 2014, https://www.hcn.org/articles/is-san-juan-countys-phil-lyman-the-new-calvin-black.

52 **federal lands into state control:** Though proponents talk about "taking back," or making the Feds "give back" the land, the Western territories were entirely owned by the U.S. from the time of the Louisiana Purchase, except for lands specifically disbursed through such policies as the Homestead and Mining Acts. In the Enabling Act that turned Montana from a territory into a state, it "forever disclaim[ed] all right and title to the unappropriated public lands within the boundaries thereof." As an editorial in the *Billings Gazette* noted, Congress has repeatedly reaffirmed the rightful ownership of these lands and resources by all Americans: the 1976 Federal Land Policy Management Act, for instance, said that "public lands should remain in federal ownership unless the disposal is in the national interest." The editors concluded, "Federal lands belong to the American people—to folks in Missouri and Iowa, New York and Texas, just as they do to Montanans and Wyomingites. Congress cannot simply give away the birthright of its citizens." See http://billingsgazette.com/news/opinion/editorial/gazette-opinion/gazette-opinion-americans-inheritance-cannot-be-given-away/article_38c4599e-2970-52f1-95f9-ef1d15f16c89.html#ixzz3CSt2Ob5R.

52 **"starving homesteaders":** An early homesteader was "the Joad of a century ago, swarming into a hostile land: duped when he started, robbed when he arrived." Howard, "Montana, High, Wide, and Handsome."

53 **1892 Johnson County War:** John W. Davis, "The Johnson County War: 1892 Invasion of Northern Wyoming," Wyoming Historical Society, http://www.wyohistory.org/essays/johnson-county-war-1892-invasion-northern-wyoming, accessed November 14, 2014.

54 **"As Ronald Reagan said in 1988":** Ronald Reagan, Message to the Congress, October 3, 1988, http://www.presidency.ucsb.edu/ws/?pid=34951.

56 **killed forty million acres:** Jason Funk et al., "Rocky Mountain Forests at

Risk: Confronting Climate-Driven Impacts from Insects, Wildfires, Heat, and Drought," Union of Concerned Scientists and the Rocky Mountain Climate Organization, September 2014.

60 **1,700 wolves:** U.S. Fish and Wildlife Service et al., "Northern Rocky Mountain Wolf Recovery Program 2013 Interagency Annual Report," edited by M. D. Jimenez and S. A. Becker (Helena, MT: USFWS, Ecological Services, 2014).

60 **"apex predators":** Though ecologists now paint a more complex picture, they still recognize the critical role of apex predators. See William J. Ripple et al., "Status and Ecological Effects of the World's Largest Carnivores," *Science* 343, no. 6167 (2014), doi 10.1126/science.1241484; and Emma Marris, "Rethinking Predators: Legend of the Wolf," *Nature* 507 (March 13, 2014): 158–60, doi:10.1038/507158a.

61 **collaborative strategies:** The collaborative approach is bearing fruit for other, less bloodthirsty but equally troubled species. Montana has the second largest population of greater sage grouse in the U.S., with two-thirds of the iconic bird's habitat in the state on private land. To avoid an Endangered Species Act listing, landowners across the bird's eleven-state range are partnering with conservation groups, oil and gas companies, and state and federal agencies to establish "habitat exchanges." Ranchers will be paid for protecting and restoring the grouse's sagebrush habitat, through the development of rigorous credits. Oil companies and others can purchase these credits to offset impacts from their rigs and roads. The same strategy is being used to protect the lesser prairie chicken, whose habitat includes Kansas. See Russell Gold, "Save a Chicken, Drill a Well," *Wall Street Journal*, July 19, 2013, http://www.wsj.com/articles/SB10001424127887324263404578612112672846 522. Landowner conservation efforts have worked for fish, too, including Montana's beautiful sail-finned trout, the nearly endangered Arctic grayling. See U.S. Fish and Wildlife Service, "Arctic Grayling Does Not Warrant Protection Under Endangered Species Act Due to Collaborative Partnerships," press release, August 19, 2014, http://www.fws.gov/mountain-prairie/pressrel/2014/08192014_ArcticGraylingDoesNotWarrantProtectionUnder ESA.php.

61 **"Calving overlaps":** Sean Wilson founded an organization called People and Carnivores, dedicated to reducing human–predator conflicts.

64 **Partners for Conservation:** Members include state and federal agencies (U.S. Fish and Wildlife Service and Natural Resources Conservation Service), other collaborative groups such as the Tallgrass Legacy Alliance of Kansas, ranchers and landowners from Washington to Florida, and environmental nonprofits. More information is available at http://partnersforconservation.org/.

64 **formalized its consultative role:** The partnership, to advance the shared goal of furthering community-based, landscape-scale conservation in the Crown, was made official in January 2013; see http://www.fs.usda.gov/ Internet/FSE_DOCUMENTS/stelprdb5441926.pdf. The departments of Agriculture and Interior are also involved in twenty-two Landscape Conservation Cooperatives (LCCs) that span the U.S. and parts of Canada and Mexico. Self-directed, these LCCs are developing the science and planning needed for conservation across jurisdictions and ownership boundaries. See National Academies of Sciences, Engineering, and Medicine, *A Review of the Landscape Conservation Cooperatives* (Washington, DC: National Academies Press, 2016).

64 **"they're people of the land":** Beginning with the Quivira coalition in 2003, collaborative conservation efforts focused on working lands have taken root throughout the arid West—creating solutions at the "radical center" that enhance ecosystem integrity at a landscape scale. For a study of these efforts see Susan Charnley, Thomas E. Sheridan and Gary P. Nabhan, eds., *Stitching the West Back Together* (Chicago: University of Chicago Press, 2014). Ranchers are also increasingly outspoken about their commitment to conservation; see http://www.hcn.org/articles/ranch-diaries-is-ranching-a-form-of-conservation?utm_source=wcn1&utm_medium=email.

71 **grass-fed beef:** Tamar Haspel lays out the many complexities of comparing grass- and grain-finished beef in "Is Grass-Fed Beef Really Better for You, the Animal, and the Planet?", *Washington Post*, February 23, 2015, http:// www.washingtonpost.com/lifestyle/food/is-grass-fed-beef-really-better-for-you-the-animal-and-the-planet/2015/02/23/92733524-b6d1-11e4-9423-f3d0a1ec335c_story.html.

71 **grasslands cover 25 percent:** Gregory P. Asner et al., "Grazing Systems, Ecosystem Responses and Global Change," *Annual Review of Environment and Resources* 29 (2004): 261–99. The most important step in reducing global emissions of meat production, according to the National Academy of Sciences, is to avoid additional "land-use change": that is, the conversion of forests or grasslands to pasture or croplands; see Petr Havlik et al., "Climate Change Mitigation Through Livestock System Transitions," *Proceedings of the National Academy of Sciences* 111, no. 10 (2014): 3709–14.

71 **emit more methane:** Doug Gurian-Sherman, "Raising the steaks: global warming and pasture-raised beef production in the United States," Union of Concerned Scientists, Cambridge, MA, 2011.

71 **vast reservoirs of carbon:** Ranchers can now be paid for the carbon benefits of their prairie preservation. In 2014, the first grassland carbon credits, issued by the American Carbon Registry, were purchased by Chevrolet, conserving 11,000 acres in critical migratory bird habitat in North Dakota.

See Sarah Jane Keller, "Chevrolet just helped bring grasslands into the carbon market: Why it matters," *High Country News,* November 26, 2014. In August 2015, the Climate Action Reserve (one of the registries developing carbon credits for California's cap-and-trade market) approved a similar protocol for paying ranchers to keep their grasslands intact. That's of particular import in California, where 90 percent of species listed as rare and endangered live within grassland ecosystems. See Environmental Defense Fund, "Why Grasslands Can Bring in the Green for Growers," January 13, 2016, http://blogs.edf.org/growingreturns/2016/01/13/why-grasslands-can-bring-in-the-green-for-growers/?utm_source=feedburner&utm_medium=email&utm_campaign=Feed%3A+GrowingReturns+%28Grow ing+Returns%29.

71 **158 million metric tons of carbon:** Lydia Olander et al., *Greenhouse Gas Mitigation Potential of Agricultural Land Management in the United States: A Synthesis of the Literature,* 3rd edition (Durham, NC: Nicholas Institute for Environmental Policy Solutions, Duke University, 2012), 38.

72 **A December 2015 study:** Insu Koh et al., "Modeling the Status, Trends, and Impacts of Wild Bee Abundance in the United States," *Proceedings of the National Academy of Sciences* 113, no. 1 (2016): 140–45.

72 **microbiotic crust:** Present in arid and semi-arid landscapes around the world, this crust may look barren but is actually millions of organisms storing moisture, discouraging weeds, reducing erosion and building nitrogen and organic matter. Its value is only beginning to be understood. See http://www.blm.gov/nstc/library/pdf/CrustManual.pdf, accessed January 11, 2016.

73 **thirty noxious weed species:** "Montana Noxious Weed Information," Montana State University, http://www.msuextension.org/invasiveplants Mangold/noxioussub.html, accessed January 13, 2016.

73 **close to 8 million acres:** "The Montana Weed Management Plan," Montana Noxious Weed Summit Advisory Council Weed Management Task Force, 2008, http://agr.mt.gov/agr/Programs/Weeds/PDF/2008weedPlan .pdf.

73 **90 percent:** "Montana Noxious Weeds," Montana Plant Life, http://montana.plant-life.org/page_weeds.htm.

73 **The invasives' names:** See "Weed ID," Montana Weed Control Association, http://www.mtweed.org/weeds/, accessed October 27, 2014. David Quammen predicts a bleak future in his essay "Planet of Weeds": a world reduced to the most successful invasives—rats, cockroaches, cheatgrass, spotted knapweed and a few tenacious human beings. "Virtually everything will live virtually everywhere, though the list of species that constitute 'everything' will be small." *Harper's,* October 1998, 57–69.

74 **Savory's idea:** While many ranchers report seeing their land and livestock thrive under Savory-style management, many rangeland scientists
remain unpersuaded that it improves water infiltration, decreases erosion
or enhances forage and livestock production. See David D. Briske, Andrew
J. Ash, Justin D. Derner, and Lynn Huntsinger, "Commentary: A Critical
Assessment of the Policy Endorsement for Holistic Management," *Agricultural Systems* 125 (March 2014): 50–53.

75 **Bud Williams's techniques:** Dusty would like to see the same practices on
public land, through bundling of individual ranchers' small grazing permits. At least one such experiment in applying adaptive, intensive grazing
is underway in southern Utah's national forests, 97 percent of which are
grazed. There, a group as diverse as Dusty's—including both woolgrowers'
and cattlemen's associations and their historic nemesis, the Grand Canyon Trust—has developed consensus recommendations to bundle national
forest leases so that they can rest meadows and vary grazing times from
one year to the next; they've also agreed on a set of indicators to monitor both ecological and social impacts, from stream biodiversity to meat
production. See Sarah Gilman, "Ranchers, Enviros and Officials Seek
a Middle Path on Public-Land Grazing," *High Country News*, February
25, 2014, http://www.hcn.org/issues/46.3/ranchers-enviros-and-officials
-seek-a-middle-path-on-public-land-grazing-in-utah. And Michele Straube
and Lorien Belton, "Collaborative Group on Sustainable Grazing for
U.S. Forest Service Lands in Southern Utah: Final Report and Consensus Recommendations," December 2012, http://ag.utah.gov/documents/
SustainableGrazingSoUtForests.pdf.

CHAPTER 2: FARMER

85 **the most severe drought on the Great Plains since 1895:** Martin Hoer-
ling, Siegfried Schubert, and Kingste Mo, "An Interpretation of the Origins
of the 2012 Central Great Plains Drought," National Oceanic and Atmospheric Administration, Drought Task Force, 2013.

86 **Kansas Hard Winter Wheat Tour:** Aaron Harries, "2013 Wheat Tour
Day Two," *The Wheat Beat* (blog), Kansas Wheat, May 1, 2013, https://
thewheatbeat.wordpress.com/2013/05/01/2013-wheat-tour-day-two/.

86 **wildly destructive downpours:** John Eligon, "After Drought, Rains
Plaguing Midwest Farms," *New York Times*, June 9, 2013, http://www
.nytimes.com/2013/06/10/us/after-drought-rains-plaguing-midwest-farms
.html?pagewanted=all.

87 **the heat returned:** Across the Western states, "temperatures on the surface
of food and forage crops hit 105 degrees, at least 10 degrees higher than

the threshold for most temperate-zone crops." Gary Paul Nabhan, "Our Coming Food Crisis," *New York Times*, July 21, 2013, http://www.nytimes.com/2013/07/22/opinion/our-coming-food-crisis.html.

87 **"died beneath blankets":** Laura Parker, "Parched: A New Dust Bowl Forms in the Heartland," *National Geographic*, May 17, 2014, http://news.nationalgeographic.com/news/2014/05/140516-dust-bowl-drought-oklahoma-panhandle-food/.

87 **That continuing volatility:** In May 2015 Justin got 4.6 inches of rain in ten days. "My soil was not able to take in near enough of that rain. It's tough when we go from so dry all through spring, then just a bunch of rain all at once. Speaks evermore to the importance of a resilient system, but difficult to be resilient with those types of extremes. A friend of mine in SE Nebraska had 11 inches over that same window. Very destructive and lots of soil loss." The summer 2015 rains hurt farmers across the Midwest. See http://blog.ucsusa.org/jason-funk/midwestern-rain-climate-change-816.

87 **most of which is paid by taxpayers:** "about 62% of total [insurance] premiums, on average, are paid by the government. In the case of catastrophic coverage, the government pays the full premium": Dennis A. Shields, "Federal Crop Insurance: Background," Congressional Research Service R40532, 2015.

88 **Wheat helped seed the Arab Spring:** Thomas L. Friedman, "The Scary Hidden Stressor," *New York Times*, March 2, 2013, http://www.nytimes.com/2013/03/03/opinion/sunday/friedman-the-scary-hidden-stressor.html.

88 **the Pentagon warned:** Coral Davenport, "Pentagon Signals Security Risks of Climate Change," *New York Times*, October 13, 2014, http://www.nytimes.com/2014/10/14/us/pentagon-says-global-warming-presents-immediate-security-threat.html. Another Pentagon report was completed in 2015 at the request of Congress: see U.S. Department of Defense, "National Security Implications of Climate-Related Risks and a Changing Climate," report to Congress, July 23, 2015, http://archive.defense.gov/pubs/150724-congressional-report-on-national-implications-of-climate-change.pdf?source=govdelivery. In September 2015 the U.N. warned that desertification could contribute to the displacement of as many as 50 million people over the next ten years. See United Nations Convention to Combat Desertification, "Water Scarcity and Desertification," Fact Sheet Series No. 2, http://www.unccd.int/Lists/SiteDocumentLibrary/Publications/Desertificationandwater.pdf, accessed January 21, 2016.

88 **the historic Dust Bowl:** 35 million acres of crop land were completely destroyed by soil erosion, 100 million acres in crops lost all of their topsoil and 125 million acres in crops were continuing to rapidly lose topsoil, leading to "the largest migration in US history." See "Surviving the Dust

Bowl: Mass Exodus from the Plains," *American Experience*, PBS, http://www.pbs.org/wgbh/americanexperience/features/general-article/dustbowl-mass-exodus-plains/, accessed April 27, 2015; David Montgomery, *Dirt: The Erosion of Civilization* (Oakland: University of California Press, 2012), 145–53; and U.S. Department of Agriculture, "Yearbook of Agriculture, 1934," edited by Milton S. Eisenhower, House Document no. 260 (Washington, DC: Government Printing Office, 1934), 78.

88 **containing a third of the planet's organisms:** Jim Robbins, "The Hidden World Under Our Feet," *New York Times*, May 11, 2013, http://www.nytimes.com/2013/05/12/opinion/sunday/the-hidden-world-of-soil-under-our-feet.html?_r=0.

88 **twenty-five horses:** From UC Berkeley biogeochemist Wendee Silver.

88 **"unprepossessing lump . . . planets":** E. O. Wilson, *Biophilia* (Cambridge, MA: Harvard University Press, 1984).

89 **90 percent . . . 60 percent:** Charles W. Rice, Distinguished Professor of Soil Microbiology, Kansas State University, in discussion with the author, February 2014.

89 ***Nature* reported:** Denise Grady, "New Antibiotic Stirs Hope Against Resistant Bacteria," *New York Times*, January 7, 2015, http://www.nytimes.com/2015/01/08/health/from-a-pile-of-dirt-hope-for-a-powerful-new-antibiotic.html?ref=health&_r=0.

90 **"Animals in a Bacterial World":** Margaret McFall-Ngai et al., "Animals in a Bacterial World: A New Imperative for the Life Sciences," *Proceedings of the National Academy of Sciences* 110, no. 9 (2013): 3229–36. In October 2015, scores of leading scientists called for a "moonshot" study of the planet's many microbiomes, from deep-sea volcanoes to Antarctic deserts. Microbial diversity is staggering: compared to the forty phyla in the animal kingdom, there are more than 1,000 microbial phyla. See Carl Zimmer, "Scientists Urge National Initiative on Microbiomes," *New York Times*, October 28, 2015, http://www.nytimes.com/2015/10/29/science/national-initiative-microbes-and-microbiomes.html.

90 **500 years:** "State Soil," Natural Resources Conservation Service, http://www.nrcs.usda.gov/wps/portal/nrcs/detail/ks/soils/?cid=nrcs142p2_033163, accessed April 10, 2015.

90 **soil loss worldwide:** Montgomery, *Dirt*, xii. And Chris Arsenault, "Only 60 Years of Farming Left If Soil Degradation Continues," *Scientific American*, December 5, 2014, http://www.scientificamerican.com/article/only-60-years-of-farming-left-if-soil-degradation-continues/. Within that loss is hidden an equally dire vanishing: the extinction of untold microbial species. In 2012, agricultural researchers from around the world founded the Global Soil Biodiversity Initiative, seeking to understand where soil life is

most critically endangered. Its chair, Diana Wall, has studied soil biodiversity in Antarctica and Kansas. See Jim Robbins, "The Hidden World Under Our Feet," *New York Times*, May 11, 2013, http://www.nytimes.com/2013/05/12/opinion/sunday/the-hidden-world-of-soil-under-our-feet.html?_r=0.

91 *doubling* **of the amount of nitrogen:** John Aber et al., "Human Alteration of the Global Nitrogen Cycle: Causes and Consequences," *Issues in Ecology* (Washington, DC: Ecological Society of America, 1997), 4.

91 **quadrupling:** Rodney M. Fujita, *Heal the Ocean: Solutions for Saving Our Seas* (Gabriola Island, BC: New Society, 2003), 56.

92 **agriculture the biggest polluter:** Agricultural sources contribute more than 70 percent of the nitrogen and phosphorus delivered to the Gulf, according to the U.S. Geological Survey National Water Quality Assessment Program; see http://water.usgs.gov/nawqa/sparrow/gulf_findings/primary_sources.html.

92 **algal blooms that deplete oxygen:** "Hypoxia," National Ocean Service, http://oceanservice.noaa.gov/hazards/hypoxia/, updated October 23, 2014.

92 **In drinking water:** Carl Zimmer, "Cyanobacteria Are Far From Just Toledo's Problem," *New York Times*, August 7, 2014, http://www.nytimes.com/2014/08/07/science/cyanobacteria-are-far-from-just-toledos-problem.html. The costs of cleaning the water fall on urban water utilities, giving rise to litigation. The city of Des Moines, Iowa, is suing to force three upstream farm counties to cut nitrogen pollution through fertilizer reductions, cover crop plantings or construction of wetland or grassland buffers. See Dan Charles, "Iowa's Largest City Sues Over Farm Fertilizer Runoff in Rivers," *The Salt,* NPR, January 12, 2015, http://www.npr.org/blogs/thesalt/2015/01/12/376139473/iowas-largest-city-sues-over-farm-fertilizer-runoff-in-rivers. Hundreds of small towns in the Midwest are now struggling with unsafe drinking water and will have to increase water rates as much as tenfold; see http://harvestpublicmedia.org/article/nitrates-costly-persistent-problem-small-towns.

94 **two hundred miles wide:** Timothy Egan, *The Worst Hard Time: The Untold Story of Those Who Survived the Great American Dust Bowl* (Boston: Houghton Mifflin, 2006), 5, 203.

94 **15 to 18 inches:** Ray Ward, remarks to the No-Till on the Plains conference, Salina, KS, February 2014.

94 **rate of loss in Iowa:** Tom Philpott, "Iowa Is Getting Sucked Into Scary Vanishing Gullies," *Mother Jones,* February 7, 2014, http://www.motherjones.com/tom-philpott/2014/02/iowas-vaunted-farms-are-losing-topsoil-alarming-rate.

94 **1.7 billion tons of topsoil:** Natural Resources Conservation Service, "Summary Report: 2012 National Resources Inventory," U.S. Department

of Agriculture and Center for Survey Statistics and Methodology, Iowa State University, http://www.nrcs.usda.gov/technical/nri/12summary.

94 **$44 billion a year:** David Pimentel et al., "Environmental and Economic Costs of Soil Erosion and Conservation Benefits," *Science-AAAS-New Series* 267, no. 5201 (1995): 1117–23.

95 **$5 an acre:** That land is worth about $5,000 an acre today.

97 **Ida Watkins claimed profits:** Egan, *The Worst Hard Time*, 44.

102 **"No-tilling," they called it:** At least one experiment in no-till began far earlier. Thomas Jefferson experimented with complex crop rotations and no-tilling, noting how well turnips succeeded "sown on stubble without hoeing. The stubble keeps the land light." He then derailed his own experiments by designing the mouldboard plow. Two centuries later, no-till pioneers were again having to improvise their own equipment. Rice recalls some yanking the heavy metal counterweights out of old sash windows to drag behind their injector knives. See Harry W. Fritz, "The Agrarian," Discovering Lewis and Clark, http://lewis-clark.org/content/content-article .asp?ArticleID=1749#, accessed April 27, 2015.

105 **more and diverse nesting birds:** Kelly R. VanBeek, Jeffrey D. Brawn and Michael P. Ward, "Does No-Till Soybean Farming Provide Any Benefits for Birds?", *Agriculture, Ecosystems & Environment* 185 (2014): 59–64. The study also reported that delaying soybean planting by just a few days would further boost nesting success, pointing again to the critical role of private lands. As one of the authors said, "there's so much land in agriculture that if only 3 or 4 percent of farmers adopted this approach, it would have a greater effect than all the land that we have in wildlife preserves in Illinois." See Diana Yates, "No-Till Soybean Fields Give (Even Some Rare) Birds a Foothold in Illinois," News Bureau of the University of Illinois, January 21, 2014, http://www.news.illinois.edu/news/14/0121no-till_JeffreyBrawn_MichaelWard.html.

106 **goo coats soil particles:** Every 1% of added organic matter builds a "bank account" of 1,000 pounds of nitrogen and 100 pounds of phosphorous per acre. Barry Fisher, Natural Resources Conservation Service, quoted in Karen Chapman, "No-Till Farming Can Reduce Input Costs and Improve Soil Health," Environmental Defense Fund *Growing Returns* blog, January 29, 2015, http://blogs.edf.org/growingreturns/2015/01/29/no-till-farming -can-reduce-input-costs-and-improve-soil-health/.

106 **desirable "aggregates":** According to Rice, 30% of prairie soils and 20% of no-till are in big aggregates, while tilled soils have almost none. No-till farmer Lance Feikert, who farms near Dodge City, KS, reported that with no-till residues "I got 3.2 inches per hour soaking into the soil. Without cover, you only get 0.22 an hour." See Tim Unruh, "No-Till Push-

ers Perplexed by Sluggish Adoption of Farming Practice," *Salina Journal,* January 29, 2015, http://www.salina.com/news/no-till-pushers-perplexed-by-sluggish-adoption-of-farming-practice/article_9a18adf3-6d55-5031-a938-0a511818925d.html. Even reduced tillage with cover cropping can increase infiltration up to 45%; see Gary Hawkins, Dana Sullivan and Clint Truman, "Water Savings Through Conservation Tillage," University of Georgia Cooperative Extension, http://www.ars.usda.gov/SP2UserFiles/Place/60480500/WaterSavingsThroughConservationTillage.pdf, accessed April 13, 2015.

106 **"Drop a chunk of healthy soil":** Natural Resources Conservation Service agronomist Chris Lawrence did this demonstration at a congressional briefing to illustrate the importance of soil organic matter to agricultural resiliency in times of drought. See Niina Heikkinen, "How to Slash CO2, Ease Drought Impacts and Curb Flooding with a Carbon-Rich Soil 'Sponge'," ClimateWire, *E & E News,* April 14, 2015, http://www.eenews.net/climatewire/2015/04/14/stories/1060016695.

107 **earthworms:** An analysis of 58 studies found earthworms boosting crop yields 25%, mainly by increasing nitrogen availability. See Jan Willem van Groenigen et al., "Earthworms Increase Plant Production: A Meta-Analysis," *Scientific Reports* 4 (2014), doi:10.1038/srep06365.

110 **"cover crop":** An alternative approach, called companion cropping, mixes cover crops in with the commercial crop, choosing species that will be dead or otherwise out of the way by harvest time (e.g., short plants amidst tall sunflowers). The companions that come up first provide nitrogen and an early canopy for the sunflower seedlings and attract beneficial insects: pollinators and predator wasps that eat pests. Justin's grandfather practiced a version of this, planting clover as a "nurse crop" for the wheat, to shelter and supply nitrogen to the young plants. See Mark Parker, "Companion Cropping Boosts No-Till Profits, Soil Health," *No-Till Farmer,* June 1, 2013, http://www.no-tillfarmer.com/articles/150-companion-cropping-boosts-no-till-profits-soil-health.

110 **yield improvements . . . of up to 10 percent:** Conservation Technology Information Center and the North Central Region Sustainable Agriculture Research and Education Program, "2013–2014 Cover Crop Survey Report," 24, http://www.northcentralsare.org/content/download/73722/1191432/2013-14_Cover_Crop_Survey_Report.pdf?inline download=1, accessed April 27, 2014.

111 **reduces runoff of nitrogen and phosphorus:** Dan Charles, "Here's How to End Iowa's Great Nitrate Fight," *The Salt,* NPR, February 2, 2015, http://www.npr.org/blogs/thesalt/2015/02/02/382475870/heres-how-to-end-iowas-great-nitrate-fight.

111 **tenfold increase in cover cropping:** Cheryl Tevis, "Clean Up the Water with Cover Crops," *Successful Farmer*, November 19, 2014, http://www.agriculture.com/crops/cover-crops/cle-up-water-with-cover-crops_568-ar46276.

111 **target set by the EPA:** Environmental Protection Agency, "States Develop New Strategies to Reduce Nutrient Levels in Mississippi River, Gulf of Mexico," news release, February 12, 2015, http://yosemite.epa.gov/opa/admpress.nsf/bd4379a92ceceeac8525735900400c27/c1feec0ba93871db85257dea005f017f!OpenDocument.

111 **chemicals:** In 2016, it appeared that chemicals would finally begin to get serious scrutiny: the House and Senate both passed important reforms of the Toxic Substances Control Act for the first time since 1976, again with important impetus from EDF. See Environmental Defense Fund, blog post, January 11, 2016, http://blogs.edf.org/health/2016/01/11/will-we-take-this-best-chance-ever-to-fix-the-law-that-helped-bring-about-duponts-pfoa-debacle/.

111 **Roundup:** In March 2015, a finding by the World Health Organization's International Agency on Cancer Risk (IARC) that glyphosate is a "probable" carcinogen sparked enormous debate, particularly because in the same year both German regulators acting on behalf of the EU and the European Food Safety Authority came to the opposite conclusion, finding glyphosate unlikely to cause cancer (as the U.S. Environmental Protection Agency has concluded in every annual review since 1991). Helpful roadmaps through the complexities are provided by Andrew Pollack, "Weed Killer, Long Cleared, Is Doubted," *New York Times*, March 27, 2015, http://www.nytimes.com/2015/03/28/business/energy-environment/decades-after-monsantos-roundup-gets-an-all-clear-a-cancer-agency-raises-concerns.html?_r=0; and Sarah Zhang, "Bacon Causes Cancer? Sort Of. Not Really. Ish," *Wired*, October 27, 2015, http://www.wired.com/2015/10/who-does-bacon-cause-cancer-sort-of-but-not-really/.

111 **"a short-term solution":** Mark Bittman, "Now This Is Natural Food," *New York Times,* October 22, 2014, http://www.nytimes.com/2013/10/23/opinion/bittman-now-this-is-natural-food.html.

111 **Chipotle's viral YouTube video:** 15.4 million views as of February 2016. "The Scarecrow," September 11, 2013, http://youtu.be/lUtnas5ScSE.

112 **poison banned nationwide:** Charles F. Wurster, *DDT Wars: Rescuing Our National Bird, Preventing Cancer, and Creating the Environmental Defense Fund* (New York: Oxford University Press, 2015).

113 **plastic mulch:** United States agriculture uses about a billion pounds of plastic annually, much of it designed for one season's use. See Elizabeth Grossman, "The Biggest Source of Plastic Trash You've Never Heard Of,"

Ensia, March 30, 2014, http://ensia.com/features/the-biggest-source-of-plastic-trash-youve-never-heard-of/.

114 **Early concerns:** See "Background Information: Bee Losses Caused by Insecticidal Seed Treatment in Germany in 2008," Federal Office of Consumer Protection and Food Safety, Germany, http://www.bvl.bund.de/EN/08_PresseInfothek_engl/01_Presse_und_Hintergrundinformationen/2008_07_15_hi_Bienensterben_en.html?nn=1414138.

115 **291-page EPA report:** Environmental Protection Agency, "Preliminary Pollinator Assessment to Support the Registration Review of Imidacloprid," January 4, 2016, http://www.regulations.gov/#!documentDetail;D=EPA-HQ-OPP-2008-0844-0140.

116 **Insects possess an astounding ability:** Dwayne Beck, remarks to the No-Till on the Plains conference, Salina, KS, February 2014.

117 **replace every feedlot:** Jim Gerrish, remarks to the No-Till on the Plains conference, Salina, KS, February 2014.

117 **Gabe Brown's 5,400-acre farm:** Justin has mulled other nonchemical options used by some organic farmers, including running a ten-foot flamethrower or dragging a steamroller through a field, but isn't persuaded that they're effective enough to merit the fossil fuel use.

118 **many times the impact:** California tomato farmer Bruce Rominger, a Big Ag hero of Mark Bittman's (and lifelong friend of the author's), argues that economies of scale are by definition measures of sustainability. "If I'm using less money that means I'm using fewer resources—fertilizer, water, diesel . . . growing more crop per drop" (remarks at the *New York Times* Food for Tomorrow conference, Pocantico Hills, NY, November 2014). Colorado State University animal science professor Temple Grandin, the reigning voice on humane treatment of livestock, sees the same dynamic playing out there. "Consumers think big is bad. But what I've found is badly managed is bad . . . I've been in big places that are really good. I've been in big places that are bad. It gets down to attitude of management." See Luke Runyon and Harvest Public Media, "For Temple Grandin, Big Farms Aren't Necessarily Bad Farms," December 28, 2015, http://www.kunc.org/post/temple-grandin-big-farms-aren-t-necessarily-bad-farms#stream/0.

118 **locally grown:** Numerous studies have demonstrated that "local food" is not inherently better for the environment. Nonlocal food may be produced far more efficiently. And transportation accounts for just 4% of the greenhouse gas emissions associated with the average American diet. See Christopher L. Weber and H. Scott Matthews, "Food-Miles and the Relative Climate Impacts of Food Choices in the United States," *Environmental Science and Technology* 42, no. 10 (2008): 3508–13. For more on this topic,

see Pierre Desrochers and Hiroko Shimizu, *The Locavore's Dilemma* (New York: Public Affairs, 2012).

118 **"whole pattern of eating":** This concept underscores the importance of native seed conservation, from small local seed sections at public libraries to organizations like Native Seed/SEARCH that focus on conserving plants adapted over millennia to particular climates and soils. See Gary Paul Nabhan, ed., *Renewing America's Food Traditions: Saving and Savoring the Continent's Most Endangered Foods* (White River Junction, VT: Chelsea Green, 2008).

118 **crops akin to the native neighbors:** Where the native ecosystem is forest, a farmer might layer fields and trees. Dusty's old ally Gloria Flora is experimenting with that kind of "agroforestry" on the two Montana acres she retired to in the mountains north of Helena. Beneath the shade of pine and spruce trees harboring native birds and mammals she has planted fruit and nut trees and, below that, layers of crops: raspberry bushes, edible flowers, grape and hop vines, medicinal yarrow and arnica. Turkeys and chickens eat the fallen fruit and return the nutrients to the soil. See Jim Robbins, "A Quiet Push to Grow Crops Under Cover of Trees," *New York Times*, November 21, 2011, http://www.nytimes.com/2011/11/22/science/quiet-push-for-agroforestry-in-us.html.

119 **deeper and more relentless:** Adding to the miseries in western Kansas, the Ogallala aquifer, which spans eight states and supplies 30% of the nation's groundwater for irrigation, has declined by a third. The water pumped out in just the past few decades would take more than a thousand years to recharge. See David R. Steward et al., "Tapping Unsustainable Groundwater Stores for Agricultural Production in the High Plains Aquifer of Kansas, Projections to 2110," *Proceedings of the National Academy of Sciences* 110, no. 37 (2013): E3477–86.

120 **"casting a vote":** Michael Pollan, "Behind the Organic-Industrial Complex," *New York Times*, May 13, 2001, http://www.nytimes.com/2001/05/13/magazine/13ORGANIC.html. The USDA's "organic" label for crops verifies only "that irradiation, sewage sludge, synthetic fertilizers, prohibited pesticides, and genetically modified organisms were not used." Though the certification says growers "must" use (and keep detailed records of) practices that improve soil fertility, including rotations, and manage nutrients and pests in ways that prevent soil and water contamination, farmers are not held to any particular performance standard or metric. See Nathanael Johnson, "What Does 'Organic' Actually Mean?", *Grist*, November 4, 2015, http://grist.org/food/what-does-organic-actually-mean/. Complete regulations of the National Organic Standards Board are

available at: http://www.ams.usda.gov/AMSv1.0/ams.fetchTemplateData
.do?template=TemplateN&navID=OrganicStandardslinkNOSBHome&
rightNav1=OrganicStandardslinkNOSBHome&topNav-&leftNav=
&page=NOPOrganicStandards&resultType=&acct=nopgeninfo, updated
April 4, 2013.

120 **"two-party" system:** Tamar Haspel, "Organic Standards Fight Over Synthetics Shows There's Room for a Third System," *Washington Post*, June 13, 2014, http://www.washingtonpost.com/lifestyle/food/organic-standards-fight-over-synthetics-shows-theres-room-for-a-third-system/2014/06/12/a509a086-eff0-11e3-bf76-447a5df6411f_story.html.

120 **"hybrid system":** A 2012 study by Jonathan Foley, then head of the University of Minnesota's Institute on the Environment, arrived at a similar conclusion: "We really need new 'hybrid' approaches, taking the best of the conventional and organic paradigms, and deploying them when and where they make the most sense." See Andrew C. Revkin, "Study Points to Roles for Industry and Organics in Agriculture," *New York Times*, April 25, 2012, http://dotearth.blogs.nytimes.com/2012/04/25/study-points-to-roles-for-industry-and-organics-in-agriculture/.

120 **the Big Ag story:** Tom Philpott, "Does 'Corporate Farming' Exist? Barely," *Mother Jones*, September 25, 2013, http://www.motherjones.com/tom-philpott/2013/09/does-corporate-farming-exist-barely.

120 **"I'm not talking about":** Mark Bittman, "A Simple Fix for Farming," *New York Times*, October 19, 2012, http://opinionator.blogs.nytimes.com/2012/10/19/a-simple-fix-for-food/.

121 **"making the world safe":** That the "agrochemical industry was going to continue being the agrochemical industry and keep promoting the same old practices," as Becky Goldburg, then senior scientist at the Environmental Defense Fund, told author Dan Charles; see *Lords of the Harvest: Biotech, Big Money, and the Future of Food* (New York: Basic Books, 2001), 97.

121 **Glickman challenged Monsanto:** Then Deputy Secretary of Agriculture Richard Rominger provided this account to the author.

121 **publicly apologize:** In his speech at the Greenpeace Business Conference, Monsanto's chairman and CEO Robert B. Shapiro acknowledged the many concerns about GMOs, including the underlying ethics, consumer safety, potential impacts on biodiversity and on traditional and organic farming practices, and the danger of concentration of power over the food supply. See http://news.bbc.co.uk/2/hi/science/nature/468147.stm.

121 **Monsanto ranked 97th:** beating out only AIG, Goldman Sachs and Dish Network. Harris Poll, 2015 Reputation Quotient Summary Report, http://skift.com/wp-content/uploads/2015/02/2015-RQ-Media-Release-Report_020415.pdf.

121 **"enlisted academics":** Eric Lipton, "Food Industry Enlisted Academics in G.M.O. Lobbying War, Emails Show," *New York Times*, September 5, 2015, http://www.nytimes.com/2015/09/06/us/food-industry-enlisted-academics -in-gmo-lobbying-war-emails-show.html. For an alternative view, see David Kroll, "What the New York Times Missed on Kevin Folta and Monsanto's Cultivation of Academic Scientists," *Forbes*, September 10, 2015, http:// www.forbes.com/sites/davidkroll/2015/09/10/what-the-new-york-times- missed-on-kevin-folta-and-monsantos-cultivation-of-academic-scientists/# 2715e4857a0b6be59bc972a0.

121 **"as an environmentalist":** Mark Lynas, "Lecture to Oxford Farming Con- ference," January 3, 2013, http://www.marklynas.org/2013/01/lecture-to- oxford-farming-conference-3-january-2013/.

122 **no evidence that eating GMOs poses risks:** Alessandro Nicolia, Alberto Manzo, Fabio Veronesi and Daniele Rosellini, "An Overview of the Last 10 Years of Genetically Engineered Crop Safety Research," *Critical Reviews in Biotechnology* 34, no. 1 (2013): 77–88; and "Statement by the American Association for Advancement of Science Board of Directors On Labeling of Genetically Modified Foods," October 2012, http://www.aaas.org/sites/ default/files/AAAS_GM_statement.pdf.

122 **public backlash and its costs:** These include the stalled release of "golden rice," engineered to remedy the vitamin A deficiency that causes blindness in children and millions of avoidable deaths each year in Asia and Africa. The humanitarian effort is being led by the nonprofit International Rice Research Institute. IRRI will price the seeds on par with other varieties (and allow farmers to save seeds), thanks to a partnership with Syngenta that secured them royalty-free access to the patents and intellectual property of several biotech companies. Washington University in St. Louis professor Glenn Davis Stone provides a nuanced history, challenging those making premature claims for the technology's readiness, but also defending the promise of genetic engineering, particularly for adding disease resistance to cassava and banana, both staples for the poor in Africa and Asia. See "Golden Rice: Glenn Davis Stone," *The Secret Ingredient*, KUT News, February 13, 2016, http://kutpodcasts.org/category/the-secret-ingredient; International Rice Research Institute, Golden Rice, "The Project," http://irri.org/golden -rice/the-project; and Amy Harmon, "Golden Rice: Lifesaver?" *New York Times*, August 24, 2013, http://www.nytimes.com/2013/08/25/sunday -review/golden-rice-lifesaver.html?pagewanted=all&_r=0.

122 **Harmon traced the environmental harm:** "A Race to Save the Orange by Altering Its DNA," *New York Times*, July 27, 2013, http://www.nytimes .com/2013/07/28/science/a-race-to-save-the-orange-by-altering-its-dna .html?pagewanted=all.

122 **Haspel has championed:** Tamar Haspel, "Unearthed: Can This GMO Save Our Oceans?" *Washington Post*, July 18, 2014, http://www.washington post.com/lifestyle/food/unearthed-can-this-gmo-save-our-oceans/2014/07/18/3562fb54-094c-11e4-a0dd-f2b22a257353_story.html.

122 **"Roundup Ready corn":** Dwayne Beck, remarks to the No-Till on the Plains conference, Salina, KS, February 2014.

122 **"pesticide treadmill":** In March 2015, the EPA announced that it will require glyphosate manufacturers to develop a weed resistance management plan; see Carey Gillam, "EPA Will Require Weed-Resistance Restrictions on Glyphosate Herbicide," Reuters, March 31, 2015, http://mobile.reuters .com/article/idUSKBN0MR2JT20150331?irpc=932.

122 **proprietary seeds:** Observers note that intellectual property claims for GMOs stir higher passions than claims for other kinds of IP because seeds are a "technology" with rich cultural and spiritual association, historically "part of the natural world that belongs to everybody and nobody, like dirt or the ocean." See Lessley Anderson, "Why Does Everyone Hate Monsanto?" *Modern Farmer,* March 4, 2014, http://modernfarmer.com/2014/03/monsantos-good-bad-pr-problem/.

123 **Public funding for seed development:** Rural Advancement Foundation International, *Proceedings of 2014 Summit on Seeds and Breeds for 21st Century Agriculture*, October 2014, http://rafiusa.org/publications/seeds/.

123 **In 1980 . . . private-sector seeds:** Jack A. Heinemann et al., "Sustainability and Innovation in Staple Crop Production in the US Midwest," *International Journal of Agricultural Sustainability* 12, no. 1 (2014): 71–88.

124 **"spatial data recording":** Tim Unruh, "Old McDonald Has a Laptop," *Salina Journal*, March 6, 2011.

124 **Justin hosted U.S. Senator Pat Roberts:** Tim Unruh, "Senator Roberts Praises Pace of Progress in Unmanned Technology for Agriculture," *Salina Journal*, November 14, 2015.

125 **"the math":** Rick Montgomery, "For Kansas Wheat, Drought and Russia Lead to a Harvest of Hard Luck," *Kansas City Star*, June 8, 2014, http://www.kansascity.com/news/business/article505048/For-Kansas-wheat-drought-and-Russia-lead-to-a-harvest-of-hard-luck.html.

126 **the Haney scale:** Named for its developer, Rick Haney of the Agricultural Research Service in Temple, TX.

126 **2014 survey by Kansas State:** Jason Bergtold, "Adoption and Intensification of Conservation Practices, Risk and Policy," paper delivered at the Risk and Profit conference, Kansas State University, August 22, 2014, http://www.agmanager.info/events/risk_profit/2014/Papers/11_Bergtold_ConservationPractices.pdf.

127 **"agroecology":** Olivier De Schutter, "Agroecology and the Right to Food" presented at the 16th Session of the United Nations Human Rights Council, http://www.srfood.org/images/stories/pdf/officialreports/20110308_a-hrc -16-49_agroecology_en.pdf.

127 **one of the planet's most vulnerable places:** P. Romero-Lankao et al., "North America," in *Climate Change 2014: Impacts, Adaptation, and Vulnerability. Part B: Regional Aspects,* edited by V. R. Barros et al. (Cambridge and New York: Cambridge University Press, 2014), 1439–98.

127 **Palmer amaranth:** Steve Karnowski, "Palmer Amaranth Weed Creeping Across Midwest from South," Associated Press, October 28, 2013, http://www.huffingtonpost.com/2013/10/28/palmer-amaranth-weed_ n_4170699.html.

127 **In 2013, Kansas State forecast:** Andrew Barkley et al., "Weather, Disease, and Wheat Breeding Effects on Kansas Wheat Varietal Yields, 1985 to 2011," *Agronomy Journal* 106, no. 1 (2014): 227–35.

128 **"How to Cope with Climate Change":** The April 2014 issue of *Midwest Producer* took a similarly straightforward tack in an article about Chuck Rice's role as lead author of the International Panel on Climate Change report; see http://m.midwestproducer.com/news/regional/k-state-professor -leads-group-making-recommendation-on-climate-change/article_53e 4b08a-c960-11e3-bb6a-0019bb2963f4.html?mobile_touch=true.

129 **"poorly served by farm groups":** An exception is Ray Gaesser, chair of the American Soybean Association, who has been outspoken about the rapid rise of extreme weather. In his five decades of farming in Iowa, he says he can't remember a period with such intense rains. See J. Gordon Arbuckle, Jr., Lois Wright Morton and Jon Hobbs, "Understanding Farmer Perspectives on Climate Change Adaptation and Mitigation: The Roles of Trust in Sources of Climate Information, Climate Change Beliefs, and Perceived Risk," *Environment and Behavior* 47, no. 2 (February 2015): 205–34; and Niina Heikkinen, "The Difficult Art of Communicating Climate Change to Farmers," ClimateWire, *E & E News,* January 28, 2015, http://www .eenews.net/stories/1060012312.

129 **Cargill's chairman Gregory Page:** Page told the *New York Times* that he doesn't know or care whether human activity causes climate change but that it would be irresponsible for Congress and farmers not to take it seriously. He noted another cost of drought and flood: the disruption of transport to market for crops. Like Kansas State, he treads lightly when addressing farmers' groups, who he says have been conditioned to think of global warming as a liberal euphemism for more regulation. "I ask simple questions: 'Would you like universities to suspend research on seeds that grow in higher temperatures?' Of course not! That's all I'm saying! . . .

You get people to acknowledge that they too have anxieties. It's a micro-acknowledgment, not a macro-acknowledgment." See Burt Helm, "Climate Change's Bottom Line," *New York Times*, January 31, 2015, http://www .nytimes.com/2015/02/01/business/energy-environment/climate-changes-bottom-line.html?_r=0.

129 **best strategies for boosting climate resiliency:** The International Food Policy Research Institute similarly identified no-till as the most important innovation for sustaining yields as the climate changes, with potential to increase wheat yields by 32%. See Mark W. Rosegrant et al., "Food Security in a World of Natural Resource Scarcity: The Role of Agricultural Technologies," International Food Policy Research Institute, 2014.

129 **carbon lost to the atmosphere through cultivation:** Agriculture has overtaken deforestation as the leading source of land-based greenhouse gas pollution. See Francesco N. Tubiello et al., "The Contribution of Agriculture, Forestry and other Land Use Activities to Global Warming, 1990–2012," *Global Change Biology* 21 (2015): 2655–60, doi:10.1111/gcb.12865.

130 **lost half or more of its stored carbon:** Rattan Lal, Ronald F. Follett, B. A. Stewart and John M. Kimble, "Soil Carbon Sequestration to Mitigate Climate Change and Advance Food Security," *Soil Science* 172, no. 12 (2007): 943–56.

130 **15 percent of the total carbon reduction:** Uta Stockmann et al., "The Knowns, Known Unknowns and Unknowns of Sequestration of Soil Organic Carbon," *Agriculture, Ecosystems and Environment* 164 (2013): 80–99. Stephen Pacala, co-director of Princeton's Carbon Mitigation Initiative, noted at a meeting in New York in September 2014 that land emissions can be reversed in a way that fossil fuel emissions cannot, at low cost, on lands that are already intensively managed, and with multiple co-benefits for productivity.

130 **Overfertilization:** Emissions from synthetic fertilizer use are a growing problem worldwide, especially in Asia. Between 2011 and 2012 total emissions from fertilizer rose 8%. "Agriculture, Forestry and Other Land Use Emissions by Sources and Removals by Sinks," March 2014, http://www .fao.org/docrep/019/i3671e/i3671e.pdf.

131 **less than 20 percent:** Brad Reagan, "Plowing Through the Confusing Data on No-Till Farming," *Wall Street Journal*, October 15, 2012, http:// www.wsj.com/news/articles/SB10000872396390443855804577602931 348705646. "'No-Till' Practices are Used on Over Half of Major Cropland Acres," USDA Economic Research Service and National Agricultural Statistics Service, http://ers.usda.gov/data-products/chart-gallery/detail .aspx?chartId=49982&ref=collection&embed=True, updated December 5, 2014. Oklahoma, where Justin's grandparents endured the worst of the

Dust Bowl, is now a leader; see Brian Bienkowski, "Putting Down the Plow in Oklahoma," *Environmental Health News*, October 15, 2015, http://www.environmentalhealthnews.org/ehs/news/2015/oct/no-till-farming-oklahoma-crop-rotation-winter-wheat-climate-chant-runoff-pesticides.

132 **prevent erosion:** As Justin told *Progressive Farmer* while planting soy: "Of what we have control of, this is the most important thing—placing the seed properly." The longer he no-tills, he added, "the deeper I feel comfortable dropping seed depth," because his fields are less crusted and more porous, so seeds can easily emerge. See Emily Unglesbee, "Planting on the Plains: Details Matter for Central Kansas Farmer," *Progressive Farmer*, May 29, 2014, http://www.dtnprogressivefarmer.com/dtnag/common/link .do;jsessionid=0F225DA23C2E33DA2ADE11BD09CDE3C7.agfree jvm2?symbolicName=/free/news/template1&product=/ag/news/bestofdtn pf&vendor.

133 **"the symbiotic relationships":** No-Till on the Plains, "About Us," http://www.notill.org/about-us, accessed January 14, 2016.

134 **Jonathan Cobb:** Cobb's story comes from his remarks to the No-Till on the Plains conference in Salina, KS, February 2014, and from Jim Steiert, "Succeeding at No-Till Without 'Chasing Acres'," *No-Till Farmer*, January 9, 2015, http://www.no-tillfarmer.com/articles/4263-succeeding-at-no-till-without-chasing-acres.

135 **network of innovators:** "Mighty Farming Microbes: Companies Harness Bacteria to Give Crops a Boost," *The Salt*, NPR, June 12, 2015, http://www.npr.org/sections/thesalt/2015/06/12/413692617/mighty-farming-microbes-companies-harness-bacteria-to-give-crops-a-boost?utm_source=facebook.com&utm_medium=social&utm_campaign=npr&utm_term=nprnews&utm_content=20150613.

135 **"biologics":** Biologics are now being enlisted to fight another of Justin's foes, invasive cheatgrass. Researchers have found that "weed-suppressive" bacteria inhibits new cheatgrass shoots. See http://www.hcn.org/articles/researchers-find-formidable-foe-for-invasive-cheatgrass.

135 **novel enzymes:** Justin's neighbor Wes Jackson has been engaged for decades in another kind of biological innovation that has brought many acolytes to his Salina farm, including author Michael Pollan, *New York Times* columnists Mark Bittman and Thomas Friedman, and Stone Barns chef Dan Barber. His aim is to turn annual grains, like wheat, into perennials, making them able to fend for themselves and build soil as robustly as native grasses. Jackson loves to show a picture of a scrawny, six-inch annual winter wheat root dwarfed by the ten-foot-long root of its perennial wheatgrass cousin, a thick, dense, beautiful dreadlock that can capture water and nutrients year-round. Justin pays close attention to Jackson's work and admires his

very long-term vision. Chuck Rice points to an inherent dilemma: a plant funneling its energy into building a big fat dreadlock root will have little energy to put into seed-making.

135 **classify produce as good, better or best:** Stephanie Strom, "Whole Foods to Rate Its Produce and Flowers for Environmental Impact," *New York Times,* October 15, 2014, http://www.nytimes.com/2014/10/16/business /whole-foods-to-rate-its-produce-and-flowers-for-environmental-impact .html?_r=0. Protests against Whole Foods' new metrics led the company to modify the ratings to give more credit and transition time to organic producers, while also continuing to "raise the bar" for things organic certification does not encompass. See http://civileats.com/2015/07/14/organic -farmers-and-whole-foods-reach-ceasefire-over-responsibly-grown-ratings/.

136 **600 member retailers:** In 2014, United Suppliers rolled out a new service for its member retailers to offer their farmer-customers: the "SUSTAIN" platform, a suite of technologies and practices for optimizing fertilizer use. A number of other consultancies and producer peer groups are also stepping up to help farmers calculate and shrink their "fieldprint." Justin has help, for instance, from consultant Sara Harper at K·Coe Isom, a firm that helps agribusinesses navigate new sustainability demands from their customers. The newly launched NutrientStar assesses the real-world performance of commercially available nutrient management tools—for different crops, regions and soil types. See http://nutrientstar .org/. For a summary of the many different agricultural sustainability initiatives, see Eileen McLellan, "An Agricultural Marriage Made in Heaven: State Programs and Private Sector Initiatives," *Growing Returns* blog, Environmental Defense Fund, January 14, 2015, http://blogs.edf .org/growingreturns/2015/01/14/an-agricultural-marriage-made-in- heaven-state-programs-private-sector-initiatives/.

136 **"Farmers often resist":** http://grist.org/food/heres-a-solution-for-those- out-of-control-toxic-algae-blooms/.

136 **"the biggest driver of farm conservation":** Though the private sector plays an increasingly important role, farmers can also now participate in public carbon markets. In June 2015, the California Air Resources Board approved the first ever crop-based carbon standards in an emissions cap-and-trade market: rice farmers can sell reductions in methane emissions into the state's market. The Environmental Defense Fund is working to achieve the same opportunity for farmers who reduce nitrous oxide emissions from fertilizer on almonds and other crops. See Sara Kroopf, "Why Almond Lovers Can Breathe Easy Again," *Growing Returns* blog, Environmental Defense Fund, October 20, 2015, http://blogs.edf.org/growingreturns/2015/10/20/why- almond-lovers-can-breathe-easy-again/?utm_source=feedburner&utm_

medium=email&utm_campaign=Feed%3A+GrowingReturns+%28Grow
ing+Returns%29.

136 **work through NRCS:** This is the approach used by a federal–state task
force created to assess the impact of farmers' conservation efforts on
water quality in the Mississippi and Gulf, relying on the trust NRCS
has built with farmers. See Mississippi River Basin Healthy Watersheds
Initiative, http://www.nrcs.usda.gov/wps/portal/nrcs/detailfull/national/
home/?cid=stelprdb1048200, accessed April 20, 2015. In April 2015 the
USDA announced a new partnership with ranchers and farmers to reduce
greenhouse gas emissions and increase carbon sequestration through voluntary measures. See U.S. Department of Agriculture, "Secretary Vilsack
Announces Partnerships with Farms and Ranchers to Address Climate
Change," blog post, April 23, 2015, http://blogs.usda.gov/2015/04/23/
secretary-vilsack-announces-partnerships-with-farmers-and-ranchers-to
-address-climate-change/.

137 **"to get insurance":** Volumes have been written about other wrongheaded
federal farm policies, like the subsidies and ethanol mandates that have
driven farmers to vast monocultures of corn: between 2008 and 2012, more
than 5 million acres of grassland were converted to cropland; a third of
those had been untouched for twenty years or more. See J. Tyler, J. Meghan
Salmon and Holly K. Gibbs, "Cropland Expansion Outpaces Agricultural
and Biofuel Policies in the United States," *Environmental Research Letters*
10, no. 4 (2015): 044003.

137 **"capture any sediments and nutrients":** Combining improvement in fertilizer management and cover crops with buffers and wetland restoration on just
1% of the region's farmland could reduce nitrogen pollution from the upper
Mississippi and Ohio River basins by 45%, the amount needed to shrink the
dead zone in the Gulf of Mexico to manageable size. See Eileen McLellan et
al., "Reducing Nitrogen Export from the Corn Belt to the Gulf of Mexico:
Agricultural Strategies for Remediating Hypoxia," *Journal of the American
Water Resources Association* 51, no. 1 (2015): 263–89, http://onlinelibrary
.wiley.com/store/10.1111/jawr.12246/asset/jawr12246.pdf?v=1&t=i8t8je
9s&s=3ee31f8cecc4fa897a24492b61b519c086b81828.

137 **"the Farm Bill cost-share":** Private water quality markets can also provide needed financial support: in the Ohio River basin, cities pay upstream
farmers to use cover crops and other practices that keep nutrients out of the
water supply. See Sara Walker and Mindy Selman, "Addressing Risk and
Uncertainty in Water Quality Trading Markets," World Resources Institute, March 2014, http://www.wri.org/sites/default/files/wri_issuebrief_
uncertainty_3-9_final.pdf, accessed April 27, 2015.

138 **butterfly "habitat exchange":** Andrew Amelinckx, "You Could Get Paid

to Save Struggling Monarchs," *Modern Farmer*, February 17, 2016, http://modernfarmer.com/2016/02/habitat-exchanges-monarchs/.

138 **double production by 2050:** David Tilman, Christian Balzer, Jason Hill and Belinda L. Befort, "Global Food Demand and the Sustainable Intensification of Agriculture," *Proceedings of the National Academy of Sciences* 108, no. 50 (2011): 20260–64.

138 **accelerating loss and degradation:** A September 2015 report estimated global losses at up to $10 trillion a year on land degraded by erosion, drought and overuse. See Economics of Land Degradation Initiative, "The Value of Land," September 2015, http://eld-initiative.org.

138 **one million acres:** "Statistics," Farmland Information Center, 2010 National Resources Inventory, http://www.farmlandinfo.org/statistics, accessed April 28, 2015.

138 **20 percent of all cultivated lands are degraded:** Olivier Dubois, "The State of the World's Land and Water Resources for Food and Agriculture: Managing Systems at Risk," U.N. Food and Agriculture Organization and Earthscan, 2011. A U.N. report found that 5,000 acres of fertile land were lost each day for the past twenty years due to salt damage. See Manzoor Qadir et al., "Economics of Salt-Induced Land Degradation and Restoration," *Natural Resources Forum* 38, no. 4 (2014): 282–95.

138 **a quarter of the world's croplands:** World Resources Institute, Aqueduct Water Risk Tool, http://www.wri.org/applications/maps/agriculture map/#x=0.00&y=-0.00&l=2&v=home&d=gmia&init=y, accessed January 14, 2016.

138 **40 percent of the global food supply:** Hugh Turral, Jacob Burke, and Jean-Marc Faurès, *Climate Change, Water and Food Security*, FAO, 2011, http://www.fao.org/docrep/014/i2096e/i2096e.pdf.

138 **nearly every one of the world's major aquifers:** J. S. Famiglietti, "The Global Groundwater Crisis," *Nature Climate Change* 4, no. 11 (2014): 945–48.

138 **The third accelerating challenge:** A fourth may be direct impacts of increased atmospheric CO2 on the nutritional value of rice and wheat, including reduced levels of zinc, iron and protein. Several billion people get most of their zinc and iron from these crops and many, especially children, are already deficient, suffering anemia, impaired cognitive development, stunted growth and weakened immunity. See Samuel S. Meyers et al., "Increasing CO2 Threatens Human Nutrition," *Nature* 510, no. 7503 (2014): 139–42.

139 **stresses will drive up food prices:** Robert Bailey, *Growing a Better Future* (Oxford: Oxfam International, 2011).

139 **230 million hectares:** M. C. Hansen et al., "High-Resolution Global Maps

of 21st-Century Forest Cover Change," *Science* 342, no. 6160 (2013): 850–53.

139 **nearly all to make way for farming:** Gabrielle Kissinger, Martin Herold, and Veronique De Sy, "Drivers of Deforestation and Forest Degradation: A Synthesis Report for REDD+ Policymakers," Lexeme Consulting, Vancouver, Canada, August 2012.

139 **"has such a large swath":** Moisés Naím, "The World is Full of Grain," *Atlantic*, October 14, 2014, http://www.theatlantic.com/international/archive/2014/10/the-world-is-full-of-grain-agriculture-economy/381413/; U.N. Food and Agriculture Organization, "The State of the World's Land and Water Resources for Food and Agriculture," 2016, http://www.fao.org/nr/solaw/main-messages/en/; and *FAO Statistical Yearbook 2013: World Food and Agriculture*, http://www.fao.org/docrep/018/i3107e/i3107e.PDF.

139 **Yields are lower in the tropics:** Paul C. West et al., "Trading Carbon for Food: Global Comparison of Carbon Stocks vs. Crop Yields on Agricultural Land," *Proceedings of the National Academy of Sciences* 107, no. 46 (2010): 19645–48.

139 **"restorative intensification":** In the fall of 2014, Kansas State launched a Sustainable Intensification Innovation Lab focused on interdisciplinary research to support "reducing global hunger, poverty and improving the nutrition of smallholder farmers" in Southeast Asia and Africa. See "Transforming Farming Systems for Smallholders," http://www.k-state.edu/siil/index.html, accessed April 22, 2015.

140 **"Global Yield Gap Atlas":** http://www.yieldgap.org/.

140 **carbon losses . . . are relatively low and the yields high:** Justin Andrew Johnson et al., "Global Agriculture and Carbon Trade-Offs," *Proceedings of the National Academy of Sciences* 111, no. 34 (2014): 12342–47.

141 **Dr. Kofi Boa:** Mike Wilson, "Africa's No-Till Revolution," *Farm Futures*, February 4, 2015, http://farmfutures.com/blogs-africas-till-revolution-9442.

142 **Genesis:** Lindsey quotes the relevant passage of Genesis in her essay on the sprayer, including its reminder that nature was not made for man's exclusive use: "To all the beasts of the earth and all the birds of the air and all the creatures that move on the ground—everything that has the breath of life in it—I give every green plant for food" (Genesis 1:30).

142 **the symbolism in John:** "Truly, truly, I say to you, unless a grain of wheat falls into the earth and dies, it remains alone; but if it dies, it bears much fruit" (John 12:24).

142 **"perspective bigger than themselves":** An increasing number of religious leaders have been echoing these themes. "This is our sin, exploiting the Earth," Pope Francis has said. "Creation is not a property, which we can rule over at will; or, even less, is the property of only a few: Creation

is a . . . wonderful gift that God has given us." That is not an abstraction, says Patrick Carolan, president of the Franciscan Action Network: "every tree, every pond, every member of every species . . . is loved individually and particularly by God." See Tara Isabella Burton, "Pope Francis's Radical Environmentalism," *Atlantic*, July 11, 2014; and Patrick Carolan, "The Catholic Perspective on the Environment," *U.S. Catholic* blog, June 1, 2011, http://www.uscatholic.org/blog/2011/05/catholic-perspective-environment.

145 **"the whisper of wind voices":** Truman Capote, *In Cold Blood* (New York: Vintage, 1994), 341.

CHAPTER 3: RIVERMAN

149 *Down the Missouri:* From *The River*, a 1938 documentary directed by Pare Lorentz for the Farm Service Administration. The film beat Leni Riefenstahl's *Olympia* to win Best Documentary at the 1938 Venice Film Festival. See Philip Kennicott, "On DVD, American Propaganda's High-Water Mark," *Washington Post*, January 28, 2007, http://www.washingtonpost.com/wp-dyn/content/article/2007/01/26/AR2007012600453.html.

150 **Carrying more than a million tons:** This history is captured in what may be the most beautiful maps ever made by an engineer: Harold Fisk's 1944 series for the Army Corps of Engineers. With various colors marking each era of deposition all the way back into prehistory, the maps look like ribbon candy or the curls of a punk Rapunzel. See "The Alluvial Valley of the Lower Mississippi River," http://www.radicalcartography.net/index.html?fisk.

150 **"the most beautiful prose":** William Grimes, "Pare Lorentz, 86, a Film Director on Socially Conscious Matters," *New York Times*, March 5, 1992, http://www.nytimes.com/1992/03/05/arts/pare-lorentz-86-a-film-director-on-socially-conscious-matters.html.

151 **most essential work of American literature:** As Ernest Hemingway said, "All American writing comes from [*Huck Finn*]. There was nothing before. There has been nothing as good since." *Green Hills of Africa* (New York: Scribner, 2003), 21.

151 **the first draft of a chapter:** Mark Twain, *Life on the Mississippi* (New York: Penguin, 1984), 51.

151 **"every tumblerful":** Twain, *Life on the Mississippi*, 170.

151 **"nutritiousness":** Ibid., 56.

152 **"the eye loves to dwell":** Ibid., 200.

153 **forcing nearly a million from their homes:** National Oceanic and Atmospheric Administration, "Flood History of Mississippi," http://www.srh

.noaa.gov/media/jan/Hydro/Flood_History_MS.pdf, accessed November 23, 2015.

153 **"an uncoiling rope":** John M. Barry, *Rising Tide: The Great Mississippi Flood of 1927 and How It Changed America* (New York: Simon and Schuster, 1997), 38.

153 **"grisly, drizzly gray mists":** Twain, *Life on the Mississippi*, 86.

153 **"blind and tangled":** Ibid., 106.

155 **far better safety record:** James Conca, "Pick Your Poison for Crude—Pipeline, Rail, Truck or Boat," *Forbes*, April 26, 2014 http://www.forbes.com/sites/jamesconca/2014/04/26/pick-your-poison-for-crude-pipeline-rail-truck-or-boat/#27951de95777. A *New York Times* short film on oil trains documented the route these "rolling bombs" take through population centers, and the growing numbers of explosions, fatalities and spills; Jon Bowermaster, "A Danger on Rails," *New York Times*, April 21, 2015, http://www.nytimes.com/2015/04/21/opinion/a-danger-on-rails.html. Merritt Lane adds, "We have quietly done our business between the levees, moving massive quantities of sensitive materials safely through major metro areas. But it's hard to get press that says another billion-ton miles transited under U.S. bridges today without incident."

155 **$405 million:** Hannah Chotin Macgowan, Canal Barge Company, email to the author, February 1, 2016.

155 **"one of the pretty girls":** Ken Otterbourg, "America's Untamable River Trade," *Fortune*, December 16, 2011, http://fortune.com/2011/12/16/americas-untamable-river-trade/.

155 **chairman of the Waterways Council:** Harry N. Cook, ed., "Capitol Currents Newsletter," *Waterways Council Newsletter*, April 27, 2015, http://waterwayscouncil.org/wp-content/uploads/2013/04/042715-mobile.htm.

156 **100 pounds:** All figures come from Captain Donnie Williams.

157 **"shooting rapids":** John McPhee, "Atchafalaya," *New Yorker*, February 23, 1987, collected in *The Control of Nature* (New York: Farrar, Straus and Giroux, 1989).

157 **Hannibal, Missouri:** Twain's boyhood home and a setting for *The Adventures of Tom Sawyer* and *The Adventures of Huckleberry Finn*.

157 **1993 and 2011 floods:** Barry Yeoman, "Life on the Mississippi, Now," *On Earth*, June 9, 2014, http://archive.onearth.org/articles/2014/06/should-we-let-the-mighty-mississippi-be-mighty-again.

158 **unprecedented Christmas floods of 2015:** Suzannah Gonzales, "Mississippi River Seen Cresting in Tennessee, Arkansas This Weekend," Reuters, January 8, 2016, http://www.reuters.com/article/us-usa-weather-idUSKBN0UM26620160108; and Kris Maher and Cameron McWhirter, "Floodwaters Begin to Crest Along Mississippi, Tributaries

Near St. Louis," *Wall Street Journal*, December 31, 2015, http://www.wsj
.com/articles/flooding-along-mississippi-river-forces-evacuations-closes
-roadways-1451568523.

158 **falling 55 feet:** Josh Sanburn, "The Not-So-Mighty Mississippi: How the
River's Low Water Levels Are Impacting the Economy," *Time.com*, July 30,
2012, http://business.time.com/2012/07/30/the-not-so-mighty-mississippi-
how-the-rivers-low-water-levels-are-impacting-the-economy/.

158 **near-record low:** Cameron McWhirter and Caroline Porter, "Barge Operators Struggle Along the Mississippi," *Wall Street Journal*, August 25, 2013,
http://www.wsj.com/articles/SB100014241278873239970045786399211
36985476.

159 **"This drifting":** Twain, *Life on the Mississippi*, 83.

159 **"economically passable":** Lynn Muench, senior VP of American Waterways Operators, says, "One inch of draft in a single barge is about 17 tons
of cargo, almost enough to fill a semi truck"; quoted in Sanburn, "The Not-
So-Mighty Mississippi."

159 **"disruption clauses":** See also McWhirter and Porter, "Barge Operations
Struggle Along the Mississippi."

160 **"We need that water":** The opposite tension had surfaced between farmers and the shipping industry in 2011, which saw the highest runoff along
the Missouri River since record-keeping began. With farms threatened by
flooding, the Army Corps opened dams to release more water, adding to
the navigation challenges in the Mississippi. See Annie Snider, "Federal
Agencies Have Made Scant Progress Since 2011 Disaster – GAO," *E &
E News PM*, June 9, 2015, http://www.eenews.net/eenewspm/2015/06/09/
stories/1060019934; and U.S. Government Accountability Office,
"Missouri River Basin: Agencies' Progress Improving Water Monitoring Is Limited," report to Congress, June 9, 2015, http://www.gao.gov/
assets/680/670715.pdf.

160 **"all one big system":** Mark Davis, director of the Tulane Institute on
Water Resources Law and Policy, calls Mississippi watershed management
"balkanized . . . it's like a child that has 800 parents." Val Marmillion,
managing director of America's Wetland Foundation, agrees: "The system
is so large, it intimidates cooperation." See David Schaper, "Mississippi River's Many 'Parents' Look to Unify," *All Things Considered*, NPR, April 19,
2013, http://www.npr.org/2013/04/25/177954961/mississippi-rivers-many-
parents-look-to-unify.

160 **a 2013 report by FEMA:** AECOM, "The Impact of Climate Change and
Population Growth on the National Flood Insurance Program through
2100," prepared for the Federal Emergency Management Agency, June

2013, http://www.aecom.ca/vgn-ext-templating/v/index.jsp?vgnextoid=e06
42ed99724e310VgnVCM100000089e1bacRCRD.

160 **increasingly concentrated rains:** Concentrated rains are expected to
increase 16% in the Missouri river drainage, 37% in the upper Mississippi
and Ohio, and 27% in the Tennessee and Cumberland rivers. See U.S.
Global Change Research Program, "National Climate Assessment 2014,"
http://nca2014.globalchange.gov/report#section-1946, accessed August 24,
2015.

160 **impacts far worse:** Wetlands capture rain and runoff, store water in the
soil and ponds, and help recharge the subterranean water table, providing
protection against both flood and drought (as well as habitat). As more
than 35 million acres of wetlands have been filled and plowed under in the
upper Mississippi River basin, sixty days of floodwater storage capacity has
been reduced to twelve. See Stacy Small-Lorenz, "Wetlands Do Triple Duty
in a Changing Climate," *National Geographic Voices*, May 1, 2014, http://
voices.nationalgeographic.com/2014/05/01/wetlands-do-triple-duty-in-a
-changing-climate/; and William Mitsch and James Gosselink, *Wetlands*,
5th edition (Hoboken, NJ: Wiley, 2015), 539.

161 **$200 billion:** Jason Plautz, "Despite White House Cuts, Appropria-
tors Look to Lock Down Some Dam Money," *National Journal*, April 8,
2015, http://www.nationaljournal.com/energy/despite-white-house-cuts-
appropriators-look-to-lock-down-some-dam-money-20150408. According
to the Big River Coalition, the Mississippi is the second most productive
river transportation system in the world, behind only China's Yangtze.

161 **12,000 miles of inland waterways:** Institute for Water Resources, U.S.
Army Corps of Engineers, "U.S. Port and Inland Waterways Moderniza-
tion: Preparing for Post-Panamax Vessels," 2012, http://www.usace.army.
mil/Missions/CivilWorks/Navigation.aspx, accessed July 19, 2015.

161 **"the heavy and still vital":** Otterburg, "America's Untamable River Trade."

161 **$11 per ton-mile:** Center for Ports and Waterways and Texas Transporta-
tion Institute, "A Modal Comparison of Domestic Freight Transportation
Effects on the General Public: 2001–2009," National Waterways Foundation, February 2012, http://www.nationalwaterwaysfoundation.org/study/
FinalReportTTI.pdf.

162 **"capable of curbing":** James Eads quoted in Barry, *Rising Tide*, 75.

162 **Mississippi River Commission:** "The Mississippi River Commission and
the Army Corps of Engineers," *American Experience*, PBS, http://www
.pbs.org/wgbh/americanexperience/features/general-article/flood-control/,
accessed November 24, 2015.

162 **Twain ridiculed:** Twain, *Life on the Mississippi*, 205–206.

162 **much venomous debate:** Those battles are central to Barry's *Rising Tide*, particularly the personal, political and intellectual rivalries among three men: James Eads, who in addition to the submersible built the first American ironclads for the Civil War, the first bridge across the Mississippi and jetties that worked to scour out the mouth of the river; Andrew Humphreys, whose ambition drove him to support a levees-only approach even though he knew it to be wrong; and the farsighted Charles Ellet, who championed multiple flood control strategies including outlets to relieve flood pressures, and foresaw that a levees-only strategy would doom Louisiana. Though Ellet was right, Humphreys largely prevailed. See *Rising Tide*, 90–91.

162 **longer than the Great Wall of China:** "There was now a Great Wall of China running up each side of the river, with the difference that while the levees were each about as long as the Great Wall they were in many places higher and in cross-section ten times as large": McPhee, "Atchafalaya."

163 **"snapping trees":** Barry, *Rising Tide*, 156.

163 **devoured by hogs:** Ibid., 279.

163 **"the myth":** Ibid., 422.

163 **"a watershed":** Ibid., 374.

163 **Bonnet Carré Spillway:** Edward Branley, "NOLA History: The Bonnet Carré Spillway," *GO NOLA*, August 17, 2011, http://gonola.com/2011/08/17/nola-history-the-bonnet-carre-spillway.html.

163 **massive hydroelectric power plants:** United States Bureau of Reclamation–Great Plains, "Lewis and Clark: Big Dam Era," http://www.usbr.gov/gp/lewisandclark/damera.html, updated July 1, 2015.

164 **Lorentz film and a parade of others:** The Lorentz film, according to the *Washington Post*'s film critic Philip Kennicott, "came as close as anything this country produced to the great collaborations between Sergei Eisenstein and Sergei Prokofiev in the Soviet Union": Kennicott, "On DVD, American Propaganda's High-Water Mark."

164 **The locks, now aging and decrepit:** Ron Nixon, "Barges Sit for Hours Behind Locks That May Take Decades to Replace," *New York Times*, February 4, 2015, http://www.nytimes.com/2015/02/05/us/barges-sit-for-hours-behind-locks-that-may-take-decades-to-replace.html?ref=todayspaper&_r=0.

164 **worsen floods:** Southern Illinois University geologist Nicholas Pinter says that dikes added 6 feet to 2008's crests; see Yeoman, "Life on the Mississippi, Now."

165 **"now they won't flood":** Among the relief valves so constrained is the New Madrid Floodway in southeast Missouri. Though it was used, finally, in 2011, the Corps's dynamiting of the levee was so delayed by a lawsuit brought by Missouri's attorney general that the levee protecting Olive Branch, Illinois, failed, filling the town "like a bathtub"; see Yeoman, "Life on the Mississippi, Now."

Also in jeopardy is a gap in the levee that currently allows the river to recharge sloughs full of bald cypress, nuttall oak, tupelo gum, golden plover, white bass and such rare fish as the golden topminnow, chain pickerel and banded pygmy sunfish. A Corps proposal to seal the gap led the advocates at American Rivers to name the middle Mississippi one of its ten most endangered rivers; see American Rivers, "America's Most Endangered Rivers for 2014: Middle Mississippi River," http://www.americanrivers.org/endangered-rivers/2014-report/middle -mississippi/#sthash.NLc7ElnV.09bdbRF6.dpuf, accessed August 3, 2015.

165 **Tigris–Euphrates:** Saddam Hussein made dark use of that significance, draining the Tigris–Euphrates marshes to drive out his Marsh Arab enemies. Iraq is now trying to recover those marshes, though contending with the same upstream pressures from dams and agriculture. See Shannon Cunniff, "From the Mississippi to the Tigris: River Restoration Lessons Travel Far," *EDF Voices* (blog), Environmental Defense Fund, June 2, 2015, https://www.edf .org/blog/2015/06/02/mississippi-tigris-river-restoration-lessons-travel-far.

166 **half the river's load:** Restore the Mississippi River Delta, "Answering 10 Fundamental Questions About the Mississippi River Delta," April 2012, http://www.mississippiriverdelta.org/files/2012/04/Mississippi RiverDeltaReport.pdf.

166 **"imprisoning the snake":** Barry, *Rising Tide*, 91.

166 **50,000 wells:** Nathaniel Rich, "The Most Ambitious Environmental Lawsuit Ever," *New York Times Magazine*, October 2, 2014, http://www .nytimes.com/interactive/2014/10/02/magazine/mag-oil-lawsuit.html? _r=0.

166 **millions of barrels of oil:** Oliver A. Houck, "The Reckoning: Oil and Gas Development in the Louisiana Coastal Zone," *Tulane Law Review* 28 no. 2 (Summer 2015): 218.

167 **greatest land loss on the planet:** Erik Kancler, "Bayou Farewell," *Mother Jones*, October 3, 2005, http://www.motherjones.com/environment/2005 /10/bayou-farewell. As Garrett Graves, who now represents Louisiana in the U.S. Congress, put it, the Corps had achieved "one of the most successful civil works in our nation's history, preventing hundreds of billions in flooding losses and protecting navigation. But it had failed at its third mandate: to protect the nation's wetlands."

167 **2,000 square miles:** Bob Marshall, Brian Jacobs and Al Shaw, "Losing Ground," *ProPublica* and *The Lens*, August 28, 2014, http://projects.propublica .org/louisiana/, accessed August 3, 2015.

168 **Leeville:** Leeville's lost beauties are captured in a painting that hangs in the South LaFourche library. See ibid.

168 **lapped all around by water:** Before LA-1 went under, it was the route to

Port Fourchon, the jumping-off point for 90% of offshore oil and gas production. Now submerged, the road has been replaced by a massive futuristic elevated expressway, a jarring counterpoint to the bedraggled remnants of the town and its drowning cemeteries.

168 **mink, otter and muskrat:** The fur industry was once a major part of the Louisiana economy; third in the world behind only Canada and Russia, the state produced more furs than the rest of the U.S. combined. Entrepreneurs tried to capitalize on that strength by importing nutria, a South American rodent, which added yet another disaster: nutria eat marsh and kill young cypress, destroying along with everything else the habitat for native fur-bearers. Bob Marshall, "Louisiana's Most Wanted," *Field and Stream*, May 1997, https://books.google.com/books?id=EXjB35Lp2_cC&pg=RA3-PA102-IA3&lpg=RA3-PA102-IA3&dq=louisiana+furs+canada+or+russia&source=bl&ots=9yu6HsYH16&sig=vmQn0Z5kHQgRwH4XqAUoj18pIAc&hl=en&sa=X&ved=0CC0Q6AEwA2oVChMIgYPwjNiryAIVRJSQCh3mLQ7C#v=onepage&q=louisiana%20furs%20canada%20or%20russia&f=false. See Restore the Mississippi River Delta, "Discover the Delta: What Went Wrong," http://www.mississippiriverdelta.org/discover-the-delta/what-went-wrong/, accessed February 19, 2016.

168 **That's all gone too:** Lafourche Parish was the river's main stem from 200 AD to the year 1200. Though abandoned when the river switched to its current course, it was nourished with Mississippi sediments through a major distributary until that was dammed in 1904. M. J. Mac et al., "Status and Trends of the Nation's Biological Resources—Vol. 1," U.S. Department of the Interior, U.S. Geological Survey, 1998, http://www.nwrc.usgs.gov/sandt/Coastal.pdf. In *Bayou Farewell* (New York: Vintage, 2004, p. 36) Mike Tidwell captured the pace of Leeville's vanishing in a chat with crabber Tim Melancon, Jr.: " 'When I was a kid,' Tee Tim says, 'I remember this canal being so much smaller that you could land your boat on solid ground right over there where that dead tree is.' He points [with crab-sorting tongs] to a rotting, limbless trunk just up the canal. It's in open water, fifty feet from the bank. More than the sight of the tree, it's those five simple words that amaze me: 'When I was a kid . . .' It makes Tee Tim sound like an old man looking back on a long lifetime. In reality he's seventeen, looking back maybe 10 years."

168 **removed thirty-five names:** "Historical Geographic Place Names Removed from NOAA Charts," National Oceanic and Atmospheric Administration, http://www.nauticalcharts.noaa.gov/history/HistoricalPlacenames_Louisiana.pdf, updated August 4, 2014.

169 **Mardi Gras Pass:** Bob Marshall, "Two-Year-Old Breach in Mississippi River Could Be Formally Named 'Mardi Gras Pass'," *The Lens*, July 11, 2014,

http://thelensnola.org/2014/07/11/two-year-old-breach-in-mississippi-river-could-be-formally-named-mardi-gras-pass/. Several local landowners are unhappy with the breach, as are some oystermen; see David Hammer, "'Natural' River Diversion at Center of Coastal Restoration Conflict," WWLV-TV, September 2, 2015, http://www.wwltv.com/story/news/local/investigations/david-hammer/2015/09/01/natural-river-diversion-at-center-of-coastal-restoration-conflict/71555912/.

169 **held it back in 1963:** U.S. Army Corps of Engineers, "Old River Control," http://www.mvn.usace.army.mil/Missions/Recreation/OldRiverControl.aspx, accessed August 24, 2015.

169 **30 percent of the river's volume:** Marshall, Jacobs and Shaw, "Losing Ground."

169 **largest river swamp:** Atchafalaya Natural Heritage Area, "Atchafalaya Basin," http://www.atchafalaya.org/page.php?name=Atchafalaya-Basin, accessed August 5, 2015.

169 **eighteen new square miles:** Marshall, Jacobs and Shaw, "Losing Ground."

170 **1,750 square miles:** Coastal Protection and Restoration Authority of Louisiana, "Louisiana's Comprehensive Master Plan for a Sustainable Coast," 2012, http://issuu.com/coastalmasterplan/docs/coastal_master_plan-v2?e=3722998/2447530.

170 **Phillip Turnipseed:** John Carey, "Architects of the Swamp," *Scientific American*, December 2013, accessed August 3, 2015, http://johncarey.biz/uploads/Wetlands-final.pdf.

172 **40 percent of all the marsh:** U.S. Geographical Service, "Louisiana Coastal Wetlands: A Resource at Risk," http://pubs.usgs.gov/fs/la-wetlands/, accessed August 3, 2015.

172 **Chinese-scale works:** Chris Mooney, "The Next Big One," *Washington Post*, August 21, 2015, http://www.washingtonpost.com/sf/national/2015/08/21/the-next-big-one/.

173 **five deep-draft Louisiana ports:** Baton Rouge, South Louisiana, New Orleans, St. Bernard and Plaquemines. See Our Coast, Our Economy, "Economics of Restoration," http://ourcoastoureconomy.org/economics-of-restoration/protecting-our-industries/navigation/, accessed November 24, 2015.

173 **nation's top producer of offshore oil:** Coastal Protection and Restoration Authority, "National Significance," http://coastal.la.gov/whats-at-stake/national-significance/, accessed December 7, 2015.

173 **Ninety percent of the country's offshore:** Coastal Protection and Restoration Authority, "Coastal Master Plans," http://coastal.la.gov/resources/library/reports/, accessed November 25, 2015. Louisiana Department of Natural Resources, "America's Energy Corridor Louisiana Serving the

Nation's Energy Needs," 2003, http://dnr.louisiana.gov/assets/docs/energy/
policypapers/AW_AmericasEnergyCorridor_Revised.pdf.

173 **fishing net thrown across the state:** U.S. Energy Information Administration, "Louisiana State Profile and Energy Estimates," http://www.eia.gov/
state/?sid=LA, updated March 27, 2014.

173 **eighty-mile stretch of river:** As of January 2015, $50 billion of new
investment is underway, as cheap U.S. natural gas attracts global companies including Germany's BASF and China's Shandong Yudang to what's
long been called "the American Ruhr." See Rachel Cernansky, "Natural Gas Boom Brings Major Growth for U.S. Chemical Plants," *Yale
Environment 360* (blog), January 29, 2015, http://e360.yale.edu/feature/
natural_gas_boom_brings_major_growth_for_us_chemical_plants
/2842/.

173 **contaminated St. Bernard's water supply:** Danny D. Reible, Charles N.
Haas, John H. Pardue and William J. Walsh, "Toxic and Contaminant
Concerns Generated by Hurricane Katrina," *The Aftermath of Katrina*,
National Academy of Engineering, Spring 2006, https://www.nae.edu/
Publications/Bridge/TheAftermathofKatrina/ToxicandContaminantConc
ernsGeneratedbyHurricaneKatrina.aspx.

173 **ten storage tanks:** Luis A. Godoy, "Performance of Storage Tanks in Oil
Facilities Damaged by Hurricanes Katrina and Rita," *Journal of Performance
of Constructed Facilities* 21, no. 6 (2007): 441–49, http://www.researchgate
.net/publication/234139064_Performance_of_Storage_Tanks_in_Oil_
Facilities_Following_Hurricanes_Katrina_and_Rita.

173 **eight million gallons:** National Marine Fisheries Service, Southeast
Regional Office, "Other Significant Oil Spills in the Gulf of Mexico,"
November 2015, http://sero.nmfs.noaa.gov/deepwater_horizon/documents/
pdfs/fact_sheets/historical_spills_gulf_of_mexico.pdf.

173 **matchless biological productivity:** Jean-Michel Cousteau, "Gulf of
Mexico Oil Spill: What Have We Learned?", *Diver*, June 2, 2012, http://
divermag.com/gulf-of-mexico-oil-spill-what-have-we-learned/.

174 **hunting here is unsurpassed:** Wildlife tourism is a $19 billion annual
industry, contributing to the region's 2.6 million tourism jobs. See Shawn
Stokes and Marcy Lowe, "Wildlife Tourism and the Gulf Coast Economy," July 9, 2013, http://www.daturesearch.com/wp-content/uploads/
WildlifeTourismReport_FINAL.pdf.

174 **nearly half of all the waterfowl:** "Flyway Conservation," *Audubon*, 2011,
http://www.audubon.org/sites/default/files/documents/ar2011-flyway
conservation.pdf.

174 **400 miles of shoreline:** Larry Handley et al., "Statewide Summary for
Louisiana," *2013 Emergent Wetlands Status and Trends in the Northern*

Gulf of Mexico: 1950–2010: USGS Scientific Investigations Report, in preparation.

174 **edge conditions:** These kinds of settings, rather than individual species, are increasingly the focus of wildlife conservationists, who see "conserving the stage"—diverse habitats such as estuaries able to support the widest variety of species—as the best way to maximize options in a fast-changing world. Critically important in this approach are physical drivers like sediment that "shape habitat and give it the dynamic ability to adapt over time." See Jim Robbins, "Resilience: A New Conservation Strategy for a Warming World," *Yale Environment 360* (blog), July 13, 2015, http:// e360.yale.edu/feature/resilience_a_new_conservation_strategy_for_a_ warming_world/2893/.

174 **bald cypress trees:** Cypress, which can last centuries without rotting, was heavily logged for boats, jetties and coffins—stripping away yet another layer of protection. Most was gone by 1925. See National Wildlife Federation, "Bald Cypress," https://www.nwf.org/Wildlife/Wildlife-Library/Plants/ Bald-Cypress.aspx. And Mollie Day, "The Unkindest Cut," Best of New Orleans, August 14, 2007, http://www.bestofneworleans.com/gambit/the -unkindest-cut/Content?oid=1248228, accessed August 26, 2015.

175 **settle along the bayou:** Their land-use planning, says Windell Curole, is based on the French "orbit" system, with houses built on the highest ground and a garden sloping to the swamp, where they get their seafood and meat.

175 **some forecast a permanent migration:** Adam Wernick, "Louisiana's Coastline Is Disappearing at the Rate of a Football Field an Hour," *Living on Earth*, Public Radio International, September 23, 2014, http://www .pri.org/stories/2014-09-23/louisianas-coastline-disappearing-rate-football- field-hour.

175 **once fifty miles from the coast:** Mike Tidwell, *Bayou Farewell: The Rich Life and Tragic Death of Louisiana's Cajun Coast* (New York: Pantheon, 2004).

176 **first billion-dollar storm:** "1965: Hurricane Betsy smashes ashore near New Orleans," *Times-Picayune*, December 8, 2011, http://www.nola.com /175years/index.ssf/2011/12/1965_hurricane_betsy_smashes_a.html.

176 **$40 million a year, first long-term proposal:** Denise J. Reed and Lee Wilson, "Coast 2050: A New Approach to Restoration of Louisiana Coastal Wetlands," *Physical Geography* 25, no. 1 (2004): 4–21, http://labs.uno.edu/ restoration/Reed%20and%20Wilson%202050.pdf.

176 **largest storm surge ever:** Jeffrey Masters, "U.S. Storm Surge Records," Weather Underground, http://www.wunderground.com/hurricane/surge_ us_records.asp, accessed August 13, 2015. 250,000 people were displaced

from New Orleans alone; 1 million were displaced throughout the gulf coast; see Tamara Houston et al., "Hurricane Katrina: A Climatological Perspective: Preliminary Report," National Climatic Data Center, 2006, http://www1.ncdc.noaa.gov/pub/data/techrpts/tr200501/tech-report-200501z.pdf.

176 **220 square miles:** Brett Israel, "5 Years After Katrina, Gulf Ecosystems on the Ropes," *livescience.com*, August 27, 2010, http://www.livescience .com/8505-5-years-katrina-gulf-ecosystems-ropes.html, accessed November 25, 2015.

176 **levees failed:** In this case, the main access for the saltwater had been the Mississippi River to the Gulf Outlet (MR-GO), built in the 1960s. During the storm MR-GO also funneled surging seawater directly into the city and St. Bernard Parish, leading in May 2015 to a federal judge finding the Corps liable for a "taking" of private property. In August 2015 a federal judge further found the Corps responsible for the full costs of restoring the wetlands destroyed by MR-GO. See U.S. Army Corps of Engineers, "History of MRGO," http://www.mvn.usace.army.mil/Missions/Environmental/ MRGOEcosystemRestoration/HistoryofMRGO.aspx, accessed August 13, 2015; John Schwartz, "U.S. Liable in New Orleans-Area Flooding," *New York Times*, May 1, 2015, http://www.nytimes.com/2015/05/02/us/ us-liable-in-new-orleans-area-hurricane-katrina-flooding.html; and Mark Schleifstein, "Judge: Corps Must Pay Full $3 Billion Cost of Restoring MR-GO Wetlands," *Times-Picayune*, August 27, 2015, http://www.nola .com/environment/index.ssf/2015/08/federal_judge_to_corps_restore .html.

181 **simulator:** The Seamen's Church Institute simulators recreate 1,000 miles of the river system including changeable weather and currents and the feel of the wheelhouse while pushing 22,000 tons of cargo, right down to the drone of the engines.

182 **"an atmosphere of song":** Twain, *Life on the Mississippi*, 229.

182 **cook gets up at 3 a.m. to prepare:** Those meals won Captain Donnie Williams a moment of television fame, when in May 2007 he hosted Food Network host Alton Brown on the M/V *Susan L. Stall*. Brown was taping a segment on towing industry cuisine for his series *Feasting on Asphalt*, and rode along from Hickman, Kentucky, Donnie's hometown, to Paducah. When he and the boat's cook made chicken, dumplings and gravy, Donnie told Brown he'd grown up on gravy: "We got up at four to milk the cows and fetch the eggs. If we was good boys, we'd have milk gravy, or water gravy if the cow didn't put out, before going off to school. We couldn't afford a lunch so we'd come home midday, and if we was good boys, if we didn't get paddlin's or start a fight, Momma'd squish tomato into the white

gravy. Man, if you've never had tomato gravy you don't know what you're missing." (After school was chocolate gravy.) When Brown told Williams he'd never heard of tomato gravy, "I reached over the ketchup bottle and squirted it in the gravy he'd just made, and he said, 'Man, what are you doin'?'"

183 **Chester, Illinois:** Local folklore in Chester, cartoonist E. C. Segar's hometown, claims that Popeye is based on Frank "Rocky" Fiegel, born in 1868, a man who was handy with his fists.

186 **Merritt can trace his maternal family history:** Much of the account here is drawn from Tulane University's Special Louisiana Research Collection; in particular, the papers of Dr. Joseph Jones, Charles Colcock Jones and the Colcock and Merrick-Jones family collections: http://specialcollections .tulane.edu/archon/index.php?p=collections/controlcard&id=117&q=jones.

186 **Major John Jones:** USGenWeb Archives, http://files.usgwarchives.net/ga/ liberty/bios/j5200005.txt, accessed August 13, 2015.

186 **award-winning histories:** *The Children of Pride: Selected Letters of the Family of the Rev. Dr. Charles Colcock Jones from the years 1860–1868*, edited by Robert Manson Myers (New Haven: Yale University Press, 1972), from which most of this account of his life is drawn, won a National Book Award in 1973. Erskine Clarke's *Dwelling Place: A Plantation Epic* (New Haven: Yale University Press, 2005) built on that earlier work, drawing on church papers and slave narratives to fill in the black families' histories.

188 **one of the first Creole cookbooks:** Natalie Vivian Scott and Caroline Merrick Jones, *The Gourmet's Guide to New Orleans* (New Orleans: Peerless Printing, 1933). Scott was a Red Cross nurse who served in both world wars and received France's highest medal for bravery for rescuing patients from a bombed building. After the war, she became, in the words of author Sherwood Anderson, "the best newspaperwoman in America." The book's foreword was by Dorothy Dix, the world's first and most widely read syndicated advice columnist. New Orleans cuisine, wrote Dix with the casual racism of the time, rested on "what an old colored cook once described as the necessary 'ingrejuns'" and was "glorified by the touch of the old Negro mammy." See Anthony Joseph Stanonis, *Dixie Emporium: Tourism, Foodways and Consumer Culture in the American South* (Athens, GA: University of Georgia Press, 2008).

189 **"the end of a life":** "In Memoriam: Joseph Merrick Jones," *Tulane Law Review* 37, no. 3 (April 1963).

190 **the first was released in 2007:** Coastal Protection and Restoration Authority, "Coastal Master Plans," p. 20, http://coastal.la.gov/resources/library/ reports/, accessed November 25, 2015.

192 **worst accidental oil spill in history:** Alan Silverleib, "The Gulf Spill:

America's Worst Environmental Disaster?" CNN, August 10, 2010, http://www.cnn.com/2010/US/08/05/gulf.worst.disaster/.

193 **"predawn of awareness":** Oliver A. Houck, "Breaking the Golden Rule: Judicial Review of Federal Water Project Planning," *Rutgers Law Review* 65 (2012): 25, http://biotech.law.lsu.edu/blog/Houck-golden-rule.pdf.

194 **a wonderful account:** Ibid., 1–57.

197 **April 2015 NOAA study:** Ariana E. Sutton-Grier, Kateryna Wowk and Holly Bamford, "Future of Our Coasts: The Potential for Natural and Hybrid Infrastructure to Enhance the Resilience of our Coastal Communities, Economies and Ecosystems," *Environmental Science and Policy* 51 (2015): 137–48, http://www.sciencedirect.com/science/article/pii/S1462901115000799.

198 **Idea Village:** Among the start-ups they've supported is the Cajun Fire Brewing Company, the only African American–owned craft brewery in the U.S. See Judith Rodin, "The Secret to New Orleans' Comeback," *Fortune*, September 3, 2015, http://fortune.com/2015/09/03/secret-new-orleans-comeback/.

198 **restoration economy:** The Environmental Defense Fund's Businesses and Coastal Restoration website maps existing and new opportunities: http://ourcoastoureconomy.org/economics-of-restoration/business-case-for-restoration/map-of-business-locations/. See also Marcy Lowe et al., "Restoring the Gulf Coast: New Markets for Established Firms," Duke Center for Globalization, Governance and Competitiveness, December 5, 2011, http://www.cggc.duke.edu/pdfs/CGGC_Gulf-Coast-Restoration.pdf.

198 **exportable equipment and expertise:** John Schwartz, "How to Save a Sinking Coast? Katrina Created a Laboratory," *New York Times*, August 7, 2015, http://www.nytimes.com/2015/08/08/science/louisiana-10-years-after-hurricane-katrina.html. In the U.S., coastal zones generate more than 42% of the nation's total economic output. Planners have been slow, however, to turn from conventional "gray" infrastructure to natural "green" infrastructure to protect those assets, in large measure because of a lack of agreed-upon engineering principles and performance metrics. Efforts are being made to fill that gap: see Shannon Cunniff, "New Report Quantifies Storm Reduction Benefits of Natural Infrastructure and Nature-Based Measures," Restore the Mississippi River Delta, blog post, September 29, 2015, http://www.mississippiriverdelta.org/blog/2015/09/29/new-report-quantifies-storm-reduction-benefits-of-natural-infrastructure-and-nature-based-measures/#sthash.Zt0TE5nz.dpuf, accessed December 6, 2015.

198 **120 million Americans:** Kristen Crossett et al., "National Coastal Population Report, Population Trends from 1970 to 2020," National Oceanic and

Atmospheric Administration and U.S. Census Bureau, March 2013, http://
stateofthecoast.noaa.gov/features/coastal-population-report.

198 **more than three billion people:** Gaia Vince, "The Rising and Sinking
Threats to Our Cities," BBC, June 13, 2013, http://www.bbc.com/future/
story/20130613-the-rising-threat-to-our-cities.

198 **One trillion dollars' worth:** Nicholas Herbert Stern, *Stern Review: The
Economics of Climate Change* (Cambridge, UK: Cambridge University
Press, 2007).

198 **the world's great deltas:** Liviu Giosan et al., "Climate Change: Protect the
World's Deltas," *Nature* 516, no. 7529 (2014): 31–33, http://www.nature.
com/news/climate-change-protect-the-world-s-deltas-1.16428.

198 **a unanimous vote:** Natalie Peyronnin et al., "Louisiana's 2012 Coastal
Master Plan: Overview of a Science-Based and Publicly Informed
Decision-Making Process," *Journal of Coastal Research* 67, special issue
(Summer 2013): 1–15, http://dx.doi.org/10.2112/SI_67_1.1.

198 **RESTORE Act:** Nikki Buskey, "Restore Act Passes Congress," *Houma
Today*, June 30, 2012, http://www.houmatoday.com/article/20120630/
articles/120639999; and U.S. Department of the Treasury, "RESTORE Act,"
http://www.treasury.gov/services/restore-act/PublishingImages/restore
_act_graph-large.jpg, updated August 10, 2015.

198 **$20.8 billion:** Environmental Law Institute Gulf Team, "BP Proposed
Consent Decree Released," ELI Oceans Program, October 13, 2015, http://
eli-ocean.org/gulf/consent-decree/. These restoration funds will need to be
defended from siphoning off by politically powerful interests, as nearly happened in a 2015 fight over funding the elevation of Highway 1 in Louisiana;
see Amy Wold, "Compromise Struck on La. 1 Elevation Project Funding;
Showdown Vote on Coastal Restoration Spending Avoided," *Advocate*,
October 21, 2015, http://theadvocate.com/news/13763970-148/showdown-
vote-on-coastal-restoration.

198 **$140 million:** Amy Wold, "A Tempting Target: Louisiana Stands to Gain
Substantial Funds for Coastal Projects, but the Governor's New Coastal
Director Warns Against Diverting Those Funds to Other Uses," *Advocate*,
July 27, 2015, http://theadvocate.com/news/13023346-123/a-tempting-
target-louisiana-stands.

199 **three winners:** Baird Team, "Delta For All," 2015, http://deltaforall.com/
wp-content/uploads/2015/08/6pg_Stakeholder_FINAL.pdf; Studio Misi-
Ziibi, "The New Misi-Ziibi Living Delta," 2015, http://www.h3studio.com/
misi-ziibi-1/; Moffatt & Nichol, "The Giving Delta," 2015, http://www
.moffattnichol.com/wp-content/uploads/ChangingCourse/MNTeam_
SummaryReportv10.pdf.

199 **"for the next big storm":** Jeff Carney, director of the LSU Coastal Sustainability Studio, quoted in Bob Marshall, "Experts: Talk Now About Drastic Changes, Or Deal With Coastal Crisis Later," *The Lens*, September 15, 2015, http://thelensnola.org/2015/09/15/coastal-planners-talk-now-about-drastic-changes-or-deal-with-crisis-later/.

201 **Corps's badly engineered levees failed:** Campbell Robertson and Jason Schwartz, "Decade After Katrina, Pointing Finger More Firmly at Army Corps," *New York Times*, May 23, 2015, http://www.nytimes.com/2015/05/24/us/decade-after-katrina-pointing-finger-more-firmly-at-army-corps.html.

201 **100,000 residents:** Emily Holden and ClimateWire, "Rising Sea Levels May Limit New Orleans Adaptation Efforts," *Scientific American,* September 10, 2015, http://www.scientificamerican.com/article/rising-sea-levels-may-limit-new-orleans-adaptation-efforts/.

202 **$150 billion:** Allison Plyer, "Facts for Features: Katrina Impact," Data Center, August 28, 2015, http://www.datacenterresearch.org/data-resources/katrina/facts-for-impact/.

202 **$10 billion investment:** These estimates were offered by Congressman Garrett Graves when he was head of the Coastal Protection and Restoration Authority. Graves's determination to prove that "principled infrastructure investments can yield exponential benefits," including significant savings in the longer term, was lauded in the conservative *National Review*. See Daniel M. Rothschild, "Saving the Bayou," January 13, 2014, http://www.nationalreview.com/article/368248/saving-bayou-daniel-m-rothschild.

202 **failure to invest now:** A 2015 LSU and RAND study found that the cost of doing nothing could reach $133 billion. See Stephen Barnes et al., "Economic Evaluation of Coastal Land Loss in Louisiana," Louisiana State University Economics & Policy Research Group and the RAND Corporation, December 2015, http://coastal.la.gov/wp-content/uploads/2015/12/LSU-Rand_Report_on_Economics_of_Land_Loss-2.pdf.

202 **more than 6.6 million homes:** Howard Botts et al., "2015 CoreLogic Storm Surge Report," June 2015, http://www.corelogic.com/research/storm-surge/corelogic-2015-storm-surge-report.pdf.

202 **largest fiscal liabilities:** Kate Sheppard, "Flood, Rebuild, Repeat: Are We Ready for a Superstorm Sandy Every Other Year?", *Mother Jones*, July/August 2013, http://www.motherjones.com/environment/2013/07/hurricane-sandy-global-warming-flooding.

202 **"ridiculous pseudo-science garbage":** Quoted in Michael O'Brien, "Vitter: Climate Change Evidence Often 'Ridiculous Pseudo-Science Garbage,'" *The Hill*, August 14, 2009, http://thehill.com/blogs/blog-briefing-room/

news/lawmaker-news/54867-vitter-climate-change-evidence-often-ridiculous-pseudo-science-garbage.

202 **letter to FEMA:** Signed by six other Republican senators, including fellow Louisianan Bill Cassidy. See Debbie Elliott, "As States Ready Disaster Plans, Feds Urge Them To Consider Climate Change," *It's All Politics*, NPR, May 19, 2015, http://www.npr.org/sections/itsallpolitics/2015/05/19/408010685/disaster-agency-to-require-states-to-consider-climate-change-in-plans.

202 **When Obama visited New Orleans:** Gov. Anthony Perrucci and CNN Wire, "Jindal: Katrina Anniversary 'Not the Time' for Climate Change Talk," WGNO, August 26, 2014, http://wgno.com/2015/08/26/gov-jindal-katrina-anniversary-not-the-time-for-climate-change-talk/.

203 **"the flag of Texaco":** John Barry, "Suing Oil and Gas Interests to Save the Coast," *Lens*, August 22, 2013, http://thelensnola.org/2013/08/22/suing-oil-and-gas-interests-to-save-the-coast-author-and-expert-john-barry-tells-why/.

203 **incorporates the leading science:** Local newspapers have also increased coverage of climate risks; see, for instance, Mark Schleifstein, "Rising Sea Level Threatens Coastal Restoration, New Orleans Levees, Scientists Say," *Times-Picayune*, August 27, 2015, http://www.nola.com/futureofneworleans/2015/08/rapidly_rising_sea_level_threa.html.

203 **seas will rise two feet:** Union of Concerned Scientists, "Causes of Sea Level Rise," 2013, http://www.ucsusa.org/sites/default/files/legacy/assets/documents/global_warming/Causes-of-Sea-Level-Rise.pdf.

203 **five feet:** Bob Marshall, "New Research: Louisiana Coast Faces Highest Rate of Sea-Level Rise Worldwide," *The Lens*, February 21, 2013, http://thelensnola.org/2013/02/21/new-research-louisiana-coast-faces-highest-rate-of-sea-level-rise-on-the-planet/.

203 **take up carbon two to four times faster:** National Oceanographic and Atmospheric Administration, Habitat Conservation, "Coastal Blue Carbon," http://www.habitat.noaa.gov/coastalbluecarbon.html, accessed August 26, 2015.

204 **fifty landowners:** Mark Schleifstein, "Carbon Credits Could Generate $1.6 Billion for Louisiana Coastal Restoration, Study Says," *Times-Picayune*, March 5, 2015, http://www.nola.com/environment/index.ssf/2015/03/carbon_credits_could_generate.html.

CHAPTER 4: SHRIMPER

209 **"the smartest people when it comes to southeast Louisiana water":** This knowledge was earned in both deltas they've called home. Levees and the destruction of wetland vegetation have left the mouth of the Mekong, like

that of the Mississippi, sinking and invaded by salt. See Jesse Hardman, "Delta Blues Part 1: The Battle To Keep Ho Chi Minh City Above Water," New Orleans Public Radio, January 21, 2015, http://wwno.org/post/delta -blues-part-1-battle-keep-ho-chi-minh-city-above-water.

210 **two million civilians:** Chase Madar, "Vietnam: A War on Civilians," *American Conservative*, July 30, 2013, http://www.theamericanconservative .com/articles/vietnam-a-war-on-civilians/.

210 **million people:** Mark Cutts, *The State of the World's Refugees, 2000: Fifty Years of Humanitarian Action* (Oxford: Oxford University Press, 2000).

212 **"where people died":** As many as 400,000 died, according to the U.N. High Commission for Refugees, lost to storms, disease and starvation. Many also suffered brutally at the hands of pirates who robbed, raped, abducted and sometimes killed refugees.

212 **boat people:** Karen J. Leong et al., "Resilient History and the Rebuilding of a Community: The Vietnamese American Community in New Orleans East," *Journal of American History* 94, no. 3 (2007): 770–79, http://www .journalofamericanhistory.org/projects/katrina/Leong.html.

212 **archbishop Philip Hannan:** Tom Wooten, *We Shall Not Be Moved: Rebuilding Home in the Wake of Katrina* (Boston: Beacon Press, 2012).

217 **breaking down on the phone:** Evacuation centers were for many a painful reminder of the refugee camps; see *New Orleans: A Village Called Versailles*, video, directed by S. Leo Chiang, New Day Films, 2008, http://www.pbs .org/frontlineworld/rough/2008/08/a_village_calle.html.

217 **"my mom put it in perspective":** Having survived first the move from north to south, then from Vietnam to America, each time leaving every-thing, building from nothing, they knew they could do it again. See Leong, "Resilient History."

217 **D & C Seafood:** Alexia Fernández Campbell and Mauro Whiteman, "Is This the End of the Line for Louisiana's Vietnamese Shrimpers," *National Journal*, October 30, 2014, http://www.nationaljournal.com/s/51997.

219 **more than 90 percent of the community had returned:** Though the city made plans as if they didn't exist: first to convert their neighborhood into a green space, drawing maps (as Father Vien Nguyen recalls) on which the community didn't even appear, then to use it as a landfill in which to dump all the contaminated debris from wrecked homes. The community fought through the city council, the courts and protests in the streets, ultimately electing the first Vietnamese member of the city council. See *New Orleans: A Village Called Versailles*, video.

220 **"the largest barrier":** Coastal Communities Consulting, mission state-ment. See Benjamin Alexander-Bloch, "USDA Awards Gretna Nonprofit a Grant for Commercial Fishers, Rural Entrepreneurs," *Times-Picayune*,

October 7, 2014, http://www.nola.com/environment/index.ssf/2014/10/usda_awards_gretna_nonprofit_a.html.

220 **three-fourths of Louisiana's shrimping fleet:** Louisiana Small Business Development Center, "LSBDC Names Top Business Consultant in Louisiana," press release, September 24, 2012, http://www.lsbdc.org/resources/?p=1199.

222 **88,000 square miles:** "BP Oil Spill: 1,000 Days," RESTORE the Mississippi River Delta, http://www.mississippiriverdelta.org/blog/2013/01/14/1000-days-later/, posted January 14, 2013.

222 **contaminated more than 1,300 miles:** National Oceanographic and Atmospheric Administration, "*Deepwater Horizon* Oil Spill: Draft Programmatic Damage Assessment and Restoration Plan and Draft Programmatic Environmental Impact Statement," Chapter 4: "Injury to Natural Resources," October 5, 2014, http://www.gulfspillrestoration.noaa.gov/wp-content/uploads/Chapter-4_Injury-to-Natural-Resources1.pdf.

223 **EPA's ultimate acquiescence:** The EPA was hamstrung, especially by the weakness of the Toxic Substances Control Act (TSCA). Kate Sheppard, "BP's Bad Breakup: How Toxic Is Corexit?," *Mother Jones,* September/October 2010, http://www.motherjones.com/environment/2010/09/bp-ocean-dispersant-corexit. But EPA has also been unconscionably slow on chemical risk assessments. See: Sam Pearson, "Obama Request Boosts Funding for Safety Reviews," *E & E News,* February 10, 2016, http://www.eenews.net/eedaily/2016/02/10/stories/1060032128.

223 **two million gallons of the dispersant Corexit:** Paul Quinlan, "More Questions Than Answers on Dispersants a Year After Spill," *E & E News,* April 22, 2011, http://www.eenews.net/special_reports/gulf_spill/stories/1059948145.

223 **increased total toxicity fifty-fold:** 2015 studies found that Corexit actually inhibits the microorganisms that naturally degrade oil, and is linked to lung damage in humans. See Sara Kleindienst et al., "Chemical dispersants can suppress the activity of natural oil-degrading microorganisms," *Proceedings of the National Academy of Sciences* 112, no. 48 (2015): 14900–05; Roberto Rico-Martínez, Terry W. Snell and Tonya L. Shearer, "Synergistic Toxicity of Macondo Crude Oil and Dispersant Corexit 9500A® to the *Brachionus plicatilis* Species Complex (Rotifera)," *Environmental Pollution* 173 (2013): 5–10; and Fu Jun Li et al., "Heme Oxygenase-1 Protects Corexit 9500A-Induced Respiratory Epithelial Injury Across Species," *PLoS ONE* 10, no. 4 (2015), http://journals.plos.org/plosone/article?id=10.1371/journal.pone.0122275.

223 **The destruction of marine and bird life:** Any ecological body count is a complex, protracted and imprecise undertaking. But because it occurred in

one of Earth's most fragile ecosystems, above the Arctic Circle, the 1989 *Exxon Valdez* spill exacted its own devastating toll, harming such species as orcas, sea otters and salmon. Twenty years later, many are still not recovered. The National Wildlife Federation, "Compare the Exxon Valdez and BP Oil Spills," https://www.nwf.org/What-We-Do/Protect-Habitat/Gulf-Restoration/Oil-Spill/Effects-on-Wildlife/Compare-Exxon-Valdez-and-BP-Oil-Spills.aspx.

223 **sea turtles:** National Oceanographic and Atmospheric Administration, "*Deepwater Horizon* Oil Spill: Draft Programmatic Damage Assessment and Restoration Plan and Draft Programmatic Environmental Impact Statement," Chapter 4: "Injury to Natural Resources," October 5, 2014, http://www.gulfspillrestoration.noaa.gov/wp-content/uploads/Chapter-4_Injury-to-Natural-Resources1.pdf.

223 **a million birds:** Martha Harbison, "More than One Million Birds Died During Deepwater Horizon Disaster," *Audubon*, May 6, 2014, https://www.audubon.org/news/more-one-million-birds-died-during-deepwater-horizon-disaster.

223 **lose the weatherproofing:** David Gessner, "The Birds of British Petroleum," *Audubon*, July–August 2015, https://www.audubon.org/magazine/july-august-2015/the-birds-british-petroleum.

223 **two trillion larvae:** National Oceanic and Atmospheric Administration, "*Deepwater Horizon* Oil Spill: Draft Programmatic Damage Assessment."

223 **lung disease and adrenal lesions:** National Wildlife Federation, "Five Years and Counting: Gulf Wildlife in the Aftermath of the *Deepwater Horizon* Disaster," March 30, 2015, http://www.nwf.org/~/media/PDFs/water/2015/Gulf-Wildlife-In-the-Aftermath-of-the-Deepwater-Horizon-Disaster_Five-Years-and-Counting.pdf.

224 **Stillbirths became commonplace:** Suzanne M. Lane et al., "Reproductive Outcome and Survival of Common Bottlenose Dolphins Sampled in Barataria Bay, Louisiana, USA, Following the *Deepwater Horizon* Oil Spill," *Proceedings of the Royal Society B* 282, no. 1818 (2015). Mark Schleifstein, "Failed Barataria Dolphin Pregnancies Linked to BP Spill," *Times-Picayune*, November 3, 2015, http://www.nola.com/environment/index.ssf/2015/11/failed_barataria_bay_dolphin_p.html.

224 **more than half of the total oiled coastline:** Jacqueline Michel et al., "Extent and Degree of Shoreline Oiling: *Deepwater Horizon* Oil Spill, Gulf of Mexico, USA," *PLoS ONE* 8, no. 6 (June 12, 2013): e65087, http://journals.plos.org/plosone/article?id=10.1371/journal.pone.0065087.

224 **rates of erosion doubled:** Brian R. Silliman et al., "Degradation and Resilience in Louisiana Salt Marshes After the BP–*Deepwater Horizon* Oil Spill,"

Proceedings of the National Academy of Sciences 109, no. 28 (2012): 11234–39, http://dx.doi.org/10.1073%2Fpnas.1204922109.

224 **Cat Island:** Emily Guidry Schatzel, "Five Years After the Oil Spill, Dead Dolphins and 25,000-Pound Tar Mat Found," *Wildlife Promise* (blog), National Wildlife Federation, March 18, 2015, http://www.mississippiriverdelta.org/blog/2015/03/18/five-years-after-the-oil-spill-dead-dolphins-and-25000-pound-tar-mat-found/#sthash.n84YcoMH.dpuf.

224 **unseen impacts:** Reductions in abundance and diversity of "benthic" (bottom-dwelling) organisms like crabs have been observed at moderate to severe levels in an area roughly 57 square miles around the wellhead. Blue crab harvests have been approximately 20 percent lower between 2011 and 2014 than they were in the ten years prior to 2010. See National Wildlife Federation, "Five Years and Counting."

225 **found in sperm whales:** National Wildlife Federation, "Five Years and Counting."

225 **eggs of migrating American white pelicans:** Mark Neuzil, "Body of Evidence: How the White Pelican Spreads Oil," *Audubon*, July–August 2015, https://www.audubon.org/magazine/july-august-2015/body-evidence-how-white-pelican-spreads-oil.

225 **2,900 square miles:** Henry Fountain, "Gulf Spill Sampling Questioned," *New York Times*, August 19, 2013, http://www.nytimes.com/2013/08/20/science/earth/new-analysis-of-gulf-oil-spill.html.

225 **oysters were still dead:** With oyster eggs, sperm and larvae exposed, scientists documented low numbers of oyster babies (called "spat") reaching maturity for several years following the spill. See National Wildlife Federation, "Five Years and Counting."

226 **lawyers scamming the Vietnamese:** Campbell Robertson and John Schwartz, "Many Hit by Spill Now Feel Caught in Claim Process," *New York Times*, April 18, 2011, http://www.nytimes.com/2011/04/19/us/19spill.html?action=click&contentCollection=U.S.&module=Related Coverage&region=Marginalia&pgtype=article.

226 **"If I hadn't met him":** Fernández Campbell and Whiteman, "Is This the End of the Line for Louisiana's Vietnamese Shrimpers?" *National Journal*.

226 **Ken Feinberg:** Feinberg was terrible, Sandy says. Without a formula for boat size and income, "some folks that had been making just $5,000 a year got twenty times that, while others whose entire income depended on the fisheries were getting nothing."

226 **"take politics out of it":** Sandy herself refuses to pull the strings she's accumulated over the years, as she recounts (referring to herself, as she sometimes does, in the third person): "I went to traffic court and that judge

is somebody that's very fond of me, but Sandy will sit there and wait two hours to get her turn. I could have just brought it to his house, but that sets a bad example for my folks. If I go into the permits department in City Hall, I know everybody, but I sit there and wait and do it right, because I'm educating my clients. You don't want to educate them to do things the wrong way, you want to educate them to do the right things."

228 **joins his deckhand:** The deckhand is an Amerasian, born of the war. They have the toughest time, says Sandy, "treated as outcasts."

232 **"one little paragraph":** The paragraph reads, *"Providing for Transitions:* As we address this crisis, sensitivity and fairness must be shown to those whose homes, lands, livelihoods and ways of life may be affected, in the near-term and long-term, by master plan projects or by continued land loss and flooding." There is in fact more. Though buried in Appendix B9 and built on an impenetrably complex formula, the Master Plan did assess how its projects, taken together, would affect the "ability for people to continue to live within their coastal community" and the "availability of natural resources determined to be culturally important for communities." They found that Sandy's west bank communities, including Buras and Venice, are in for a rough time with or without the Master Plan. Byron's east bank communities, by contrast, including Pointe a la Hache, will see their chances of survival significantly increased by the Master Plan. See http://coastal .la.gov/a-common-vision/2012-coastal-master-plan/cmp-appendices/.

234 **bycatch of other species:** National Marine Fisheries Service, "U.S. National Bycatch Report," edited by William A. Karp, Lisa L. Desfosse and Samantha G. Brooke, NOAA Technical Memo, 2011, http://www.nmfs .noaa.gov/by_catch/bycatch_nationalreport.htm.

234 **80 percent of a shrimper's catch:** John McQuaid, "Gulf Shrimpers Take a Pounding," *Times-Picayune*, March 28, 1996, http://www.nola.com/ environment/index.ssf/1996/03/gulf_shrimpers_take_a_pounding.html.

234 **"fish-eye":** "Definition and History: Bycatch Reduction Device (BRD)," Louisiana's Fisheries, Louisiana State University Agricultural Center, http://www.lsu.edu/seagrantfish/management/TEDs&BRDs/brds_specs. htm, accessed October 15, 2015.

234 **his own branded business:** Anna Maria Shrimp; http://annamarieshrimp. com/.

235 **Seafood Watch:** Benjamin Alexander-Bloch, "Seafood Watch Removes Louisiana Shrimp from 'Avoid' List," *Times-Picayune*, July 2, 2015, http://www.nola.com/environment/index.ssf/2015/07/seafood_watch_ removes_louisian.html; and SeaWeb, "Louisiana Gulf Shrimp," http:// seaweb.org/news/SeaWebSuccessStories_GulfShrimp.php, accessed October 15, 2015.

235 **Mid-Barataria Diversion:** Chuck Perrodin, "The CPRA Recommends Advancing Two Mississippi River Sediment Diversions," Coastal Protection and Restoration Authority, October 21, 2015, http://coastal.la.gov /wp-content/uploads/2015/10/FINAL-SedimentDiversionsRecommen dation-2.pdf.

235 **as much water as . . . the Missouri River:** J. C. Kammerer, "Largest Rivers in the United States," U.S. Geological Survey, http://pubs.usgs.gov/of/1987/ ofr87-242/, updated September 1, 2005.

237 **"intimidated or fearful":** At one point a girl chimes in, translating for the Cambodians what a new member of Sandy's staff has said: "This is the first time that they come to listen to us, so Auntie, Uncle and all, do not lose hope and please trust them."

238 **made matters worse:** This diversion, called West Bay, was built by the Corps in 2003. It failed to build land until 2009, when on the advice of local rancher Earl Armstrong the Corps added islands to slow the outflow and allow the silt to settle. It has since built 1,000 acres but is almost certainly a lost cause, given the high rates of subsidence and wave action this far downriver. Future projects will be built well upriver, where they stand a chance of surviving. See John Snell, "Cattle Rancher Crusades to Save Coastal Project," *Fox 8 Live*, December 27, 2012, http://www.fox8live.com/ story/19981026/cattle-rancher-crusades-to-save-a-coastal-project, accessed October 19, 2015.

239 **stunningly unhelpful:** In tallying the reasons for land loss, she also omitted entirely the Corps's role: "Ninety percent of what we've lost is due to factors you guys know already: sea level rise, subsidence, salt water intrusion when the oil and gas companies came in and drilled wells." A fisherman promptly caught her out, noting that parts of New Orleans are several feet further below sea level than Buras: "so we can talk about whether it's a Gulf problem or a levee problem."

239 **"vital to the state's economy":** Shrimp is one of the highest-value fisheries in the Southeast, bringing in more than $400 million a year. See National Marine Fisheries Service, "Regional Summary Gulf of Mexico Region Fisheries," 2011, https://www.st.nmfs.noaa.gov/Assets/economics/documents/ feus/2011/FEUS2011%20-%20Gulf%20of%20Mexico.pdf.

240 **Grand Liard Ridge:** The Grand Liard bayou used to flow out of the river between high, tree-lined banks, nourishing miles of marsh. When levees severed the bayou's connection to the river, it was like cutting an umbilical cord, leaving the banks and marsh to waste away.

242 **one dredge would take 300 to 450 days:** Natalie Peyronnin, former senior coastal resource scientist for Louisiana's Coastal Protection and Restoration Authority, interview with the author.

242 **Pumping sediments:** Dumping sediment through a pipe would also kill marsh, and would never be permitted. Doug Rader, chief marine scientist, Environmental Defense Fund, interview with the author.

243 **hundreds of millions of dollars:** U.S. Army Corps of Engineers, "USACE Navigation—Meeting America's Maritime Transportation Needs," February 2, 2015, http://www.usace.army.mil/Portals/2/docs/civilworks/budget/strongpt/fy16sp_navigation.pdf.

243 **Community Focus Group:** Coastal Protection and Restoration Authority, "Resilience Working Groups," http://coastal.la.gov/project-content/ccrp/resilience-working-groups/, accessed January 19, 2016.

245 **conference on urban resilience:** "Field Lessons from the Ground: Building Urban Resilience Upon the Occasion of the 10th Anniversary of Hurricanes Katrina and Rita," MAS Cities Global Resilience Network, https://bellagiomas.wordpress.com/about/, accessed October 20, 2015.

245 **Caernarvon Freshwater Diversion:** U.S. Army Corps of Engineers, "Modification of Caernarvon Diversion," January 2013, http://www.mvn.usace.army.mil/Portals/56/docs/environmental/LCA/Near-Term%20Projects/ModofCaernarvonFactSheetJanuary2013PAO.pdf.

246 **"birth of a new delta":** In homage to Tidwell's *Bayou Farewell*, Lopez and team call the bayou that winds through here Bayou Bonjour.

247 **"contrary to plan, running it wide open":** Sammy Nunez, then president of the Louisiana state senate, overrode the operation scheme that had been negotiated. See Jeffrey Meitrodt and Aaron Kuriloff, "Shell Game," *Times-Picayune*, May 11, 2003, http://www.nola.com/speced/shellgame/index.ssf?/speced/shellgame/grounds11.html.

247 **"we got destroyed":** Byron Encalade contends that the Caernarvon Diversion actually resulted in net land loss. Todd Masson, "Caernarvon Diversion Moving Fish and Killing Marsh, Anglers Say," *Times-Picayune*, March 26, 2013, http://www.nola.com/outdoors/index.ssf/2013/03/caernarvon_diversion_moving_fi.html. Encalade is also opposed to John Lopez's proposal to keep Mardi Gras Pass open even though it's not in the Master Plan because, he says, the crevasse has accelerated the loss of oysters, and because it sets a precedent for the Master Plan to be ignored.

248 **"Open Letter to the Citizens of Louisiana":** Available at http://www.mississippiriverdelta.org/files/2015/05/AnOpenLettertotheCitizensofLouisiana.pdf.

249 **edges of the marsh:** See also: Lawrence P. Rozas and Denise J. Reed, "Nekton Use of Marsh-Surface Habitats in Louisiana (USA) Deltaic Salt Marshes Undergoing Submergence," *Marine Ecology: Progress Series* 96 (1993): 147–57.

250 **clean the water flowing into the Gulf:** Debate continues about the relative

value of river-built freshwater marsh versus saltwater marsh. The former is more easily damaged but also rebounds; the latter is more robust but once gone is gone for good.

250 **a quarter of total annual nutrient pollution:** Victor H. Rivera-Monroy et al., "Landscape-Level Estimation of Nitrogen Removal in Coastal Louisiana Wetlands: Potential Sinks under Different Restoration Scenarios," *Journal of Coastal Research*, special issue 67: "Louisiana's 2012 Coastal Master Plan Technical Analysis" (2013): 75–87, doi: http://dx.doi.org/10.2112/SI_67_6.

251 **Asian imports:** In 1990 the U.S. imported 500 million pounds of shrimp. By 2010, the figure was 1.2 billion pounds, 75% of it from Asia. *Consumer Reports* tested 342 samples of frozen shrimp from large retail chains in twenty-seven U.S. cities, most of it farm-raised imports, and found harmful bacteria and illegal antibiotic residues. Fraud is also rampant; even in the Gulf, more than one-third of the shrimp labeled "Gulf" were farmed imports.

251 **85 cents in 2015:** Jacob Batte, "Targeting Imports: Legislation Could Protect Domestic Shrimpers," *Houma Today*, June 29, 2015, http://m.houmatoday.com/article/20150629/ARTICLES/150629680?Title=Targeting-imports-Legislation-could-protect-domestic-shrimpers.

251 **reduced U.S. Gulf shrimping by about 35 percent:** Lydia Mulvany, "Asian Shrimp Imports Are Chewing Up U.S. Suppliers," *Bloomberg News*, September 7, 2015, http://www.bloomberg.com/news/articles/2015-09-07/all-you-can-eat-shrimp-imports-chew-up-u-s-suppliers-amid-slump.

251 **destruction of . . . mangroves:** George Black, "Hu Tieu, A Vietnamese Dish Spiced with Prosperity and Climate Change," *Guardian*, December 14, 2014, http://www.theguardian.com/vital-signs/2014/dec/14/vietnam-dish-environment-fish-farm-shrimp-climate-aquaculture.

251 **water pollution:** Tom Philpott, "Today's Seafood Special: Pig Manure, Antibiotics, and Diarrhea Bugs," *Mother Jones*, January/February 2013, http://www.motherjones.com/environment/2013/01/imported-seafood-shrimp-fda.

251 **labor conditions:** Tom Philpott, "Did a Slave Process the Shrimp in Your Scampi?", *Mother Jones*, June 17, 2013, http://www.motherjones.com/tom-philpott/2013/06/did-slave-process-shrimp-your-scampi.

251 **greatest help they could get:** A top priority is passage of Louisiana congressman Charles Boustany's PROTECT bill, designed to stop illegal imports. See "Boustany's Bill to PROTECT Louisiana Seafood Passes House," website of Congressman Charles W. Boustany, June 12, 2015, https://boustany.house.gov/114th-congress/boustanys-bill-to-protect-louisiana-seafood-passes-house/. In April 2015, the U.S. Court of International Trade rubbed salt in local wounds with its ruling that "no domestic industry is being

harmed by imports of frozen warm water shrimp from China, Ecuador, India, Malaysia and Vietnam," because they were merely filling the void left by the BP spill. See "US Court: BP Oil Spill Hurt US Shrimp Industry, Not Farmed Imports," *Undercurrent News*, April 6, 2015, http://www .undercurrentnews.com/2015/04/06/us-court-bp-oil-spill-hurt-us-shrimp -industry-not-farmed-imports/.

252 **"speckled trout":** Quoted in Marshall, "Louisiana Coastal Experts Beginning Study."

CHAPTER 5: FISHERMAN

257 **By 2001 federal fishery managers:** National Marine Fisheries Service, Southeast Regional Office, "How Has the Red Snapper Fishery Changed Over Time?", http://sero.nmfs.noaa.gov/sustainable_fisheries/gulf_fisheries /red_snapper/overview/index.html, accessed November 4, 2015.

258 **Until 1990, Wayne could bring in:** The first fishery management plan, passed in 1984, set only a few gear restrictions and a 13-inch minimum size limit. See Gulf of Mexico Fishery Management Council, "Reef Fish Management Plans—Archives," http://gulfcouncil.org/fishery_management_ plans/reef_fish_management_archives.php, accessed November 4, 2015.

259 **the Gulf's iconic fish:** In official terms it was "overfished," meaning there were too few fish in the water to replenish the population and maximize harvest over the long term. And fishermen were still overfishing: removing fish at too high a rate for it to ever recover. See National Marine Fisheries Service, Southeast Regional Office, "How Has the Red Snapper Fishery Changed Over Time?"

259 **fifty pounds over fifty years:** Ellen Bolen, "10 Key Facts About Red Snapper," *Oceans Currents* (blog), Ocean Conservancy, May 20, 2013, http:// blog.oceanconservancy.org/2013/05/20/10-key-facts-about-red-snapper/.

259 **a ten-year-old:** National Marine Fisheries Service, Southeast Regional Office, "Rebuilding Red Snapper: What Have Fishery Managers and Fishermen Done to Rebuild Red Snapper?", http://sero.nmfs.noaa.gov/ sustainable_fisheries/gulf_fisheries/red_snapper/overview/rebuilding/ index.html.

260 **"big old fat fertile female":** C. E. Porch et al., "Estimating the Dependence of Spawning Frequency on Size and Age in Gulf of Mexico Red Snapper," *Marine and Coastal Fisheries* 7, no. 1 (2015): 233–45.

260 **2.6 percent in 1990:** National Marine Fisheries Service, Southeast Regional Office, "How Has the Red Snapper Fishery Changed Over Time?"

260 **Magnuson Fishery Conservation and Management Act:** Magnuson followed close on the heels of the National Environmental Protection Act, Clean Water Act, Clean Air Act, Marine Mammal Protection Act and

Endangered Species Act. See Eric C. Schwaab, "The Road to End Overfishing: 35 Years of Magnuson Act," National Marine Fisheries Service,
April 11, 2011, http://www.nmfs.noaa.gov/stories/2011/20110411roadendo
verfishing.htm, accessed December 9, 2015.

260 **two hundred nautical miles:** Or to an international border in cases where
one EEZ bumps up against another. The 200-nautical-mile Exclusive Economic Zone was codified internationally in the 1982 U.N. Convention on
the Law of the Sea. The U.S. controls more ocean than any other country
in the world. See Paul Greenberg quoted in Adam Wernick, "How American Seafood Goes Almost Everywhere Except America," Public Radio
International, August 7, 2014, http://www.pri.org/stories/2014-08-07/how-
american-seafood-goes-almost-everywhere-except-america.

260 **jurisdiction over near-shore waters:** Texas and Florida state waters
on the Gulf Coast extend nine miles; the others extend three miles. See
National Oceanographic and Atmospheric Administration, Office of General Counsel, "Maritime Zones and Boundaries," http://www.gc.noaa.gov/
gcil_maritime.html#eez, accessed November 6, 2015.

260 **Gulf of Mexico Fishery Management Council:** The Gulf Council's seventeen members include the directors of the five state fish and wildlife
agencies, eleven members nominated by the governor and chosen by the
commerce secretary, and the National Marine Fisheries Service southeast
administrator. Its charge is to develop management plans for fish stocks
that meet legal standards and are based on the best available science as
well as input from fishermen, scientists and the public. The Gulf Council
currently has management plans for seven fisheries: reef fish (snapper and
groupers), shrimp, spiny lobster, corals, pelagics, drum and aquaculture.
It also has an essential habitat plan applicable to all the rest. See Gulf of
Mexico Fishery Management Council, "About Us," http://gulfcouncil.org/
about/index.php, accessed November 9, 2015.

261 **51 percent to commercial fishermen:** Gulf of Mexico Fishery Manage-
ment Council, "Red Snapper Individual Fishing Quota Program 5-Year
Review," April 2013, http://gulfcouncil.org/docs/amendments/Red%20
Snapper%205-year%20Review%20FINAL.pdf.

261 **February 22, February 16:** Gulf of Mexico Fisheries Management Council, "Reef Fish Management Plans—Archives," http://gulfcouncil.org/
fishery_management_plans/reef_fish_management_archives.php, accessed
November 9, 2015.

261 **ever more stringent rules:** Using "amendments" to management plans,
close to forty in all; ibid.

261 **"mini" and then a "micro" derby:** Gulf of Mexico Fisheries Management
Council, "Red Snapper Individual Fishing Quota Program 5-Year Review."

261 **more than half the seventeen derby years:** Ibid.

262 **more than a million fish each year:** Ibid.

262 **released fish died:** D. L. Nieland et al., "Red Snapper in the Northern Gulf of Mexico: Age and Size Composition of the Commercial Harvest and Mortality of Regulatory Discards," *American Fisheries Society Symposium* 60, 2007.

263 **John Breaux:** Senator Breaux found Wayne impressive: "[I] always have great respect for those of you who are out there every day fishing . . . And then to have to come in and talk about all this stuff. It's like alphabet soup . . . This is my job, and I find it horribly confusing. To find a fishing boat captain out there who can also understand this stuff is truly amazing to me." See statement of Wayne Werner, Commercial Fisherman, Galliano, Louisiana, "Reauthorization of the Magnuson–Stevens Fishery Conservation and Management Act," Field Hearing Before the Subcommittee on Oceans and Fisheries of the Committee on Commerce, Science, and Transportation, U.S. Senate, 106th Cong., 1st sess., 1999, http://www.gpo.gov/fdsys/pkg/CHRG-106shrg77584/html/CHRG-106shrg77584.htm.

267 **"rights-based management":** The new EU initiative to move toward rights-based management calls them "Transferable Fishing Concessions."

267 **Catch shares:** The ideas for governing common resources that underlie catch shares—including exclusive access and community participation in designing the rules—have deep roots. Economist Elinor Ostrom found villages in Switzerland, Japan, Spain and the Philippines successfully sharing their common grazing, forest and irrigation resources over many centuries. Francis Christy proposed the modern fisheries variation in the early 1970s. See Elinor Ostrom, *Governing the Commons: The Evolution of Institutions for Collective Action* (Cambridge, UK: Cambridge University Press, 1990); and Francis Christy, "Fishermen's Quotas: A Tentative Suggestion for Domestic Management," Law of the Sea Institute, occasional paper no. 19, 1973.

268 **Coastal Conservation Association:** CCA was founded in 1977 to combat "commercial overfishing along the Texas coast, [which had] decimated redfish and speckled trout populations." See Coastal Conservation Association, "About CCA," http://www.joincca.org/about, accessed December 13, 2015.

268 **Senate subcommittee field hearing:** Hearing on the Reauthorization of the Magnuson Fishery Conservation and Management Act: Before the Subcommittee on Oceans and Fisheries of the Committee on Commerce, Science, and Transportation, United States Senate, 104th Cong., 1st sess., March 4, 1995.

268 **"Well I'm one of the I's":** His exact words were: "I am one of the individuals . . . who had my life voted on the other day."

269 **"moratorium on *any* new IFQs":** The National Marine Fisheries Service

had gotten so far as to approve a snapper IFQ in 1995, but Congress first blocked funding and then erected the moratorium. See Pam Baker, Felix G. Cox and Peter M. Emerson, "Managing the Gulf of Mexico Commercial Red Snapper Fishery," presented to the Committee to Review Individual Fishing Quotas, Ocean Studies Board, National Research Council, New Orleans, Louisiana, January 27, 1998, http://media.law.stanford.edu/organizations/programs-and-centers/enrlp/doc/slspublic/redsnapper2_ex.pdf.

269 **until 2000:** The moratorium was originally until 2000 but later extended until 2002. See Donna R. Christie and Richard G. Hildreth, *Coastal and Ocean Management Law in a Nutshell* (St. Paul, MN: West Academic, 2007), 245.

270 **"probably 80 percent":** According to EDF's Pam Baker, who played a central role in advancing the IFQ, "The fishermen were the first to say that 80 percent of released fish were dying and research eventually concluded the same."

271 **how many fish are really out there:** A comprehensive description of stock assessment methods can be found in Shannon Cass-Calay et al., "SEDAR Red Snapper 2014 Update Assessment: September 7, 2015," http://sedarweb.org/docs/suar/SEDARUpdateRedSnapper2014_FINAL_9.15.2015.pdf.

272 **not a property right:** A critical difference between an access right and a property right is that when an access right, such as a grazing lease, is revoked, the government does not have to pay compensation for a "taking."

274 **"Ad Hoc Red Snapper Advisory Panel":** Plus four non-voting members representing economics, biology, law enforcement and environmental interests.

274 **protection for their small businesses:** These protections are the norm for most US catch share systems, though each design choice entails tradeoffs among social, economic and biological outcomes, as the Gulf Council's 2015 review of the IFQ makes clear. And though consolidation is occurring in the fish supply chain, it is driven primarily by globalization. Every dock in the South Atlantic used to have a fish house; now there are almost none—even where no catch shares exist. See Corbett A. Grainger, "Defining and Designing Property Rights in Marine Fisheries," Conservation Leadership Council, 2013, http://www.leadingwithconservation.org/wp-content/uploads/2013/07/clc-grainger_04.pdf.

275 **the most ardent and public champion:** Buddy is the star of *Big Fish, Texas*, a reality show on the National Geographic Channel, http://channel.nationalgeographic.com/big-fish-texas/; he was also featured in Isobel Yeung, "Countdown to Extinction," *VICE*, episode 25, HBO, March 20, 2015.

275 **3.3 million, 7.3 million:** Gulf of Mexico Fishery Management Council,

"Red Snapper Quotas for 2015–2017+," March 2015, http://gulfcouncil
.org/docs/amendments/Final%20Red%20Snapper%20Framework%20
Action%20Set%202015-2017%20Quotas.pdf.

276 **Gulf Wild:** http://www.gulfwild.com/index.php.

277 **more than forty fishermen:** Alex Dropkin, "The Gulf's Red Snapper
Fishery Makes a Comeback," *Texas Observer*, September 24, 2014, http://
www.texasobserver.org/red-snapper-fishery-texas-gulf-sustainable, accessed
December 1, 2015.

277 **Recent tests:** By Oceana, between 2010 and 2012 at 674 retailers. In fact,
only 7 of the 120 samples of red snapper purchased nationwide were actually red snapper. See Kimberly Warner et al., "Oceana Study Reveals Seafood Fraud Nationwide," Oceana, February 2013, http://oceana.org/sites/
default/files/reports/National_Seafood_Fraud_Testing_Results_FINAL
.pdf, accessed November 12, 2015.

277 **more than halfway to its rebuilding goal:** National Marine Fisheries
Service, Southeast Regional Office, "How Has the Red Snapper Fishery
Changed Over Time?"

277 **Seafood Watch:** "Seafood Watch Removes Gulf of Mexico Commercial
Red Snapper Fishery From 'Red List,'" Environmental Defense Fund, September 17, 2013, https://www.edf.org/media/seafood-watch-removes-gulf-
mexico-commercial-red-snapper-fishery-"red-list".

278 **"tragedy of the commons":** Garrett Hardin, "The Tragedy of the Commons," *Science* 162, no. 3859 (1968): 1243–48.

279 **"marine protected areas":** Felicia C. Coleman, Pamela B. Baker and
Christopher C. Koenig, "A Review of Gulf of Mexico Marine Protected
Areas: Successes, Failures, and Lessons Learned," *Fisheries* 29, no. 2 (2004):
10–21.

279 **strict accountability:** Efforts are underway to move fisheries to electronic
logbooks, which enable a fisherman to record the species, number, weight
and size of his catch, including discards, stamped by time, location and
depth, directly into National Marine Fisheries Service databases. Working
with its "study fleet"—fishermen who collaborate on research—NOAA
has found digital data-gathering to be nearly as accurate as putting observers on each boat. With onboard cameras, it would become verifiable. See
Emily Yehle, "Technology Buoys Fishermen Devastated By Cod's Collapse," GreenWire, *E & E News*, August 10, 2015, http://www.eenews.net/
stories/1060023209.

280 **Between 1991 and 2013:** Gulf of Mexico Fishery Management Council,
"Red Snapper Quotas for 2015–2017+".

280 **Testifying to Congress:** Oversight Hearing Before the Committee on
Natural Resources, U.S. House of Representatives, 113th Cong, 1st sess.,

2013, http://www.gpo.gov/fdsys/pkg/CHRG-113hhrg81806/html/CHRG
-113hhrg81806.htm.

281 **failing recreational fishermen:** Pamela Baker, "The Red Snapper Management System 'Stinks and Punishes Everyone,'" Saving Seafood, March
1, 2013, http://www.savingseafood.org/opinion/edfs-pamela-baker-the-red-
snapper-management-system-stinks-and-punishes-everyone/.

281 **recreational harvest is tracked only sporadically:** by the federal Marine
Recreational Information Program (MRIP) and state wildlife agencies;
see National Marine Fisheries Service, Southeast Regional Office, "2015
Gulf of Mexico Red Snapper Recreational Season Length Estimates,"
April 20, 2015, http://sero.nmfs.noaa.gov/sustainable_fisheries/lapp_dm/
documents/pdfs/2015/rs_2015_rec_quota_projection.pdf.

282 **"fishing in less than 100 feet":** "If fishing farther offshore to get big fish,"
Louisiana State University Professor James Cowan wrote in an August 2015
letter to the Gulf Council, recreational "discard mortality rates are likely to
be much higher than the 10% used by the Council to set quota." James H.
Cowan Jr., letter to the Gulf Council, August 8, 2015, http://gulfcouncil
.org/docs/minority%20reports/A28_Minority%20Report_FINAL%
20w%20Appendices.pdf. The Millennium Ecosystem Assessment found
average fishing depths increasing from 175 meters in 1950 to 300 meters
in 2001. See Millennium Ecosystem Assessment, *Ecosystems and Human
Well-Being: Synthesis* (Washington DC: Island Press, 2005).

282 **no limits on who can participate:** In 2013, 3.4 million recreational fishermen took more than 26 million fishing trips in the Gulf and caught 192
million fish. See National Marine Fisheries Service, "Fisheries of the United
States, 2014," Current Fishery Statistics no. 2014, http://www.st.nmfs.noaa
.gov/Assets/commercial/fus/fus14/documents/FUS%202014%20FINAL
.pdf.

282 *doubled*: Gulf of Mexico Fishery Management Council, "Red Snapper
Quotas for 2015–2017+".

282 *six-fold*: National Marine Fisheries Service, Southeast Regional Office,
"Framing the Red Snapper Issue in the Gulf of Mexico (March 2013),"
http://sero.nmfs.noaa.gov/sustainable_fisheries/gulf_fisheries/red_snapper/
documents/pdfs/gulf_red_snapper_web_article.pdf.

282 **ever faster boats:** Capable of up to 65 knots, compared to Wayne's top
speed of 11 knots.

283 **"the way they're running the fishery":** As LSU scientist James Cowan
wrote in his 2015 letter to the Gulf Council, "We have demonstrated over
and over again that there is sufficient fishing capacity in the US Gulf to
deplete red snapper stocks."

283 **"elite group of snapper barons":** The attacks have been unremitting, often

as part of membership or fundraising appeals. See, for instance, "What's Your Freedom to Fish Worth?" Recreational Fishing Alliance, http://joinrfa .org/whats-your-freedom-to-fish-worth/, accessed December 10, 2015; and Captain Tom Hilton's blog, http://www.theonlinefisherman.com/599-wrapped-external-conservation-sites/5876-steve-southerland.

281 **tried every policy move:** See Thad Altman and Frank Artiles, "States Have Remedy for Red Snapper Mess," Coastal Conservation Association, December 4, 2014, http://www.joincca.org/articles/769; "Recreational Fishing And Boating Community Underwhelmed by House Magnuson–Stevens Act Reauthorization Bill," Coastal Conservation Association, May 29, 2014, http://www.joincca.org/articles/666; and "Amendment 28—Red Snapper Reallocation Briefing Document," Coastal Conservation Association, February 25, 2014, http://s3.amazonaws.com/assets.clients/ cca/ckeditor_assets/attachments/507/snapper_briefing02282014_web pdf?1393623815.

282 **State management . . . is a bad idea:** Other species held up as models of successful state management, including striped bass and Dungeness crab, live close to shore, not in deep federal waters. See Monica Goldberg, "Solutions for Recreational Red Snapper Not Found in Other Fisheries," *EDFish* (blog), Environmental Defense Fund, October 21, 2015, http://blogs.edf .org/edfish/2015/10/21/solutions-for-recreational-red-snapper-not-found -in-other-fisheries/.

284 **"cannibalistic on each other":** Already, several states have staked out positions favoring their own fishermen, motivated in part by lucrative sales of angler licenses. (Florida, for instance, earned $36 million selling 1.6 million saltwater licenses in the 2013–14 season.) Texas led the way: for years it has defied the federal rule requiring state fishing seasons to mirror the federal season, and instead set a 365-day season. Other states followed: in 2014, Louisiana's state waters were open for 286 days, Florida's for 52, Mississippi's and Alabama's for 21. After accounting for all the landings made in state waters during those extended seasons, the resulting federal recreational red snapper season had to be cut to just 9 days. See Florida Fish and Wildlife Conservation Commission, "The Economic Impact of Saltwater Fishing in Florida," http://myfwc.com/conservation/value/saltwater -fishing/, accessed November 16, 2015.

284 **noncompliant:** Gulf of Mexico Fishery Management Council, "Recreational Red Snapper Sector Separation," December 2014, http://gulfcouncil .org/docs/amendments/RF%2040%20-%20Final%2012-17-2014.pdf.

284 **half-abundance might look like a lot:** The red snapper also appear more abundant than they are because they've been intentionally lured to congre-

gate by the artificial reefs, including decommissioned oil rigs, that Alabama and others have been building for four decades. Though some scientists claim those reefs grow the population, their impact may be the reverse: giving both fishermen and predators the opportunity to essentially shoot fish in a barrel. See J. H. Cowan Jr. et al., "Red Snapper Management in the Gulf of Mexico: Science- or Faith-Based?", *Reviews in Fish Biology and Fisheries* 21, no. 2 (2011): 187–204.

285 **"Administrative discretion":** *Guindon vs. Pritzker*, 2014 WL 1274076 (D.D.C., March 26, 2014), http://cases.justia.com/federal/district-courts/district-of-columbia/dcdce/1:2013cv00988/160778/61/0.pdf?ts=1411527224; and Alex Dropkin, "The Gulf's Red Snapper Fishery Makes a Comeback," *Texas Observer*, September 24, 2014, http://www.texasobserver.org/red-snapper-fishery-texas-gulf-sustainable.

285 **allocate *more* fish:** "Gulf Council Shifts Red Snapper Quota From Commercial to Recreational Sector," *Gulf Seafood News*, August 25, 2015, http://gulfseafoodnews.com/2015/08/25/23191gulf-council-shifts-red-snapper-quota/, accessed November 17, 2015.

286 **renowned chefs:** Including Haley Bitterman of New Orleans's famous Brennan's.

286 **coalition of conservationists and businesses:** http://sharethegulf.org/.

286 **Gulf Headboat Collaborative Pilot:** A customer on a headboat pays per person, rather than chartering the vessel. See National Marine Fisheries Service, Southeast Regional Office, "Gulf of Mexico Headboat Collaborative Pilot Program Frequently Asked Questions," http://sero.nmfs.noaa.gov/sustainable_fisheries/ifq/documents/pdfs/Headboat%20FAQs_20140110144135.pdf, accessed November 17, 2015.

286 **Angler Action Program:** Captain Ron Presley, "iAngler," Snook and Gamefish Foundation, http://www.snookfoundation.org/news/research/561-iangler.html, accessed November 17, 2015.

286 **"iSnapper":** Harte Research Institute for Gulf of Mexico Studies, Texas A & M University—Corpus Christi, "iSnapper: Design, Testing, and Analysis of an iPhone-Based Application as an Electronic Logbook in the For-Hire Gulf of Mexico Red Snapper Fishery," January 2014, http://www.sefsc.noaa.gov/P_QryLDS/download/CR954_Stunz_2014.pdf?id=LDS, accessed November 17, 2015.

287 **modeled on duck stamps:** Limited in number and distributed via lottery or otherwise, each stamp would allow a fisherman to catch one or several fish. They would be overseen by public or private angler management organizations. Snook and Gamefish Foundation, "License and Stamp," http://www.snookfoundation.org/info/license.html, accessed December 10, 2015.

287 **two-thirds of all fish:** Rod Griffin, "Fish Forever: Can Small-Scale Fisherman Help Save the Oceans," *EDF Solutions,* Summer 2013, https://www.edf.org/sites/default/fileds/edfSolutions-summer2013.pdf.

287 **Alaskan pollock:** "What type of fish do you use in the Filet-O-Fish?" FAQs, McDonald's, http://www.mcdonalds.com/us/en/your_questions/our_food/what-type-of-fish-do-you-use-in-the-filet-o-fish.html.

287 **quota is community-owned:** The community's purchase of quota was financed by the city and EDF's revolving California Fisheries Fund. See Phoebe Higgins, "CFF Supports New Fishing Community-Led Nonprofit in Monterey," *EDFish* (blog), Environmental Defense Fund, September 14, 2015, http://blogs.edf.org/edfish/2015/09/14/cff-supports-new-fishing-community-led-nonprofit-in-monterey/.

287 **Fisheries Trust:** Monterey Bay Fisheries Trust, "Background," http://www.montereybayfisheriestrust.org/our-story, accessed November 18, 2015.

287 **NOAA's 2014 annual report:** National Marine Fisheries Service, "Status of the Stocks 2014," April 2015, http://www.nmfs.noaa.gov/sfa/fisheries_eco/status_of_fisheries/archive/2014/2014_status_of_stocks_final_web.pdf.

287 **the agency's 2015 analysis:** Eric Thunberg et al., "Measuring Changes in Multi-Factor Productivity in US Catch Share Fisheries," *Marine Policy* 62 (2015): 294–301.

287 **ninety species:** National Marine Fisheries Service, "The West Coast Groundfish IFQ Fishery: Results from the First Year of Catch Shares 2011," June 2012, http://www.nmfs.noaa.gov/stories/2012/07/docs/catch_sharesyear1_report.pdf.

287 **declared a federal disaster:** Brian Gorman and Mike Fergus, "Commerce Secretary Daley Announces West Coast Groundfish Fishery Failure," National Oceanic and Atmospheric Administration, January 9, 2000, http://www.publicaffairs.noaa.gov/releases2000/jan00/noaa00r103.html, accessed November 18, 2015.

288 **canary rockfish and petrale sole had recovered:** Shems Jud, "Canary Rockfish Rebound Dramatically: 40 years Ahead of Schedule," *EDFish* (blog), Environmental Defense Fund, June 18, 2015, http://blogs.edf.org/edfish/2015/06/18/canary-rockfish-rebound-dramatically-40-years-ahead-of-schedule/. Bearing out Wayne's argument that fishermen with a secure stake become champions of conservation practices, many of the Pacific fishermen signed agreements to establish voluntary "no-take zones" to avoid accidental catch of severely depleted species, and adopted new technologies to track and share information about overfished areas and bycatch. Thanks to those efforts, bycatch is down 80 percent. See Shems Jud, "New and Innovative Management Practices Revitalize West Coast Fish-

ery," *EDF Voices* (blog), Environmental Defense Fund, September 4, 2014, https://www.edf.org/blog/2014/09/04/new-and-innovative-management -practices-revitalize-west-coast-fishery.

288 **25,000 pounds of fish:** Rod Griffin, "Restoring Oceans to Abundance," *EDF Solutions,* Fall 2014, https://www.edf.org/sites/default/files/content/ solutions_fall2014.pdf.

288 **Needless deaths . . . reduced:** Jonathan H. Adler and Nathaniel Stewart, "Learning How to Fish," *Regulation,* Spring 2014, http://object.cato.org/ sites/cato.org/files/serials/files/regulation/2014/4/regulation-v37n1-1.pdf. A study published in the journal *Marine Policy* found on average safety nearly tripled in fisheries in the U.S. and British Columbia that switched to catch shares. Dietmar Grimm et al., "Assessing Catch Shares' Effects, Evidence from Federal United States and Associated British Columbian fisheries," *Marine Policy* 36, no. 3 (2012): 644–57.

288 **IFQs are held up as a model:** Jonathan H. Adler, "Conservative Principles for Environmental Reform," *Duke Environmental Law and Policy Forum* 23 (Spring 2013): 253–80, http://www.perc.org/sites/default/files/pdfs/ Adler,%20Conservative%20Principles%20for%20Environmental%20 Reform.pdf.

288 **directly involved in their design:** Donald Leal, "Fencing the Fishery," Political Economy Research Center, June 1, 2002, http://www.perc.org/ sites/default/files/guide_fish.pdf.

288 **the gathering rush for seabed minerals:** The Gulf oil boom has resumed, with more activity than before the explosion. Oil companies have built a dozen new multibillion-dollar platforms, including Shell's *Olympus,* 130 miles off the Louisiana coast and larger than a New York City block, and another designed to tap an oil field under 9,500 feet of water—almost twice the depth of the Macondo. Even BP plans to spend $4 billion a year in the Gulf for the next decade. See Daniel Gilbert, Amy Harder and Justin Scheck, "Oil Boom Returns to U.S. Gulf After *Deepwater Horizon* Disaster," *Wall Street Journal,* November 21, 2014, http://www.wsj.com/articles/ oil-rigs-return-to-gulf-after-deepwater-horizon-disaster-1416599464; and David McFadden, "Deep-Sea Mining Looms on Horizon as UN Body Issues Contracts," Associated Press, July 25, 2015, http://bigstory.ap.org/article/ f8119a26f1034e7487f67725a2ffde80/deep-sea-mining-looms-horizon-un -body-issues-contracts.

289 **lowest numbers of juvenile red snapper:** This would have set the fishery back except that closure of the Gulf gave the big fish a compensating rest.

289 **lesions or rotting fins:** Lesions were most common in bottom-dwellers, including snapper, particularly north of the wellhead. See Steven A.

Murawski et al., "Prevalence of External Skin Lesions and Polycyclic Aromatic Hydrocarbon Concentrations in Gulf of Mexico Fishes, Post-Deepwater Horizon," *Transactions of the American Fisheries Society* 43, no. 4 (2014): 1084–97.

289 **"hearsaying":** "Fish near BP oil spill site are sick," *Star Democrat*, April 20, 2012, http://www.stardem.com/news/national/article_0137d3d3-5371 -57e6-9834-c429bcdba091.html.

289 **reductions in females' reproductive success:** National Oceanographic and Atmospheric Administration, "*Deepwater Horizon* Oil Spill: Draft Programmatic Damage Assessment and Restoration Plan and Draft Pro-grammatic Environmental Impact Statement," Chapter 4: "Injury to Natu-ral Resources," October 5, 2014, http://www.gulfspillrestoration.noaa.gov/ wp-content/uploads/Chapter-4_Injury-to-Natural-Resources1.pdf.

290 **growing acidity of the ocean:** According to the National Oceanographic and Atmospheric Administration, acidity has increased 30 percent since the dawn of the Industrial Revolution. See Malavika Vyawahare, "Sci-entists Warn That 2 C Target is Not Enough to Protect Corals, Marine Life," ClimateWire, *E & E News*, August 18, 2015, http://www.eenews.net/ climatewire/2015/08/18/stories/1060023581.

290 **100,000 years before the ocean recovered:** Sebastian Hennige, J. Mur-ray Roberts and Phillip Williamson, eds., "An Updated Synthesis of the Impacts of Ocean Acidification on Marine Biodiversity," Technical Series no. 75, Secretariat of the Convention on Biological Diversity, U.N. Envi-ronment Program, 2014, http://web.mit.edu/12.000/www/m2018/pdfs/ acidification-biodiversity.pdf.

290 **Acidification compromises:** According to NOAA, half of the minuscule sea snails called pteropods, a prime food source for pink salmon, mack-erel and herring in coastal Pacific waters, are losing their shells to human-induced ocean acidification. See Elspeth Dehnert, "Acidity is Dissolving the Shells of a Critical Component in the Ocean Food System," Climate-Wire, *E & E News*, May 1, 2014, http://www.eenews.net/climatewire/ stories/1059998804/search?keyword=+ocean+acidification+losing+shells; and N. Bednaršek et al., "Limacina helicina shell dissolution as an indicator of declining habitat suitability owing to ocean acidification in the California Current Ecosystem," *Proceedings of the Royal Society B* 281, no. 1785 (June 2014), http://rspb.royalsocietypublishing.org/content/281/1785/20140123.

290 **Shrimp react:** Woods Hole Oceanographic Institution, "In CO2-Rich Environment, Some Ocean Dwellers Increase Shell Production," December 1, 2009, http://www.whoi.edu/page.do?pid=7545&tid=3622&cid=63809.

291 **Three billion:** U.N. Food and Agriculture Organization, "The State of

World Fisheries and Aquaculture," 2014, http://www.fao.org/3/a-i3720e .pdf.

291 **38 million, 90 percent:** Christophe Béné, Graeme Macfadyen and Edward Hugh Allison, "Increasing the Contribution of Small-Scale Fisheries to Poverty Alleviation and Food Security," U.N. Food and Agriculture Organization, 2007, http://www.fao.org/3/a-a0965e.pdf. Often subsistence fisherman have few options. Some Somali pirates were once fishermen who now have no fish; see Ishaan Tharoor, "How Somalia's Fishermen Became Pirates," *Time*, April 18, 2009, http://content.time.com/time/world/article/0,8599,1892376,00.html.

291 **overexploited or collapsed:** Rainer Froese et al., "What Catch Data Can Tell Us About the Status of Global Fisheries," *Marine Biology* 159, no. 6 (2012): 1283–92. Nearly all of this depletion has occurred since 1950; compared to the 1950s, just 5% of bluefin tuna, 10% of sharks and 5% of Atlantic cod remain. See Sylvia Earle, "My Wish: Protect Our Oceans," TED Long Beach California, February 2009, http://ideas.ted.com/4-gifs -that-show-what-happened-to-the-oceans/.

291 **catch underestimated:** Daniel Pauly and Dirk Zeller, "Catch Reconstructions Reveal that Global Marine Fisheries Catches Are Higher than Reported and Declining," *Nature Communications* 7 (2016), doi 10.1038/ ncomms10244. "Finding the Missing Fish," *Pew Charitable Trusts,* January 9, 2015, http://www.pewtrusts.org/en/research-and-analysis/analysis /2015/01/09/finding-the-missing-fish.

291 **with a secure access right:** Nicolás L. Gutiérrez, Ray Hilborn and Omar Defeo, "Leadership, Social Capital and Incentives Promote Successful Fisheries," *Nature* 470, no. 7334 (2011): 386–89.

292 **In 2015, Belize:** Prototypes update: 50 in 10, "Making Sustainable Fisheries a Reality around the Globe," http://www.50in10.org/home-featured /prototypes-update-making-sustainable-fisheries-a-reality-around-the -globe-2/, accessed December 13, 2015.

292 **a study of almost 5,000 fisheries:** Christopher Costello et al., "Global Fishery Prospects Under Contrasting Management Regimes," *Proceedings of the National Academy of Sciences* 113, no. 18 (2016): 5125–29. Food and profit projections are compared to a business as usual scenario.

293 **it is not . . . too late:** Douglas J. McCauley et al., "Marine Defaunation: Animal Loss in the Global Ocean," *Science* 347, no. 6219 (January 16, 2015), http://science.sciencemag.org/content/347/6219/1255641; Fen Montaigne, "Why Ocean Health Is Better and Worse Than You Think," *Yale Environment 360* (blog), February 18, 2015, http://e360.yale.edu/feature/ why_ocean_health_is_better_and_worse_than_you_think/2848/; and Carl

Zimmer, "Ocean Life Faces Mass Extinction, Broad Study Says," *New York Times*, January 15, 2015, http://www.nytimes.com/2015/01/16/science/earth/study-raises-alarm-for-health-of-ocean-life.html.

296 **fresh-caught grouper:** Wayne reports every fish he catches, including those he eats for lunch.

298 **self-propelled jack-up:** Floated into place with their legs in the air, the rigs jack their legs down to the seafloor and use bow thrusters to walk sideways.

298 **interdependencies:** In October 2015, Cuba and the U.S. signed an historic agreement to work together to protect marine life in their shared seas. See Erica Goode, "Crown Jewel of Cuba's Coral Reefs," *New York Times*, July 13, 2015, http://www.nytimes.com/2015/07/14/science/crown-jewel-of-cubas-coral-reefs.html; and Victoria Burnett, "Cuba and U.S. Agree to Work Together to Protect Marine Life," *New York Times*, October 6, 2015, http://www.nytimes.com/2015/10/06/world/americas/cuba-and-us-agree-to-work-together-to-protect-marine-life.html.

ACKNOWLEDGMENTS

For their support, patience and indispensable help, I wish to thank:

My present and former colleagues: Fred Krupp, Eric Pooley, David Festa, Nick Nicholas, Amanda Leland, Cynthia Hampton, Dominique Browning, Eric Holst, Suzy Friedman, Steve Cochran, Brian Jackson, Natalie Peyronnin, Doug Rader, Pam Baker, Kristen McConnell, Robert Jones, Daniel Willard, Matt Tinning, James Workman, Joel Plagenz, Lisa Moore, Tony Kreindler, Paul Harrison, Nicole Possin, Julie Benson, Elizabeth Van Cleve, Anne Marie Borch, David Bjerklie, and my tireless researcher, Ashley Rood.

My subtle and graceful editors at Norton: Starling Lawrence, Ryan Harrington and Allegra Huston, and insightful agents Gail Ross and Howard Yoon.

My filmmaking colleagues, close and encouraging readers at every turn: John Hoffman, Susan Froemke, Alexandra Moss, Jon Bardin and Beth Aala.

My earliest and lifelong teachers, lepidopterist Don MacNeill, California master gardener Rose Horn, forester Stuart Saucke and Central Valley farmers Rich, Evelyn, Rick, Charlie, Ruth and Bruce Rominger.

My friends and family, all of them, but especially Frank and Barby Harvey, David and Karen Horn, Aaron Steinberg, Corrie Yackulic, Sindri Anderson, Quinn Moss, Katie Mark, Shari Levine, Ruth

Friedman, Kathy Jo Vincent, Randi Hopkins, Ilan Averbuch, Alka Mansukhani, Patti Cohen, Eddie Sutton, Guillaume and Christina Malle, Denise Ribera and—our lifeline always—Eva Ansley.

I wish to honor the memory of my brother Martin.

Above all, I thank everyone in this book—for opening up their lives to me and the world and giving us all cause for hope.

INDEX

Note: Page numbers after 298 refer to notes.

Folklore
of Canada